Mathematical Engineering

More information about this series at http://www.springer.com/series/8445

Igor V. Shevchuk

Modelling of Convective Heat and Mass Transfer in Rotating Flows

Springer

Igor V. Shevchuk
MBtech Group GmbH and Co. KGaA,
Powertrain Solutions
Fellbach-Schmiden
Germany

ISSN 2192-4732 ISSN 2192-4740 (electronic)
Mathematical Engineering
ISBN 978-3-319-20960-9 ISBN 978-3-319-20961-6 (eBook)
DOI 10.1007/978-3-319-20961-6

Library of Congress Control Number: 2015944450

Springer Cham Heidelberg New York Dordrecht London

Printed on acid-free paper

Springer International Publishing AG Switzerland is part of Springer Science+Business Media (www.springer.com)

To my wife Nataliya and sons Vladimir,
Aleksandr and Nikolay

Preface

This monograph was conceived as an overview of the potential and achievements of the analytical and numerical modelling of convective heat and mass transfer in different types of rotating flows. Flow rotation can be a consequence of (i) *system rotation*, (ii) *flow swirl* imposed by so-called swirl generators and (iii) *curvature of surfaces* or larger segments of the geometry such as turns, bends, curved connections, etc. Rotation, swirl or curvature-induced volume forces often referred to as centrifugal and Coriolis forces can significantly affect the flow pattern, as well as the heat and mass transfer rate.

Rotating flows arise in numerous scientific and engineering applications. As practical examples one can mention turbomachinery, energy systems, automotive engineering, aerospace engineering, medical equipment, processing engineering and many others. One of the important scientific applications is a rotating disk electrode involved in experimental determination of the diffusion coefficient in electrolytes. A cone-plate or a cone-disk device, which includes a fixed disk and a rotating cone that touches the disk by its apex, is widely known in measurements of the viscosity coefficient of liquids.

One can mention several books, which elucidate many aspects of the physics and provide quantification for many parameters of rotating flows. These are classical books of L.A. Dorfman "Hydrodynamic Resistance and the Heat Loss of Rotating Solids" (Oliver and Boyd, Edinburgh, UK, 1963) and V.G. Levich "Physicochemical Hydrodynamics" (Prentice-Hall, Inc., Englewood Cliffs, N.J., 1962), which for many decades were desktop books for specialists in the fields of convective heat and mass transfer in rotating disk systems. The fundamental review monograph of J.M. Owen & R.H. Rogers "Flow and Heat Transfer in Rotating-Disc Systems" (Research Studies Press Ltd., UK, 1989 & 1995) summarized results of experimental investigations and theoretical modelling in the area of secondary air cooling systems of gas turbines, including rotor−stator systems and rotating cavities formed by parallel co-rotating disks. The book "Heat Transfer and Fluid Flow in Rotating Coolant Channels" (Research Studies Press, J. Wiley and Sons, 1981) by W.D. Morris is devoted mostly to experimental investigations of the hydraulic resistance and average heat transfer in channels rotating about a parallel axis. In the recent book "Rotating flow" (Elsevier Inc., Amsterdam etc., 2011), P.R.N. Childs shed light on the basic theory of rotating flows and contributed to the development of the integral methods for rotating-disk systems, including rotor-stator configurations and rotating cavities.

During the past decades, methods of experimental and theoretical investigations of scientific and practical problems of convective heat and mass transfer, which the aforementioned books deal with, have made considerable progress. This resulted in obtaining new accurate experimental and numerical results for a series of rotating geometries studied before and emerging during the past years. As a result, the analysis and generalizations of the experimental and numerical data provided in the aforementioned books are often insufficient. Integral methods have been rather successfully applied to several rotating-disk geometries. However, theoretical model assumptions underlying the known integral methods demonstrated their restricted capabilities in light of the newly obtained experimental and numerical data. A powerful modelling technique based on the exact self-similar solutions of the Navier–Stokes and energy equations appeared to be insufficiently developed for a few rotating disk geometries, where appropriate self-similar forms of the solutions have not been derived. In addition, a few important scientific and practical problems were not touched in the above-mentioned books: (a) transient conjugate heat transfer; (b) uniform orthogonal flow impingement onto a disk; (c) fluid flow, heat and mass transfer in a small gap between a rotating disk and/or a cone that touches the disk by its apex; (d) convective heat and mass transfer at Prandtl and Schmidt numbers, both moderately larger than unity with the application to the experimental technique based on naphthalene sublimation in air, and much larger than unity with the application in electrochemistry.

All said, the above became an incentive for me to undertake investigations that were summarized in the form of a book by Shevchuk I.V. "Convective Heat and Mass Transfer in Rotating Disk Systems (Springer Verlag, Berlin, Heidelberg, 2009). Since then, I have conducted new studies on the subject of rotating flows published as a series of research papers. In the same time, the international scientific community has also contributed much to this research area, which provided valuable material for validation and corroboration of the models and numerical results presented in my book.

In comparison with the previous book, my new monograph outlines the further progress in the integral methods, self-similar and analytical solutions for the problems of convective heat and mass transfer in rotating-disk systems validated through extensive comparisons with the experimental data including those that have been published during the past six years. Most part of the new monograph is devoted to system rotation-induced fluid flows and deals with several rotating-disk geometries. Swirl flows were also modelled in some of these geometries. In addition, the scope of the new monograph was extended to cover other types of rotating flows such as those in (a) the channels rotating around a parallel axis, and (b) the two-pass ribbed channels with 180° bends. These studies provide examples of design optimization of air cooling systems of the rotors of electrical motors and gas turbine blades, respectively.

The present book consists of nine chapters. The book is mainly focused on convective heat transfer in air flow, with the exception of Chap. 6, which deals with heat and mass transfer at Prandtl or Schmidt numbers larger than unity.

Chapter 1 depicts geometries studied in this book, outlines forces influencing the flow and presents a general mathematical description in the form of momentum, continuity, energy and convective diffusion equations written in a vector form, Cartesian and cylindrical polar coordinate systems.

In Chap. 2, the general mathematical description is adapted to rotating disk systems. The chapter contains also an overview of the existing methods of solution, the integral method developed by the author, and a general analytical solution for the turbulent boundary layer flow and heat transfer in rotating-disk systems obtained using this method.

Chapter 3 is devoted to steady-state and unsteady heat transfer of a single rotating disk. As demonstrated here, the present integral method is significantly more accurate and incorporates a wider variety of thermal boundary conditions than other integral methods. Chapter 3 critically overviews the most reliable experimental data for transitional flow, provides recommendations for the calculation of average heat transfer of an *entire* disk and briefly outlines some important aspects of laminar transient heat transfer.

In Chap. 4, results of the analytical and numerical modelling of external flow over a rotating disk and outward flow between parallel co-rotating disks are described and compared with experimental data. In particular, Chap. 4 presents solutions for the cases of (a) disk rotation in a fluid subject to solid-body rotation, (b) accelerating non-rotating radial flow and (c) centrifugal swirling radial flow in a gap between parallel co-rotating disks.

Chapter 5 focuses on laminar flow, heat and mass transfer between a disk and a cone that touches the disk with its apex. It comprises such geometries as "rotating cone—stationary disk", "rotating disk—stationary cone", "co-rotating or contra-rotating disk and cone" and "non-rotating conical diffuser". Novel is the section describing effects of the Prandtl and Schmidt numbers, as well as a review of the relevant recently published works.

In Chap. 6, results of different authors for the problems of convective heat and mass transfer for the Prandtl and Schmidt numbers larger than unity are critically analysed and generalized. Chapter 6 presents original theoretical models of the author developed for naphthalene sublimation in air and electrochemical problems. In the integral method of the author, effects of large Prandtl and Schmidt numbers are taken into account.

Chapter 7 describes results of the CDF modelling of convective heat transfer in pipes rotating around a parallel axis including effects of the flow angle of attack at the inlet to the pipe, as well as influence of the cross-section geometry (circular or elliptic pipes).

In Chap. 8, original results of the simulation and optimization of convective heat transfer in the varying aspect ratio two-pass internal ribbed cooling channels with 180° bends are outlined and analysed from a single viewpoint.

Chapter 9 presents overall conclusions to the material presented in the book.

The present book is an official publication of my habilitation thesis (Habilitationsschrift) prepared during my work as a university lecturer at the Institute of Aerospace Thermodynamics (ITLR), University of Stuttgart, Germany and successfully presented at the Faculty of Aerospace Engineering and Geodesy, University of Stuttgart on 24 April 2015. I would like to deeply thank the main referee of my habilitation thesis, Director of the ITLR Prof. Dr.-Ing. habil. Bernhard Weigand for his strong support of my aspiration to successfully accomplish this work, as well as for his numerous valuable advices and fruitful discussions. I would also like to warmly thank the co-referees, Professor D.Phil. Peter R.N. Childs (Imperial College London, UK), Professor Dr. Andrey V. Kuznetsov (North Carolina State University, U.S.A.) and Professor Dr.-Ing. habil. Yuri B. Zudin (National Research Center "Kurchatov Institute", Moscow, Russia) for reviewing my habilitation thesis and for participation in the habilitation process.

I have obtained results included in this work during two decades of research studies conducted in collaboration with different research institutions in Germany, USA, UK, France, Sweden and Ukraine. I would like to thank all my colleagues, with whom I collaborated during that time, for their contribution, useful advices and friendly discussions. The Research Fellowship of Alexander von Humboldt Foundation (Germany), which enabled my research stay at Technische Universität Dresden in 2003–2005 and summarizing a large part of the material used in the present book, is also gratefully acknowledged.

I am very thankful to my family for their invaluable constant support and understanding during my preparation of the habilitation thesis.

Stuttgart, Germany Igor V. Shevchuk

Contents

Nomenclature

a	Thermal diffusivity; (m^2/s)
a	Radial velocity gradient on the outer boundary of the boundary layer, Eq. (2.27); (1/s)
a	Semi-major axis of an ellipse; (m)
$A = ad_j/V_j$	Non-dimensional radial velocity gradient on the outer boundary of the boundary layer; (–)
$AR = H/W$	Aspect ratio; (–)
b	Outer radius of a disk; (m)
b	Semi-minor axis of an ellipse; (m)
$Bi_1 = \alpha_1 b/\lambda_w$	Biot number at a cylindrical surface of a disk; (–)
$Bi = 0.5\alpha s/\lambda_w$	Biot number at a flat surface of a disk; (–)
$Bi_2 = 0.5\alpha_2 s/\lambda_w$	Biot number at a flat surface of a disk; (–)
C	Concentration; (mol/m^3)
$c_f/2 = \tau_w/(\rho V_*)^2$	Surface friction coefficient; (–)
$C_M = 4M/(\rho\omega^2 b^5)$	Moment coefficient of two flat sides of a rotating disk; (–)
c_p	Isobaric specific heat; (J/(kg K))
$C_w = \dot{m}/(\mu b)$	Non-dimensional radial mass flowrate through a cavity between two rotating disks; (–)
$c_{0^*} = (T_{w,l} - T_\infty)_{n_*=0}$	Constant temperature difference on the surface with $T_{w,l}$ = const. and T_∞ = const.; (K)
D	Disk diameter; (m)
D	Diameter of the circular pipe; (m)
D_j	Nozzle diameter; (m)
D_m	Diffusion coefficient; (m^2/s)
$D_h = 4S/P_e$	Hydraulic diameter (arbitrary cross-section); (m)
$D_h = \dfrac{2HW}{H+W}$	Hydraulic diameter (rectangular channel); (m)
$D_e = \sqrt{4S/\pi}$	Equivalent diameter; (m)
e	Rib height; (m)

$\boldsymbol{F}$	Mass force per unit volume (boldface denotes vector parameter); (N/m^3)
F_x, F_y, F_z	Mass force components in Cartesian coordinates (per unit volume); (N/m^3)
F_r, F_φ, F_z	Mass force components in cylindrical polar coordinates (per unit volume); (N/m^3)
F, G, H, P	Self-similar functions, Eq. (2.26); (–)
F_0, G_0, H_0	Self-similar velocity components, free rotating disk (subscript "0"), Eq. (2.26) at $\beta = 0$, $N = 0$; (–)
$\mathrm{Fo} = 4a_w t/s^2$	Fourier number; (–)
$F_\mathrm{t}(t) = \dfrac{T_w(t,r) - T_\infty}{T_{w,i}(r) - T_\infty}$	Non-dimensional disk surface temperature in the unsteady heat transfer problem; (–)
$f = \Delta p D_h/(0.5\rho \bar{V}^2 L)$	Friction factor in a single channel/pipe; (–)
$f = \Delta p D_{hi}/(0.5\rho U_i^2 L)$	Friction factor in a two-pass channel; (–)
f_0	Friction factor, Blasius Eq. (7.11) or McAdams Eq. (8.1) (–)
g	Acceleration of gravity; (m/s^2)
$h = r\ \mathrm{tg}\gamma$	Height of a conical gap; (m)
h_j	Nozzle-to-disk distance; (m)
$H = b/(0.5s)$	Parameter in Eq. (4.21); (–)
H	Eccentricity; (m)
H	Height of a rectangular channel, m; (m)
I	Turbulence intensity; (–)
$\boldsymbol{j}$	Acceleration of a mass force; (m/s^2)
k	Turbulent kinetic energy per unit mass; (m^2/s^2)
$K_H = \dfrac{\int_0^{\delta_T} v_r (T - T_\infty)\,\mathrm{d}z}{(T_w - T_\infty)\int_0^{\delta_T} v_r \mathrm{d}z}$	Shape-factor of the temperature profile; (–)
$K_m = \delta^{-1}\int_0^{\delta} \frac{v_r}{\omega r}\mathrm{d}z$	Non-dimensional radial mass flow rate through the boundary layer; (–)
$K_V = \dfrac{\int_0^{\infty} v_r (v_\varphi - v_{\varphi,\infty})\mathrm{d}z}{(\omega r - v_{\varphi,\infty})\int_0^{\infty} v_r \mathrm{d}z}$	Shape-factor of the velocity profile; (–)
L	Length of a single channel/pipe; (m)
L	Characteristic length in a two-pass channel; (m)
$M = -2\pi\int_0^b r^2 \tau_{w\varphi}\mathrm{d}r$	Moment of one side of a rotating disk; (Pa m^3)
$\dot{m}$	Total radial mass flowrate through the cavity between two rotating disks; (kg/s)
$\dot{m}$	Mass flowrate through a pipe/channel; (kg/s)
$\dot{m}_\mathrm{d} = 2\pi r\rho\int_0^{\delta} v_r \mathrm{d}z$	Mass flowrate through the momentum boundary layer over a rotating disk; (kg/s)
$\dot{m}_\mathrm{d,T} = 2\pi r\rho\int_0^{\delta_T} v_r \mathrm{d}z$	Mass flowrate through the thermal boundary layer over a rotating disk; (kg/s)

n	Exponent in the power-law approximation of the velocity profiles; (–)
n_{T}	Exponent in the power-law approximation of the temperature profiles; (–)
n_*	Exponent in the power-law approximation of the surface temperature, Eqs. (2.29)–(2.31); (–)
$N = v_{r,\infty}/(\omega r)$	Non-dimensional radial velocity in potential flow outside of the boundary layer; (–)
$Nu = \dfrac{q_w r}{\lambda(T_w - T_\infty)}$	Local Nusselt number for a rotating disk; (–)
$Nu = \alpha D_{\mathrm{h}}/\lambda$	Local Nusselt number, single pipe/channel; (–)
$Nu = \alpha D_{\mathrm{e}}/\lambda$	Local Nusselt number, single pipe/channel; (–)
$Nu = \alpha D_{\mathrm{hi}}/\lambda$	Local Nusselt number in a two-pass channel; (–)
$\overline{Nu} = \bar{\alpha} D_{\mathrm{h}}/\lambda$	Average Nusselt number, single pipe/channel; (–)
$\overline{Nu} = \bar{\alpha} D_{\mathrm{e}}/\lambda$	Average Nusselt number, single pipe/channel; (–)
$\overline{Nu}_0$	Average Nusselt number, standard conditions (smooth straight pipe/channel, no rotation); (–)
$\overline{Nu} = \bar{\alpha} D_{\mathrm{hi}}/\lambda$	Average Nusselt number, two-pass channel; (–)
$\overline{Nu}_{\mathrm{st}}$	Average Nusselt number, straight smooth channel, CFD simulations; (–)
Nu_0	Nusselt number, Dittus-Boelter Eq. (7.10); (–)
Nu_1	Nusselt number, Eq. (8.2); (–)
$Nu_{\mathrm{b}} = \dfrac{q_w b}{\lambda(T_w - T_\infty)}$	Nusselt number based on the outer radius of a rotating disk; (–)
$Nu_D = \dfrac{q_w D}{\lambda(T_w - T_\infty)}$	Local Nusselt number based on the diameter of a rotating disk; (–)
$Nu_{Dj} = \dfrac{q_w D_j}{\lambda(T_w - T_\infty)}$	Local Nusselt number at flow impingement onto a rotating disk; (–)
$Nu_{\mathrm{av}} = \dfrac{q_{w.\mathrm{av}} b}{\lambda(T_w - T_\infty)_{\mathrm{av}}}$	Average Nusselt number for a rotating disk; (–)
$Nu_{\mathrm{av}} = \dfrac{b\int_0^b Nu(T_w - T_\infty)\mathrm{d}r}{\int_0^b (T_w - T_\infty) r\mathrm{d}r}$	Average Nusselt number for a rotating disk; (–)
p	Static pressure; (Pa)
p	Pitch between ribs; (m)
P_e	Perimeter of the pipe/channel; (m)
$Pr = \mu c_p/\lambda$	Prandtl number; (–)
q	Heat flux per unit area; ($\mathrm{W/m^2}$)
$q_w = -\lambda\left(\dfrac{\mathrm{d}T}{\mathrm{d}z}\right)_{z=0}$	Wall heat flux per unit area; ($\mathrm{W/m^2}$)
$q_{w,\mathrm{av}} = \int_0^b q_w r\mathrm{d}r / \int_0^b r\mathrm{d}r$	Surface-averaged wall value of the heat flux per unit area; ($\mathrm{W/m^2}$)
$Ra_H = \dfrac{\omega^2 H^3 D_{\mathrm{h}} \beta \Delta T}{2\lambda v}$	Rayleigh number for a rotating pipe; (–)
$Re_a = aD^2/v$	Reynolds number in radial flow over a disk; (–)

$Re_\omega = \omega r^2/\nu$	Local rotational Reynolds number for a rotating disk; (–)
$Re_\omega = \omega D_h^2/\nu$	Rotational Reynolds number for a rotating pipe; (–)
$Re_{\omega D} = \omega D^2/\nu$	Rotational Reynolds number based on the disk diameter; (–)
$Re_{\omega j} = \omega D_j^2/\nu$	Rotational Reynolds number based on the nozzle diameter of an impinging jet; (–)
$Re_\Omega = \Omega r^2/\nu$	Local rotational Reynolds number for rotating cones in cone-disk systems; (–)
$Re_\varphi = \omega b^2/\nu$	Rotational Reynolds number at the outer radius of a disk; (–)
$Re_j = V_j D_j/\nu$	Reynolds number based on an impingement velocity; (–)
$Re_{V*} = V_* \delta/\nu$	Reynolds number based on the velocity V_*; (–)
$Re_T^{**} = \dfrac{\omega r \delta_T^{**}}{\nu}$	Enthalpy Reynolds number; (–)
$Re = Re_\Omega \eta_1^2/12,$	Reynolds number, cone-and-plate systems; (–)
$Re = Re_\omega \eta_1^2/12$	Reynolds number, cone-disk systems; (–)
$Re = \bar{V} D_h/\nu$	Axial Reynolds number in a pipe; (–)
$Re = U_i D_{hi}/\nu$	Axial Reynolds number, two-pass channel; (–)
$Ro = \omega D/\bar{V}$	Rossby number in a rotating pipe; (–)
r, φ, z	Cylindrical polar coordinates; (m or rad)
s	Spacing (height) between rotating disks; (m)
s	Thickness of a disk in the problem of unsteady conjugate heat transfer; (m)
S	Cross-section area of a pipe/channel; (m^2)
S	Contact surface; (m^2)
$Sc = \nu/D_m$	Schmidt number; (–)
$Sh = \alpha_m r/D_m$	Sherwood number for a rotating disk; (–)
$Sh_{av} = \alpha_{m,av} b/D_m$	Average Sherwood number for a rotating disk; (–)
$St = \dfrac{q_w}{\rho c_p V_* (T_w - T_\infty)}$	Stanton number; (–)
t	Time; (s)
T	Temperature; (K)
T_B	Bulk fluid temperature; (K)
T_i	Inlet temperature in a pipe/channel; (K)
T_w	Stationary/instantaneous wall temperature; (K)
$T_{w,l}$	Initial value of the wall temperature in unsteady heat transfer; (K)
T_∞	Temperature in potential flow outside of the boundary layer; (K)
T_{ref}	Reference temperature; (K)
$T_{m-out} = \dfrac{1}{\dot{m}} \int_S T d\dot{m}$	Temperature of mixing at the pipe outlet; (K)

$T^+ = (T_w - T)\rho_\infty V_\tau / q_w$	Local temperature in wall coordinates; (K)
$\tan\varphi = \dfrac{v_r}{\omega r - v_\varphi}$	Tangent of the flow swirl angle; (–)
$V = \left[v_r^2 + (v_\varphi - \omega r)^2\right]^{1/2}$	Total velocity; (m/s)
$V^+ = V/V_\tau$	Total velocity in wall coordinates; (m/s)
$V_\tau = (\tau_w/\rho)^{1/2}$	Friction velocity; (m/s)
v_r, v_φ, v_z	Velocity components in cylindrical coordinates; (m/s)
$\overline{v}_r = \dfrac{v_r}{\omega r - v_{\varphi,\infty}}$	Non-dimensional radial velocity; (–)
$\overline{v}_\varphi = \dfrac{v_\varphi - \omega r}{v_{\varphi,\infty} - \omega r}$	Non-dimensional tangential velocity; (–)
$V_* = \omega r\lvert\beta - 1\rvert(1 + \alpha^2)^{1/2}$	Characteristic velocity; (m/s)
V_j	Axial flow velocity at infinity or at the outlet of a nozzle; (m/s)
$\bar{V}$	Mean axial velocity in the pipe; (m/s)
u, v, w	Velocity components in Cartesian coordinates; (m/s)
U	Velocity in a two-pass channel; (m/s)
U_b	Bulk-averaged velocity, two-pass channel; (m/s)
U_i	Channel mean axial velocity at the inlet of a two-pass channel; (m/s)
W	Width of a rectangular channel; (m)
W_{in}	Width of the inlet pass, two-pass channel; (m)
W_{out}	Width of the outlet pass, two-pass channel; (m)
W_{el}	Tip wall distance from the divider wall; (m)
W_{web}	Divider wall thickness; (m)
x, y, z	Cartesian coordinates; (m)
$x = r/b$	Non-dimensional radial coordinate; (–)
$y = z/(0.5s)$	Non-dimensional axial coordinate in the problem of heat conduction inside a disk; (–)
$\tilde{z} = z/h$	Non-dimensional coordinate in conical gaps; (–)
$z^+ = zV_\tau/\nu$	Wall-law coordinate; (–)
$z^* = z/D_h$	Dimensionless axial coordinate in a pipe; (–)
$\alpha = -\tau_{wr}/\tau_{w\varphi}$	Tangent of the flow swirl angle on the wall; (–)
α	Heat transfer coefficient; (W/(m^2 K))
$\alpha\, S$	Overall cooling efficiency; (W/K)
α_m	Mass transfer coefficient; (m/s)
$\alpha_{m,\text{av}} = \dfrac{2}{b^2}\displaystyle\int_0^b \alpha_m r dr$	Surface-averaged mass transfer coefficient; (m/s)

$\beta = v_{\varphi,\infty}/(\omega r)$	Parameter of flow swirl, i.e. Dimensionless tangential velocity component in potential flow outside of the boundary layer; (–)
β	Angle of attack at the inlet in a rotating pipe; (°)
γ	Angle of conicity between a cone and a disk; (°)
δ	Thickness of a momentum boundary layer; (m)
$\bar{\delta} = \delta/b$	Non-dimensional thickness of a momentum boundary layer; (–)
δ_T	Thickness of a thermal boundary layer; (m)
$\Delta = \delta_T/\delta$	Relative thickness of a thermal/diffusion boundary layer; (–)
ε	Dissipation rate of k per unit mass: (m^2/s^3)
δ^*	Displacement thickness; (m)
δ^{**}	Momentum thickness; (m)
$\delta_T^{**} = \int_0^{\delta_T} \frac{v_r}{\omega r} \frac{T - T_\infty}{T_w - T_\infty} dz$	Enthalpy thickness; (m)
$\bar{\delta}_T^{**} = \delta_T^{**}/\delta$	Non-dimensional enthalpy thickness; (–)
$\Delta P^* = \Delta p/(0.5\rho U_i^2)$	Relative pressure drop; (–)
$\Delta T = T_w - T_\infty$	Temperature difference on a surface; (K)
$\Delta T_{av} = \frac{\int_0^b (T_w - T_\infty) r dr}{\int_0^b r dr}$	Surface-averaged temperature difference; (K)
$\Delta T_i = T_{w,i}(r) - T_\infty$	Surface temperature difference at the initial moment of time $t = 0$ in unsteady conditions; (K)
$\Delta T_t(t, r) = T_w(t, r) - T_\infty$	Instantaneous temperature difference on a surface in unsteady heat transfer; (K)
$\Delta T_{x=1}$	Temperature difference on a surface at $x = 1$; (K)
$\overline{\Delta T} = \Delta T/\Delta T_{x=1}$	Relative non-dimensional temperature difference on a surface; (–)
ζ	Self-similar variable, Eq. (2.26); (–)
$\Theta = \frac{T - T_w}{T_\infty - T_w}$	Non-dimensional temperature; (–)
$\theta = \frac{T - T_\infty}{T_w - T_\infty}$	Non-dimensional temperature; (–)
$\vartheta(t, r, z) = (T - T_\infty)/c_{0^*}$	Non-dimensional temperature inside a disk for unsteady heat transfer; (–)
$\kappa = \tan \varphi_\infty = \frac{v_{r,\infty}}{\omega r - v_{\varphi,\infty}}$	Tangent of the flow swirl angle, potential flow; (–)
λ	Thermal conductivity; (W/(m K))
μ	Dynamic viscosity; (Pa s)
ν	Kinematic viscosity; (m^2/s)
ν_T	Turbulent kinematic viscosity; (m^2/s)
$\xi = z/\delta$	Non-dimensional coordinate; (–)

$\xi_T = z/\delta_T$	Non-dimensional coordinate; (–)
ρ	Density; (kg/m^3)
τ	Shear stress; (Pa)
$\tau_w = (\tau_{wr}^2 + \tau_{w\varphi}^2)^{1/2}$	Total shear stress on the wall; (Pa)
$\tau_{wr} = \mu(dv_r/\mathrm{d}z)_{z=0}$	Radial shear stress on the wall; (Pa)
$\tau_{w\varphi} = \mu(dv_\varphi/\mathrm{d}z)_{z=0}$	Tangential shear stress on the wall; (Pa)
$\varphi_w = \arctan\left[\frac{v_r}{\omega r - v_\varphi}\right]_{z=0}$	Swirl angle at the wall of a rotating disk; (Pa)
$\varphi_w = \arctan\left[\frac{v_r}{\Omega r - v_\varphi}\right]_{z=0}$	Swirl angle at the wall of a stationary disk; (Pa)
χ	Reynolds analogy parameter, Eq. (2.52); (–)
ω	Angular velocity of rotation of a disk (or co-rotating disks); (1/s)
ω	Angular velocity of rotation of a pipe; (1/s)
ω	Specific dissipation rate of k (1/s)
Ω	Angular velocity of rotation of a fluid in rotating-disk systems; (1/s)
Ω	Angular velocity of rotation of a cone in cone-disk systems. (1/s)

Subscripts

av	Average value
c	Centrifugal forces (accelerations)
Cor	Coriolis forces (accelerations)
E	Ekman layers
i	Initial moment of time
i	Inlet to a cavity
in	Inlet to a channel
j	Impinging jet
lam	Laminar flow
max	Value at a point of maximum
ref	Reference value
t	Turbulent parameters
t	Transient/instantaneous value of a parameter
turb	Turbulent flow
T	Parameters of a thermal boundary layer
tr	Parameters at the point of abrupt transition from laminar to turbulent flow
tr1	Parameters at the point of the beginning of transition from laminar to turbulent flow
tr2	Parameters at the point of the end of transition from laminar to turbulent flow

tran	Transitional flow
w	Wall value (at $z = 0$)
w	Thermophysical properties of the wall material
0	Standard conditions
	(a) free rotating disk at $v_{r,\infty} = 0$ and $v_{\varphi,\infty} = 0$
	(b) smooth non-rotating pipe/channel of a circular cross-section
1	Boundary of the viscous/heat conduction layer
1	Outer cylindrical surface of a disk
2	Flat surface of a disk in unsteady heat transfer
∞	Potential flow outside of a boundary layer
$\overline{u'v'}$	Time-averaged pulsation turbulent values

Mathematical Symbols

$$\nabla^2 = \left(\frac{\partial^2}{\partial r^2} + \frac{1}{r}\frac{\partial}{\partial r} + \frac{1}{r^2}\frac{\partial^2}{\partial \varphi^2} + \frac{\partial^2}{\partial z^2}\right)$$ Operator Nabla

Acronyms

CFD	Computational fluid dynamics
DES	Detached-Eddy simulation
LES	Large-Eddy simulation
RANS	Reynolds-averaged Navier–Stokes (equations)
RSM	Reynolds stress model
TLC	Thermochromic liquid crystals
1D	One-dimensional
2D	Two-dimensional
3D	Three-dimensional

Chapter 1
Overview of Rotating Flows

1.1 Applications of Rotating Flows

Rotating flows can be often met in different industrial applications. Rotating flows include [1] (i) flows caused by *system rotation*, (ii) *swirl flows* caused by swirl generators, and (iii) *curvilinear flows* in turns and bends. Common for these rotating flows is the emergence of volume forces (i.e., centrifugal and Coriolis forces) affecting the flow patterns, though the nature of these forces is different.

A large part of this work is devoted to rotating flows caused by system rotation. These include several types of rotating-disk systems, as well as straight pipes rotating about a parallel axis. Swirl flows are considered as applied to some of the rotating configurations mentioned above. Curvilinear flow effects are studied in detail while investigating different geometries of two-pass ribbed and smooth channels with 180° bends.

Rotating-disk systems are typically employed in gas turbine design, electrochemistry (rotating-disk electrodes), bio- and chemical reactors, transport engineering (automobile breaks), rotating-disk cleaners, etc. This work incorporates results of investigations for free rotating disks or disks placed in a fluid subjected to radial acceleration or rotating as a solid body (Fig. 1.1a), as well as impingement cooling of a rotating disk (Fig. 1.1b).

Flows *between a cone and a disk* with the cone apex touching the disk surface are used in medical equipment, viscosimetry, etc. Results of simulations for configurations where the disk and the cone rotate independently (Fig. 1.2a) and stationary conical diffusers with swirl flows in them (Fig. 1.2b) are outlined in this work.

Straight pipes rotating about a parallel axis (Fig. 1.3a) represent air cooling channels of electric motors. Configuration and location of such pipes and rotation effects on fluid flow and heat transfer in them are investigated in the present work.

Two-pass ribbed and smooth channels with 180° bends (Fig. 1.3b) are typical geometries of internal cooling channels of gas turbine blades. Effect of the aspect

I.V. Shevchuk, *Modelling of Convective Heat and Mass Transfer in Rotating Flows*, Mathematical Engineering, DOI 10.1007/978-3-319-20961-6_1

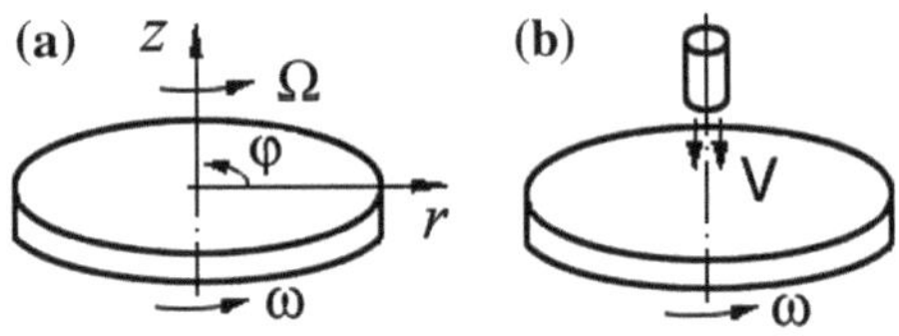

Fig. 1.1 Rotating-disk systems: **a** a free rotating disk and a disk placed in a rotating fluid, **b** impingement cooling of a rotating disk [2]

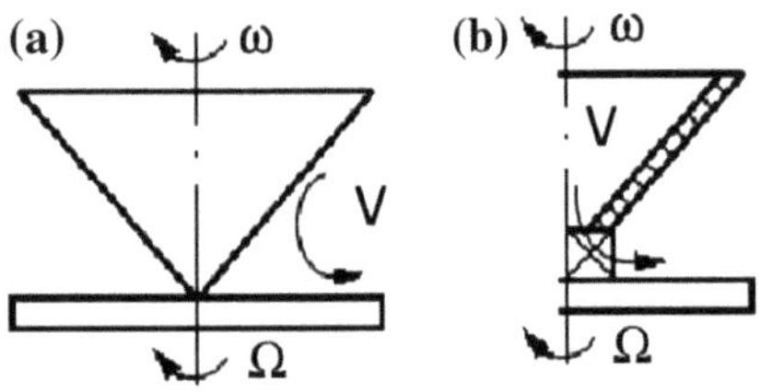

Fig. 1.2 Rotating flow between a rotating disk and/or a cone (**a**), and **b** swirl flow in a stationary conical diffuser [2]

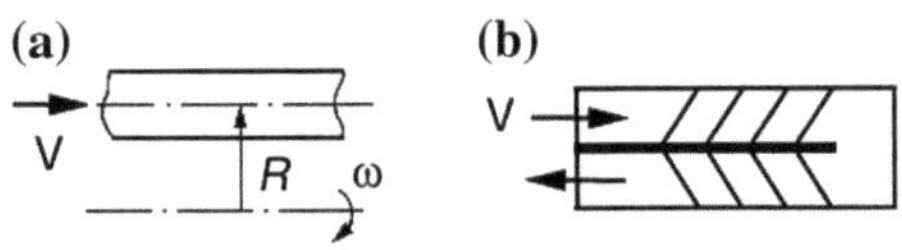

Fig. 1.3 Pipe rotating about a parallel axis (**a**); two-pass ribbed channel with a 180° bend (**b**)

ratios of the inlet and outlet passes of such channels, design of the bend region itself, location and height of the ribs on fluid flow, and heat transfer are also studied in detail in this work.

1.2 Volume Forces and Their Description

Fluid particles are subject to effects of volume forces and surface forces [3]. Volume forces affect each elementary volume of fluid (fluid particle). Inertia, gravity, magnetic, electric forces, etc. are classified as volume forces. Surface forces act on elementary surface segments. Pressure and frictional (viscous) forces are examples of surface forces.

Global gravity of the Earth engenders gravitational forces. Acceleration/deceleration of the configuration, where fluid flows, causes inertia forces. For instance, system rotation, as well as rotation of a fluid in a stationary configuration, also results in inertia forces. A study of effects of electromagnetic forces is not an objective of the present work.

In rotating systems, inertia forces are caused by the motion of the configuration itself and peculiarities of fluid flow. Inertia forces can also result from streamline curvature in rotating flows in a stationary configuration (for instance, swirl or curvilinear flows). In this case, the direction and strength of the inertia forces are determined by counteraction of the velocity field with pressure and viscous forces.

Inertia and gravity forces can be described by an expression

$$\boldsymbol{F} = \boldsymbol{j}\rho, \tag{1.1}$$

where $\boldsymbol{j}$ is a volume force acceleration. For example, on the Earth surface, $\boldsymbol{j} = \boldsymbol{g}$ for gravity force. Volume forces are reduced here (and throughout this chapter) to a unit of volume; boldface symbols denote vectors.

Centrifugal forces are caused by system rotation or streamline curvature and are directed outwards and orthogonal to the axis of rotation

$$\boldsymbol{F}_{\mathrm{c}} = \rho\boldsymbol{\omega} \times (\boldsymbol{\omega} \times \boldsymbol{R}) = -\rho\boldsymbol{\omega} \cdot (\boldsymbol{R} \cdot \boldsymbol{\omega}) + \rho\boldsymbol{R}\omega^2. \tag{1.2}$$

Here the local radius vector $\boldsymbol{R}$ of a fluid particle is counted relative to the rotation axis; the symbol × stands for a vector product. Vectors $\boldsymbol{R}$ and $\boldsymbol{\omega}$ are perpendicular to each other, therefore the scalar product $\boldsymbol{R} \cdot \boldsymbol{\omega}$ is equal to zero. System rotation is absent in curvilinear or swirl flows, hence a local rotation velocity at each location can be written as $\boldsymbol{\omega} = \boldsymbol{V}/\boldsymbol{R}$, and Eq. (1.2) reduces to

$$\boldsymbol{F}_{\mathrm{c}} = \rho\boldsymbol{R}(\boldsymbol{V}/\boldsymbol{R})^2 = \rho\boldsymbol{V}^2/\boldsymbol{R}, \tag{1.3}$$

where $\boldsymbol{V}$ is the relative velocity, i.e., fluid velocity relative to the considered configuration.

Coriolis forces emerge in rotating systems, where the vectors of the relative velocity $\boldsymbol{V}$ and angular velocity of rotation $\boldsymbol{\omega}$ are not parallel to each other. In a rotating coordinate system [2, 4, 5], Coriolis force is written as

$$\boldsymbol{F}_{\mathrm{Cor}} = -2\rho\boldsymbol{\omega} \times \boldsymbol{V}. \tag{1.4}$$

Coriolis force is orthogonal to a surface, where the vectors $\boldsymbol{\omega}$ and $\boldsymbol{V}$ are located. If the origins of the vectors $\boldsymbol{F}_{\mathrm{Cor}}$, $\boldsymbol{\omega}$, and $\boldsymbol{V}$ are matched, the Coriolis force is directed toward the point from which the shortest turn from $\boldsymbol{\omega}$ to $\boldsymbol{V}$ would be seen counterclockwise.

Volume forces being often a main cause of a particular type of flow can also (a) engender secondary flows (in form of vortex or recirculation flows) or (b) render a stabilizing/destabilizing effect on the fluid. The latter effects can exhibit themselves only if volume forces undergo spatial variation in frames of the given configuration. *An excessive volume force* is a difference of the volume forces between two points of the configuration

$$\Delta\boldsymbol{F} = \boldsymbol{F}_2 - \boldsymbol{F}_1 = \rho_2\boldsymbol{j}_2 - \rho_1\boldsymbol{j}_1. \tag{1.5}$$

Nonuniformity in the distribution of density and/or volume force acceleration in the configuration engender the excessive volume force.

Shchukin [5] wrote that "the character of fluid flow can be affected only by the volume forces, whose value is different from the pressure gradient caused by these volume forces and counteracting with them." In other words, the difference between the volume force and pressure gradient simultaneously acting in the configuration is in reality the excessive volume force affecting the flow.

All physical processes on the Earth are subject to the gravitational field said to be simple. If volume forces of different nature act simultaneously in the system, the field of volume forces is complex. As compared to inertia forces, the gravitational force is very often insignificant being therefore neglected in physical models.

The surface restricting the flow pattern can be located at different angles with respect to the volume force vector. Also, volume force fields can be steady and unsteady. If the inequality $\mathrm{grad}|\boldsymbol{F}| > 0$ characterizes the volume force field, its effects on fluid flow are conservative (stabilizing the flow, suppressing turbulence, or sporadic perturbations). Once $\mathrm{grad}|\boldsymbol{F}| < 0$, effects of the volume forces on fluid flow are active (disturbing it, causing secondary flows and increasing turbulence) [2, 5].

1.3 Differential Equations of Continuity, Momentum, and Heat Transfer

Mathematical modeling of any physical process requires stating a boundary problem. In frames of the methodology used in the present work, this means writing differential equations describing momentum and heat/mass transfer, continuity equation, equation of state, as well as proper boundary and initial conditions.

In a rotating coordinate system, for incompressible subsonic flow of a fluid with constant physical properties and negligible viscous dissipation effects, the equations of momentum transfer and continuity can be written in a vector form [2, 4, 6]

$$\rho\frac{\mathrm{D}\boldsymbol{V}}{\mathrm{D}t} = \rho\Bigg[\underbrace{\frac{\partial \boldsymbol{V}}{\partial t}}_{\mathrm{I}} + \underbrace{(\boldsymbol{V}\mathrm{grad})\boldsymbol{V}}_{\mathrm{II}}\Bigg] = \underbrace{\rho\boldsymbol{F}}_{\mathrm{III}} \underbrace{-\,\mathrm{grad}\,\boldsymbol{p}}_{\mathrm{IV}} + \underbrace{\mathrm{div}\,\boldsymbol{\Pi}}_{\mathrm{V}} \underbrace{-\,2\rho\boldsymbol{\omega}\times\boldsymbol{V}}_{\mathrm{VI}} \underbrace{-\,\rho\boldsymbol{\omega}\times(\boldsymbol{\omega}\times\boldsymbol{R})}_{\mathrm{VII}}, \quad (1.6)$$

$$\mathrm{div}(\rho\boldsymbol{V}) = 0. \quad (1.7)$$

The value D/Dt in Eq. (1.6) is the total derivative incorporating local and convective derivatives, i.e., terms I and II, respectively (where term I is zero for steady state processes). Term III stands for the *volume forces except for centrifugal and Coriolis forces*. Terms IV and V represent pressure and friction effects, respectively. Here $\boldsymbol{\Pi}$ is the stress tensor including viscous and turbulent stresses. Terms VI and VII denote Coriolis $\boldsymbol{F}_{\mathrm{Cor}}$ and centrifugal $\boldsymbol{F}_{\mathrm{C}}$ forces, respectively. As above, the relative velocity vector $\boldsymbol{V}$ denotes flow velocity with respect to the coordinate system associated with the rotating configuration.

The tensor of stresses $\boldsymbol{\Pi}$ in Eqs. (1.1) and (1.6) has the following form:

$$\boldsymbol{\Pi} = \begin{pmatrix} \tau_{11} & \tau_{21} & \tau_{31} \\ \tau_{21} & \tau_{22} & \tau_{23} \\ \tau_{31} & \tau_{23} & \tau_{33} \end{pmatrix}, \tag{1.8}$$

where τ_{ik} are stress components including viscous and turbulent stresses.

In a nonrotating coordinate system, the vector $\boldsymbol{V}$ in Eqs. (1.6) and (1.7) denotes the absolute velocity, and terms VI and VII are neglected. In doing so, Lamé coefficients in a curvilinear coordinate system and turbulent viscosity model account for the centrifugal force effects.

For the volume force field $\boldsymbol{F}$ = grad A representing the potential A, one can introduce a so-called modified (reduced) pressure

$$\boldsymbol{p}^* = \boldsymbol{p} + \rho A - \frac{1}{2}\rho(\omega \times \boldsymbol{R})(\omega \times \boldsymbol{R})^2, \tag{1.9}$$

and Eq. (1.6) can be rewritten as

$$\frac{\partial \boldsymbol{V}}{\partial t} + (\boldsymbol{V}\mathrm{grad})\boldsymbol{V} + 2\rho\omega \times \boldsymbol{V} = \rho \boldsymbol{F} - \frac{1}{\rho}\mathrm{grad}\,\boldsymbol{p}^* - \frac{1}{\rho}\mathrm{div}\,\boldsymbol{\Pi}. \tag{1.10}$$

In a *Cartesian coordinate system*, Navier–Stokes, continuity and energy equations for incompressible turbulent flow with constant fluid properties with account for the volume forces can be written as [6]

$$\rho\left(\frac{\partial u}{\partial t} + u\frac{\partial u}{\partial x} + v\frac{\partial u}{\partial y} + w\frac{\partial u}{\partial z}\right) = F_x - \frac{\partial p}{\partial x} + \mu\left(\frac{\partial^2 u}{\partial x^2} + \frac{\partial^2 u}{\partial y^2} + \frac{\partial^2 u}{\partial z^2}\right) - \rho\left(\frac{\partial \overline{u'^2}}{\partial x} + \frac{\partial \overline{u'v'}}{\partial y} + \frac{\partial \overline{u'w'}}{\partial z}\right), \tag{1.11}$$

$$\rho\left(\frac{\partial v}{\partial t} + u\frac{\partial v}{\partial x} + v\frac{\partial v}{\partial y} + w\frac{\partial v}{\partial z}\right) = F_y - \frac{\partial p}{\partial y} + \mu\left(\frac{\partial^2 v}{\partial x^2} + \frac{\partial^2 v}{\partial y^2} + \frac{\partial^2 v}{\partial z^2}\right) - \rho\left(\frac{\partial \overline{v'^2}}{\partial y} + \frac{\partial \overline{u'v'}}{\partial x} + \frac{\partial \overline{v'w'}}{\partial z}\right), \tag{1.12}$$

$$\rho\left(\frac{\partial w}{\partial t}+u\frac{\partial w}{\partial x}+v\frac{\partial w}{\partial y}+w\frac{\partial w}{\partial z}\right)=F_z-\frac{\partial p}{\partial z}+\mu\left(\frac{\partial^2 w}{\partial x^2}+\frac{\partial^2 w}{\partial y^2}+\frac{\partial^2 w}{\partial z^2}\right)$$
$$-\rho\left(\frac{\partial\overline{w'^2}}{\partial z}+\frac{\partial\overline{u'w'}}{\partial x}+\frac{\partial\overline{v'w'}}{\partial y}\right), \quad (1.13)$$

$$\frac{\partial u}{\partial x}+\frac{\partial v}{\partial y}+\frac{\partial w}{\partial z}=0, \quad (1.14)$$

$$\frac{\partial T}{\partial t}+u\frac{\partial T}{\partial x}+v\frac{\partial T}{\partial y}+w\frac{\partial T}{\partial z}=a\left(\frac{\partial^2 T}{\partial x^2}+\frac{\partial^2 T}{\partial y^2}+\frac{\partial^2 T}{\partial z^2}\right)$$
$$-\left(\frac{\partial\overline{u'T'}}{\partial x}+\frac{\partial\overline{v'T'}}{\partial y}+\frac{\partial\overline{w'T'}}{\partial z}\right). \quad (1.15)$$

In a *cylindrical polar coordinate system*, Navier–Stokes, continuity and energy equations for incompressible turbulent flow with constant fluid properties including volume forces can be written as [2, 7, 8]

$$\rho\left(\frac{\partial v_r}{\partial t}+v_r\frac{\partial v_r}{\partial r}+\frac{v_\varphi}{r}\frac{\partial v_r}{\partial\varphi}-\frac{v_\varphi^2}{r}+v_z\frac{\partial v_r}{\partial z}\right)=F_r-\frac{\partial p}{\partial r}+\mu\left(\nabla^2 v_r-\frac{v_r}{r^2}-\frac{2}{r^2}\frac{\partial v_\varphi}{\partial\varphi}\right)$$
$$-\rho\left[\frac{1}{r}\frac{\partial}{\partial r}\left(r\overline{v_r'^2}\right)+\frac{1}{r}\frac{\partial}{\partial\varphi}\left(\overline{v_r'v_\varphi'}\right)\right.$$
$$\left.+\frac{\partial}{\partial z}\left(\rho\overline{v_r'v_z'}\right)-\frac{1}{r}\left(\overline{v_\varphi'^2}\right)\right], \quad (1.16)$$

$$\rho\left(\frac{\partial v_\varphi}{\partial t}+v_r\frac{\partial v_\varphi}{\partial r}+\frac{v_\varphi}{r}\frac{\partial v_\varphi}{\partial\phi}+\frac{v_r v_\varphi}{r}+v_z\frac{\partial v_\varphi}{\partial z}\right)=F_\varphi-\frac{\partial p}{\partial\varphi}+\mu\left(\nabla^2 v_\varphi+\frac{2}{r^2}\frac{\partial v_r}{\partial\varphi}-\frac{v_\varphi}{r^2}\right)$$
$$-\rho\left[\frac{1}{r^2}\frac{\partial}{\partial r}\left(r^2\overline{v_r'v_\varphi'}\right)\right.$$
$$\left.+\frac{1}{r}\frac{\partial}{\partial\varphi}\left(\overline{v_\varphi'^2}\right)+\frac{\partial}{\partial z}\left(\overline{v_\varphi'v_z'}\right)\right], \quad (1.17)$$

$$\rho\left(\frac{\partial v_z}{\partial t}+v_r\frac{\partial v_z}{\partial r}+\frac{v_\varphi}{r}\frac{\partial v_z}{\partial\varphi}+v_z\frac{\partial v_z}{\partial z}\right)=F_z-\frac{\partial p}{\partial z}+\mu(\nabla^2 v_z)$$
$$-\rho\left[\frac{1}{r}\frac{\partial}{\partial r}\left(r\overline{v_r'v_z'}\right)\right.$$
$$\left.+\frac{1}{r}\frac{\partial}{\partial\varphi}\left(\overline{v_\varphi'v_z'}\right)+\frac{\partial}{\partial z}\left(\overline{v_z'^2}\right)\right], \quad (1.18)$$

$$\frac{\partial(rv_r)}{\partial r}+\frac{1}{r}\frac{\partial(rv_\varphi)}{\partial\varphi}+\frac{\partial(rv_z)}{\partial z}=0, \tag{1.19}$$

$$\begin{aligned}\frac{\partial T}{\partial t}+v_r\frac{\partial T}{\partial r}+\frac{v_\varphi}{r}\frac{\partial T}{\partial\varphi}+v_z\frac{\partial T}{\partial z}=\frac{1}{r}\frac{\partial}{\partial r}\left[r\left(a\frac{\partial T}{\partial r}-\overline{v_r'T'}\right)\right]\\+\frac{1}{r^2}\frac{\partial}{\partial\varphi}\left(a\frac{\partial T}{\partial\varphi}-r\overline{v_\varphi'T'}\right)+\frac{\partial}{\partial z}\left(a\frac{\partial T}{\partial z}-\overline{v_z'T'}\right).\end{aligned} \tag{1.20}$$

If the fluid flow is steady state and axisymmetric, while heat transfer is unsteady, all derivatives with respect to the φ-coordinate, as well as derivatives with respect to time in Eqs. (1.16)–(1.18) are equal to zero: $\partial/\partial\varphi\equiv\partial/\partial t\equiv 0$. As a consequence, Eqs. (1.16)–(1.20) can be rewritten as

$$\begin{aligned}v_r\frac{\partial v_r}{\partial r}-\frac{v_\varphi^2}{r}+v_z\frac{\partial v_r}{\partial z}=\frac{1}{\rho}F_r-\frac{1}{\rho}\frac{\partial p}{\partial r}+\nu\left(\frac{\partial^2 v_r}{\partial r^2}+\frac{1}{r}\frac{\partial v_r}{\partial r}-\frac{v_r}{r^2}+\frac{\partial^2 v_r}{\partial z^2}\right)\\-\left[\frac{1}{r}\frac{\partial}{\partial r}\left(r\overline{v_r'^2}\right)+\frac{\partial}{\partial z}\left(\rho\overline{v_r'v_z'}\right)-\frac{1}{r}\left(\overline{v_\varphi'^2}\right)\right],\end{aligned} \tag{1.21}$$

$$\begin{aligned}v_r\frac{\partial v_\varphi}{\partial r}+\frac{v_rv_\varphi}{r}+v_z\frac{\partial v_\varphi}{\partial z}=\frac{1}{\rho}F_\varphi+\nu\left(\frac{\partial^2 v_\varphi}{\partial r^2}+\frac{1}{r}\frac{\partial v_\varphi}{\partial r}-\frac{v_\varphi}{r^2}+\frac{\partial^2 v_\varphi}{\partial z^2}\right)\\-\left[\frac{1}{r^2}\frac{\partial}{\partial r}\left(r^2\overline{v_r'v_\varphi'}\right)+\frac{\partial}{\partial z}\left(\overline{v_\varphi'v_z'}\right)\right],\end{aligned} \tag{1.22}$$

$$\begin{aligned}v_r\frac{\partial v_z}{\partial r}+v_z\frac{\partial v_z}{\partial z}=\frac{1}{\rho}F_z-\frac{1}{\rho}\frac{\partial p}{\partial z}+\nu\left(\frac{\partial^2 v_z}{\partial r^2}+\frac{1}{r}\frac{\partial v_z}{\partial r}+\frac{\partial^2 v_z}{\partial z^2}\right)\\-\left[\frac{1}{r}\frac{\partial}{\partial r}\left(r\overline{v_r'v_z'}\right)+\frac{\partial}{\partial z}\left(\overline{v_z'^2}\right)\right],\end{aligned} \tag{1.23}$$

$$\frac{\partial(rv_r)}{\partial r}+\frac{\partial(rv_z)}{\partial z}=0, \tag{1.24}$$

$$\begin{aligned}\frac{\partial T}{\partial t}+v_r\frac{\partial T}{\partial r}+v_z\frac{\partial T}{\partial z}=\frac{1}{r}\frac{\partial}{\partial r}\left[r\left(a\frac{\partial T}{\partial r}-\overline{v_r'T'}\right)\right]\\+\frac{\partial}{\partial z}\left(a\frac{\partial T}{\partial z}-\overline{v_z'T'}\right).\end{aligned} \tag{1.25}$$

For laminar flow, all terms including fluctuating velocity/temperature components in Eqs. (1.11)–(1.25) are zero.

Equations (1.11)–(1.25) will be transformed to a rotating coordinate system in the subsequent chapters individually for each rotating configuration.

1.4 Differential Equation of Convective Diffusion

The differential equation describing convective diffusion in a fluid looks analogous to the energy equation. The difference consists in that the concentration C in the equation substitutes the temperature, while the diffusion coefficient D_m replaces the thermal diffusivity coefficient. Convective diffusion equations in Cartesian and cylindrical polar coordinate systems are written below.

In the *Cartesian coordinate system*, the convective diffusion equation for incompressible turbulent flow with constant physical properties of the substance taking part in the convective diffusion process has the following form:

$$\frac{\partial C}{\partial t}+u\frac{\partial C}{\partial x}+v\frac{\partial C}{\partial y}+w\frac{\partial C}{\partial z}=D_m\left(\frac{\partial^2 C}{\partial x^2}+\frac{\partial^2 C}{\partial y^2}+\frac{\partial^2 C}{\partial z^2}\right)-\left(\frac{\partial\overline{u'C'}}{\partial x}+\frac{\partial\overline{v'C'}}{\partial y}+\frac{\partial\overline{w'C'}}{\partial z}\right). \quad (1.26)$$

In a *cylindrical polar coordinate system*, the convective diffusion equation for incompressible turbulent flow with constant physical properties of the substance looks as follows:

$$\frac{\partial C}{\partial t}+v_r\frac{\partial C}{\partial r}+\frac{v_\varphi}{r}\frac{\partial C}{\partial \varphi}+v_z\frac{\partial C}{\partial z}=\frac{1}{r}\frac{\partial}{\partial r}\left[r\left(D_m\frac{\partial C}{\partial r}-\overline{v_r'C'}\right)\right]+\frac{1}{r^2}\frac{\partial}{\partial\varphi}\left(D_m\frac{\partial C}{\partial\varphi}-r\overline{v_\varphi'C'}\right)+\frac{\partial}{\partial z}\left(D_m\frac{\partial C}{\partial z}-\overline{v_z'C'}\right). \quad (1.27)$$

Once the fluid flow is steady state and axisymmetric, whereas mass transfer is unsteady, all φ-derivatives in Eq. (1.27) are equal to zero.

$$\frac{\partial C}{\partial t}+v_r\frac{\partial C}{\partial r}+v_z\frac{\partial C}{\partial z}=\frac{1}{r}\frac{\partial}{\partial r}\left[r\left(D_m\frac{\partial C}{\partial r}-\overline{v_r'C'}\right)\right]+\frac{\partial}{\partial z}\left(D_m\frac{\partial C}{\partial z}-\overline{v_z'C'}\right). \quad (1.28)$$

Again, for laminar flow, terms containing only fluctuating velocity/concentration components in Eqs. (1.26)–(1.28) are zero.

The convective diffusion equation is employed in Chap. 6 while modeling convective heat/mass transfer for the Prandtl or Schmidt numbers larger than unity.

References

1. Childs PRN (2011) Rotating flow. Elsevier Inc, Amsterdam
2. Shevchuk IV (2009) Convective heat and mass transfer in rotating disk systems. Springer, Berlin, Heidelberg

3. Loitsyanskii LG (1966) Mechanics of liquids and gases. Pergamon, Oxford, UK
4. Filippov IF (1986) Heat transfer in electrical machines. Energoatomizdat, Leningrad, USSR (in Russian)
5. Shchukin VK (1980) Heat transfer and hydrodynamics of pipe flows in mass-force fields. Izdatel'stvo Mashinostroenie, Moscow, USSR (in Russian)
6. Schlichting G (1968) Boundary-layer theory. McGraw-Hill Book Company, New York
7. Hinze JO (1975) Turbulence. McGraw-Hill Inc, New York
8. Munson BR, Young DF, Okiishi TH, Huebsch WW (2009) Fundamentals of fluid mechanics. Wiley, New York

Chapter 2
Mathematical Modeling of Convective Heat Transfer in Rotating-Disk Systems

2.1 Differential and Integral Equations

2.1.1 Navier–Stokes and Energy Equations in Differential Form

A schematic of a stationary axisymmetric problem of convective heat transfer over rotating disks, whose axis of symmetry serves as the axis z of a stationary cylindrical coordinate system with the point $z = 0$ placed on the disk surface, is depicted in Fig. 2.1. The angular velocity is high, so that gravitational effects are negligible, i.e., $F_r = F_\varphi = F_z = 0$.

Thus, Eqs. (1.21)–(1.25) are reduced [1–3] to

$$v_r \frac{\partial v_r}{\partial r} + v_z \frac{\partial v_r}{\partial z} - \frac{v_\varphi^2}{r} = -\frac{1}{\rho}\frac{\partial p}{\partial r} + \nu\left(\nabla^2 v_r - \frac{v_r}{r^2}\right) - \left[\frac{1}{r}\frac{\partial}{\partial r}\left(r\overline{v_r'^2}\right) + \frac{\partial}{\partial z}\left(\rho\overline{v_r' v_z'}\right) - \frac{1}{r}\left(\overline{v_\varphi'^2}\right)\right], \tag{2.1}$$

$$v_r \frac{\partial v_\varphi}{\partial r} + \frac{v_r v_\varphi}{r} + v_z \frac{\partial v_\varphi}{\partial z} = \nu\left(\frac{\partial^2 v_\varphi}{\partial r^2} + \frac{1}{r}\frac{\partial v_\varphi}{\partial r} - \frac{v_\varphi}{r^2} + \frac{\partial^2 v_\varphi}{\partial z^2}\right) - \left[\frac{1}{r^2}\frac{\partial}{\partial r}\left(r^2\overline{v_r' v_\varphi'}\right) + \frac{\partial}{\partial z}\left(\overline{v_\varphi' v_z'}\right)\right], \tag{2.2}$$

$$v_r \frac{\partial v_z}{\partial r} + v_z \frac{\partial v_z}{\partial z} = -\frac{1}{\rho}\frac{\partial p}{\partial z} + \nu\left(\frac{\partial^2 v_z}{\partial r^2} + \frac{1}{r}\frac{\partial v_z}{\partial r} + \frac{\partial^2 v_z}{\partial z^2}\right) - \left[\frac{1}{r}\frac{\partial}{\partial r}\left(r\overline{v_r' v_z'}\right) + \frac{\partial}{\partial z}\left(\overline{v_z'^2}\right)\right], \tag{2.3}$$

I.V. Shevchuk, *Modelling of Convective Heat and Mass Transfer in Rotating Flows*, Mathematical Engineering, DOI 10.1007/978-3-319-20961-6_2

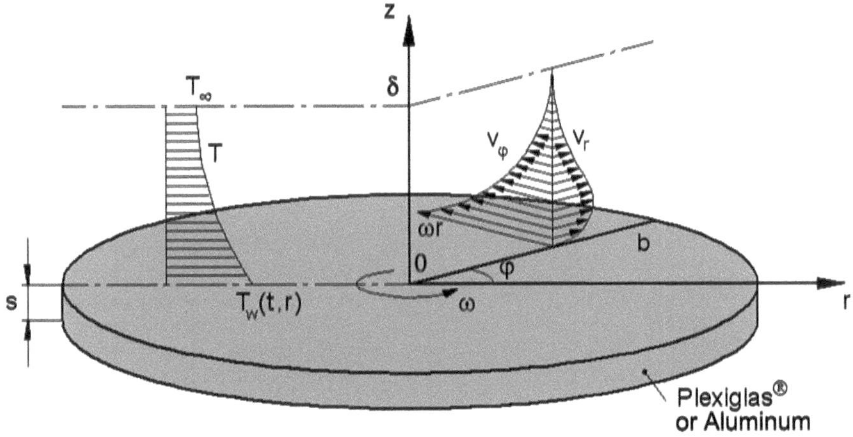

Fig. 2.1 Geometrical arrangement and main parameters of the problem of fluid flow and heat transfer over a rotating disk in still air [3]

$$\frac{\partial v_r}{\partial r} + \frac{v_r}{r} + \frac{\partial v_z}{\partial z} = 0, \tag{2.4}$$

$$\frac{\partial T}{\partial t} + v_r \frac{\partial T}{\partial r} + v_z \frac{\partial T}{\partial z} = \frac{1}{r}\frac{\partial}{\partial r}\left[r \left(a\frac{\partial T}{\partial r} - \overline{v_r' T'} \right) \right] + \frac{\partial}{\partial z}\left(a \frac{\partial T}{\partial z} - \overline{v_z' T'} \right). \tag{2.5}$$

One can assign the coordinate system in Fig. 2.1 to be rotating together with the disk. In doing so, Eqs. (2.1)–(2.3) for laminar flow can be re-written as [1–3]

$$v_r \frac{\partial v_r}{\partial r} + v_z \frac{\partial v_r}{\partial z} - \frac{v_\varphi^2}{r} - 2\omega v_\varphi - \omega^2 r = -\frac{1}{\rho}\frac{\partial p}{\partial r} + \nu \left(\frac{\partial^2 v_r}{\partial r^2} + \frac{1}{r}\frac{\partial v_r}{\partial r} - \frac{v_r}{r^2} + \frac{\partial^2 v_r}{\partial z^2} \right), \tag{2.6}$$

$$v_r \frac{\partial v_\varphi}{\partial r} + v_z \frac{\partial v_\varphi}{\partial z} + \frac{v_r v_\varphi}{r} + 2\omega v_r = \nu \left(\frac{\partial^2 v_\varphi}{\partial r^2} + \frac{1}{r}\frac{\partial v_\varphi}{\partial r} - \frac{v_\varphi}{r^2} + \frac{\partial^2 v_\varphi}{\partial z^2} \right), \tag{2.7}$$

$$v_r \frac{\partial v_z}{\partial r} + v_z \frac{\partial v_z}{\partial z} = -\frac{1}{\rho}\frac{\partial p}{\partial z} + \nu \left(\frac{\partial^2 v_z}{\partial r^2} + \frac{1}{r}\frac{\partial v_z}{\partial r} + \frac{\partial^2 v_z}{\partial z^2} \right). \tag{2.8}$$

The terms $2\omega v_\varphi$ and $2\omega v_r$ stand for the r- and φ-components of the Coriolis force, respectively. The term $\omega^2 r$ is the r-component of the centrifugal force (all divided by ρ). Equations (2.1)–(2.2) for turbulent flow can be derived in analogy to Eqs. (2.6)–(2.8) [1].

2.1.2 Differential Equations of the Boundary Layer

To simplify Eqs. (2.1)–(2.5) for boundary layers, the following assumptions are made [1, 2, 4]:

(a) velocity components v_r and v_φ are an order of magnitude larger than the v_z-velocity;
(b) velocity and temperature vary in the z-direction much more significantly than they do in the r-direction; and
(c) variation of the static pressure in z-direction is negligible.

The equation of continuity, Eq. (2.4), does not undergo any change. As a result, Eqs. (2.1)–(2.5) reduce to the following final form [1, 2, 4]:

$$v_r\frac{\partial v_r}{\partial r}+v_z\frac{\partial v_r}{\partial z}-\frac{v_\varphi^2}{r}=-\frac{1}{\rho}\frac{\partial p}{\partial r}+\frac{1}{\rho}\frac{\partial \tau_r}{\partial z}, \tag{2.9}$$

$$v_r\frac{\partial v_\varphi}{\partial r}+v_z\frac{\partial v_\varphi}{\partial z}+\frac{v_r v_\varphi}{r}=\frac{1}{\rho}\frac{\partial \tau_\varphi}{\partial z}, \tag{2.10}$$

$$\frac{1}{\rho}\frac{\partial p}{\partial z}=0, \tag{2.11}$$

$$\frac{\partial T}{\partial t}+v_r\frac{\partial T}{\partial r}+v_z\frac{\partial T}{\partial z}=-\frac{1}{\rho c_p}\frac{\partial q}{\partial z}, \tag{2.12}$$

$$\tau_r=\mu\frac{\partial v_r}{\partial z}-\rho\overline{v_r' v_z'}, \tag{2.13}$$

$$\tau_\varphi=\mu\frac{\partial v_\varphi}{\partial z}-\rho\overline{v_\varphi' v_z'}, \tag{2.14}$$

$$q=-\left(\lambda\frac{\partial T}{\partial z}-\rho c_p\overline{T' v_z'}\right). \tag{2.15}$$

The pressure across the boundary layer is constant and equal to the pressure in the potential flow region, i.e., $p = p_\infty$. Equations (2.13)–(2.15) include only the most significant turbulent shear stress and heat flux components.

For a stationary thermal boundary layer, the term $\partial T/\partial t$ in Eq. (2.12) vanishes.

Equations (2.9)–(2.15) are closed with an equation of potential flow, where functions $v_{r,\infty}$, $v_{\varphi,\infty}$, and p_∞ do not vary in the z-direction:

$$\frac{1}{2}\frac{\mathrm{d}v_{r,\infty}^2}{\mathrm{d}r}-\frac{v_{\varphi,\infty}^2}{r}=-\frac{1}{\rho}\frac{\mathrm{d}p_\infty}{\mathrm{d}r}. \tag{2.16}$$

2.1.3 Integral Equations of the Boundary Layer

For steady-state conditions, Eqs. (2.9)–(2.11), (2.13)–(2.20) with allowance for Eqs. (2.4) and (2.16) can be re-written in an integral form [1, 2, 4]:

$$\frac{\mathrm{d}}{\mathrm{d}r}\left[r\int\limits_0^{\delta} v_r\left(v_{r,\infty}-v_r\right)\mathrm{d}z\right]+r\frac{\mathrm{d}v_{r,\infty}}{\mathrm{d}r}\int\limits_0^{\delta}\left(v_{r,\infty}-v_r\right)\mathrm{d}z-\int\limits_0^{\delta}\left(v_{\varphi,\infty}^2-v_{\varphi}^2\right)\mathrm{d}z = r\tau_{wr}/\rho, \tag{2.17}$$

$$\frac{\mathrm{d}}{\mathrm{d}r}\left[r^2\int\limits_0^{\delta} v_r\left(v_{\varphi}-v_{\varphi,\infty}\right)\mathrm{d}z\right]+\frac{\dot{m}_d}{2\pi\rho}\frac{\mathrm{d}}{\mathrm{d}r}\left(rv_{\varphi,\infty}\right)=-r^2\tau_{w\varphi}/\rho, \tag{2.18}$$

or

$$\frac{\mathrm{d}}{\mathrm{d}r}\left[r^2\int\limits_0^{\delta} v_r v_{\varphi}\mathrm{d}z\right]+rv_{\varphi,\infty}\frac{\mathrm{d}}{\mathrm{d}r}\left(\frac{\dot{m}_d}{2\pi\rho}\right)=-r^2\tau_{w\varphi}/\rho, \tag{2.19}$$

$$\frac{\mathrm{d}}{\mathrm{d}r}\left[r\int\limits_0^{\delta_T} v_r(T-T_{\infty})\mathrm{d}z\right]+\frac{\mathrm{d}T_{\infty}}{\mathrm{d}r}\cdot\frac{\dot{m}_{d,T}}{2\pi\rho}=rq_w/(\rho c_p). \tag{2.20}$$

Another notation of Eqs. (2.17), (2.18) and (2.20) looks as [1, 2, 4, 5]

$$\frac{\mathrm{d}}{\mathrm{d}r}\left(v_{r,\infty}^2 r\delta\overline{\delta}_r^{**}\right)+v_{r,\infty}r\delta\frac{\mathrm{d}v_{r,\infty}}{\mathrm{d}r}\overline{\delta}_r^{*}-v_{\varphi,\infty}^2\delta\overline{\delta}_{\varphi}^{**}=r\tau_{wr}/\rho, \tag{2.21}$$

$$\frac{\mathrm{d}}{\mathrm{d}r}\left[\delta r^2(\omega r)^2\overline{\delta}_{\varphi r}^{**}\right]+\frac{\dot{m}_d}{2\pi\rho}\frac{\mathrm{d}}{\mathrm{d}r}\left(rv_{\varphi,\infty}\right)=-r^2\tau_{w\varphi}/\rho, \tag{2.22}$$

$$\frac{\mathrm{d}}{\mathrm{d}r}\left[\omega r^2\delta\overline{\delta}_T^{**}(T_w-T_{\infty})\right]+\frac{\mathrm{d}T_{\infty}}{\mathrm{d}r}\cdot\frac{\dot{m}_{d,T}}{2\pi\rho}=rq_w/(\rho c_p), \tag{2.23}$$

where

$$\overline{\delta}_r^{*}=\int\limits_0^1(1-\tilde{v}_r)\mathrm{d}\xi,\quad \overline{\delta}_r^{**}=\int\limits_0^1\tilde{v}_r(1-\tilde{v}_r)\mathrm{d}\xi,\quad \overline{\delta}_{\varphi}^{**}=\int\limits_0^1\left(1-\frac{v_{\varphi}^2}{v_{\varphi,\infty}^2}\right)\mathrm{d}\xi, \tag{2.24}$$

$$\overline{\delta}_{\varphi r}^{**}=\int\limits_0^1\frac{v_r\left(v_{\varphi}-v_{\varphi,\infty}\right)}{(\omega r)^2}\mathrm{d}\xi,\quad \tilde{v}_r=v_r/v_{r,\infty}. \tag{2.25}$$

2.2 Methods of Solution

2.2.1 Self-similar Solution

Exact solutions of the Navier–Stokes and energy equations were found for a free rotating disk subject to laminar flow [1, 2, 4, 6–12]. For this purpose, self-similar variables F, G, H, P, and ζ were employed:

$$\begin{array}{lll} v_r = (a+\omega)rF(\zeta), & v_\varphi = (a+\omega)rG(\zeta), & v_z = \sqrt{(a+\omega)\nu}H(\zeta), \\ p = -\rho\nu\omega P(\zeta), & \theta = (T - T_\infty)/(T_w - T_\infty), & \zeta = z\sqrt{(a+\omega)/\nu}. \end{array} \tag{2.26}$$

The respective boundary conditions had the following form:

$$\zeta \to \infty:\ v_{r,\infty} = ar, \quad v_{z,\infty} = -2az, \quad v_{\varphi,\infty} = \Omega r, \quad \beta = \Omega/\omega = \text{const.}, \quad \theta = 0, \tag{2.27}$$

$$\zeta = 0:\ F = H = 0, \quad G = 1, \quad \theta = 1, \tag{2.28}$$

$$\zeta = 0:\ T_w = T_{\text{ref}} + c_{0w}r^{n_*}, \quad T_\infty = T_{\text{ref}} + c_{0\infty}r^{n_*} \text{ or } T_\infty = T_{\text{ref}} + \beta c_{0w}r^{n_*}. \tag{2.29}$$

Here, c_0, c_{0w}, $c_{0\infty}$, and n_* are the empirical constants. Equation (2.29) can be re-written as

$$\Delta T = T_w - T_\infty = c_0 r^{n_*} \quad (\text{for } c_0 = c_{0w} - c_{0\infty}), \tag{2.30}$$

$$\text{or} \quad \Delta T = c_{0w}(1-\beta)r^{n_*}. \tag{2.31}$$

Equations (2.1)–(2.4) and (2.12) (for $\partial T/\partial t = 0$), with allowance for Eq. (2.16), reduce to a self-similar form:

$$F^2 - G^2 + F'H = \frac{N^2 - \beta^2}{(1+N)^2} + F'', \tag{2.32}$$

$$2FG + G'H = G'', \tag{2.33}$$

$$HH' = P' + H'', \tag{2.34}$$

$$2F + H' = 0, \tag{2.35}$$

$$\theta'' - Pr(n_* F\theta + H\theta') = 0. \tag{2.36}$$

Here, $N = a/\omega$ = const. A solution of Eqs. (2.32)–(2.35) for simultaneously non-zero values of β and N does not exist. However, such a solution can be found either for $N \neq 0$ and $\beta = 0$, or for $\beta \neq 0$ and $N = 0$.

Equations (2.32)–(2.36) have been often solved with the help of so-called in-house computer codes using a spectral collocation method based on the Chebyshev polynomials [13–18], Keller box [19] or quasi-linearization method [20], expansions in power/exponential series [9, 21], finite difference schemes [22], shooting methods [1, 8, 10, 12, 23], etc. Computer mathematics softwares like Mathcad, Mathlab, Mathematica, etc. enable solving Eqs. (2.32)–(2.36) via user interface programing options [3, 20].

A self-similar energy equation involving dissipation terms allows using only one value of the exponent $n_* = 2$ in the boundary conditions (2.29)–(2.31) [1, 2, 4]. At subsonic flow of air, dissipation effects, as well as radial heat conduction, are negligible. Therefore, we neglected the respective terms in Eq. (2.36) of the thermal boundary layer, which enabled us using arbitrary values of the parameter n_*.

Exact solutions of Eqs. (2.32)–(2.36) serve as benchmark datasets used in validations of experiments or CFD models developed for more complicated problems. Based on the self-similar solutions, it is also possible to develop approximate analytical solutions of problems, whose boundary conditions differ from Eqs. (2.27)–(2.31).

2.2.2 *Approximate Analytical Methods for Laminar Flow*

Laminar impingement flow over a single rotating disk at N = const. and $\beta = 0$ was simulated using an approximate mathematical method of Slezkin-Targ in [4]. Velocity components were approximated by sixth-order polynomials. A polynomial of third order resulted in an inaccuracy in the surface friction of up to $\sim$25 % at $N = 5$. This inaccuracy increases fast for higher values of N. Should the author [4] extend this method to model heat transfer? This would yield a cumbersome solution for the Nusselt number.

A complex combination of exponential and logarithmic functions resulted in an approximate solution for laminar flow over a single rotating disk [24]. The heat transfer problem was not solved. Such an extension of the method [24] would, however, yield even more inconvenient and cumbersome relations for the Nusselt number than in [4].

For porous injection through a rotating disk, an approximate solution was presented as a combined expansion in power and exponential series. It is obvious that this approach has the same deficiencies as the aforementioned methods [4, 24].

Analytical solutions [19] were obtained for a stretching disk for (a) a case of no rotation and (b) infinitely large stretching rate. Both situations have very limiting application; a general analytical solution for a stretching rotating disk does not exist.

Based on the above, one can conclude that a search for an exact analytical solution for the velocity, pressure, and temperature profiles in laminar flow over a rotating disk is a very complicated and inexpedient mathematical task. Alternatively, as demonstrated below, a match of an integral method and a self-similar solution yields a transparent and accurate approximate analytical solution for fluid flow and heat transfer characteristics.

2.2.3 *Numerical Methods*

At early stages, finite difference methods implemented in in-house codes were used by different authors [25–38] to simulate laminar/turbulent fluid flow and heat transfer in rotating cavities formed by parallel co-rotating disks using algebraic [39] or low-Reynolds-number k-ε turbulence models [40–42]. A finite difference method was employed by the author [43] to simulate a 3D air flow in a rotating-disk grinder of solid particles with the RANS approach with a k-ε turbulence model [44].

Commercial CFD codes (e.g., FLUENT, CFX, Phoenics, etc.) using RANS approaches have been widely used by different authors to simulate fluid flow in rotating-disk systems [33, 36, 45–52]. Turbulence was modeled using closure with standard and realizable k-ε models, RNG k-ε model, k-ω SST model, Spalart–Allmaras model, and others.

The LES approach was employed by [53] to simulate a stationary turbulent flow over a rotating disk. The LES approach was also used in [54–58] to simulate turbulent flow and heat transfer over a single disk in air flow parallel to the disk surface.

Numerical simulation using in-house or commercial CFD codes is the most widely used universal tool for problems with arbitrary geometry and boundary conditions to be performed in academic and especially applied/industrial research. Given a proper mesh, accuracy of results depends here on the selection of the turbulence model, which is to be performed individually for each problem to be solved.

A disadvantage of CFD modeling is that it provides only an array of numerical data, which is often an inconvenience in comparison with analytical solutions. Therefore, methods delivering exact or approximate analytical solutions are advantageous for relatively simple geometries and boundary conditions.

2.3 Integral Methods

2.3.1 *Momentum Boundary Layer*

In frames of an integral method, Eqs. (2.17)–(2.23) are solved accompanied with models for (a) velocity/temperature profiles (or enthalpy thickness), as well as (b) shear stresses on the wall (velocity boundary layer) and wall heat flux (thermal boundary layer).

To briefly outline a history of the integral methods for rotating-disk systems, fundamentals of them were laid by von Karman [9] and Dorfman [4]. Further development of model assumptions for integral methods was done in the works [1, 2, 48, 59]. An important feature of the method [1, 2] further elaborated in the present work consists in the use of the same mathematical form of the models for laminar or turbulent flow, which differ from each other only by numerical values of certain parameters. In fact, this confirms the idea of Loytsyanskiy [60], who said that there exists "an analogy between basic characteristics of laminar and turbulent boundary layers."

The radial v_r and tangential v_φ velocity components in the boundary layer are interrelated in accordance with the equation [61]

$$\bar{v}_r = \bar{v}_\varphi \tan\varphi. \tag{2.37}$$

In case where potential flow in the r-direction is negligible, i.e., $v_{r,\infty} = 0$, approximations of the velocity profiles were written by the authors [1, 2] in the following form:

$$\bar{v}_\varphi = 1 - g(\xi), \quad \bar{v}_r = \alpha f(\xi), \tag{2.38}$$

where the functions $g(\xi)$ and $f(\xi)$ of the variable $\xi = z/\delta$ were set to be independent of the coordinate r. For laminar flow,

$$g(\xi) = G_0(\xi), \quad f(\xi) = F_0(\xi)/\alpha_0. \tag{2.39}$$

The functions $G_0(\xi)$ and $F_0(\xi)$ are a solution of Eqs. (2.32)–(2.35) for a free rotating disk, i.e., for $N = 0$ and $\beta = 0$ [1, 2, 4].

For turbulent flow, power-law profiles were employed:

$$g(\xi) = 1 - \xi^n, \tag{2.40}$$

$$f(\xi) = \xi^n(1 - \xi), \quad \tan\varphi = \alpha(1 - \xi), \tag{2.41}$$

where n = 1/5–1/10 [1, 2, 4, 9, 48, 59, 62–64]. Approximations (2.40) and (2.41) were formulated for the first time by von Karman [9]. The characteristic Reynolds number determines the value of the exponent n (see Figs. 2.2, 2.3 and 2.4).

A more accurate approximation for $f(\xi)$ in turbulent flow is [65–68]

$$f(\xi) = \xi^n(1 - \xi)^2, \quad \tan\varphi = \alpha(1 - \xi)^2. \tag{2.42}$$

Nevertheless, Eq. (2.42) has been rarely used apparently due to the somewhat more complicated form of expressions resulting from the integration of Eqs. (2.17)–(2.19).

More elaborate power-law profiles were used by the authors [69]

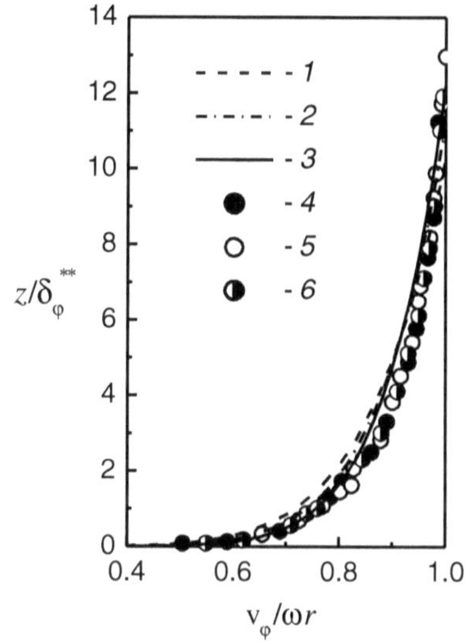

Fig. 2.2 Profiles of the non-dimensional tangential velocity in the turbulent boundary layer over a free rotating disk [3]. Calculation by Eq. (2.40) [61]: *1*—n = 1/7, *2*—1/8, *3*—1/9. Experiments: *4*—Re_ω = (0.4–1.6) × 10^6 [70], *5*—(0.6–1.0) × 10^6 [71], *6*—(0.65–1.0) × 10^6 [72]. Here $\delta_\varphi^{**} = \int_0^b \frac{v_\varphi}{\omega r}(1 - \frac{v_\varphi}{\omega r})\mathrm{d}z$ (definition of [70–74])

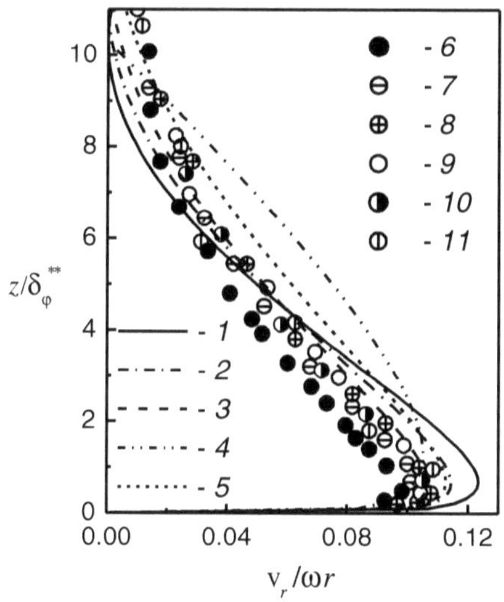

Fig. 2.3 Profiles of the non-dimensional radial velocity in the turbulent boundary layer over a free rotating disk [3]. Calculation by Eq. (2.42) or (2.58) [61]: *1*—n = 1/7, *2*—1/8, *3*—1/9. Equation (2.41), [9]: *4*—n = 1/7. Equation (2.44): *5*—n = 1/7, b = 0.7, c = 1.2, α = 0.2003. Experiments: *6*—Re_ω = 0.4 × 10^6, *7*—0.65 × 10^6, *8*—0.94 × 10^6, *9*—1.6 × 10^6 [70], *10*—0.6 × 10^6, *11*—1.0 × 10^6 [71]

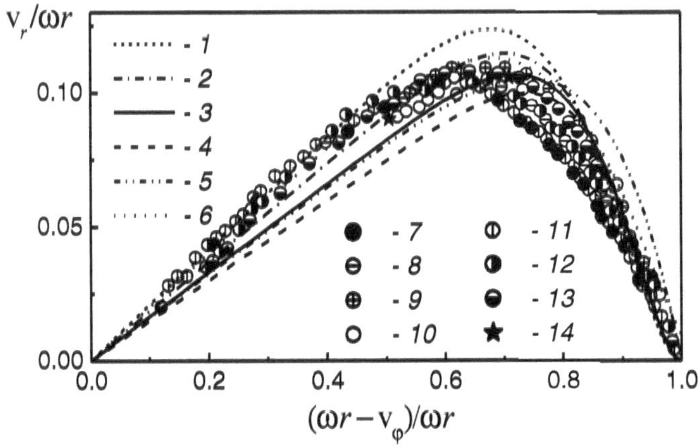

Fig. 2.4 Correlation between the radial and tangential velocity components in the boundary layer [3]. Calculation by Eq. (2.59) at $L = 2$ (curves *1–4*) or $L = 1$ (curve *5*) [61]: *1*—$n = 1/7$, *2*—1/8, *3*—1/9, *4*—1/10, *5*—1/7, von Karman's method [9], *6*—1/7, Eq. (2.44) for $b = 0.7$, $c = 1.2$, $\alpha = 0.2003$. Experiments: *7*—$Re_\omega = 0.4 \times 10^6$, *8*—$0.65 \times 10^6$, *9*—$0.94 \times 10^6$, *10*—$1.6 \times 10^6$ [70], *11*—0.4×10^6, *12*—0.6×10^6, *13*—1.0×10^6 [71], *14*—2.0×10^6 [75]

$$f(\xi) = \xi^n(1 - \xi^{n/m}), \quad \tan\varphi = \alpha(1 - \xi^{n/m}), \tag{2.43}$$

with exponents n and m independent from each other. The authors [69] have not further developed their model, apparently because of its excessively complicated structure.

A trigonometric function approximating the function $\tan\varphi$ in Eq. (2.37)

$$\tan\varphi = \alpha[(1 - \sin^b(c\xi)] \tag{2.44}$$

was used by [76, 77]. The values of the constants $b = 0.7$, $c = 0.12$ at $n = 1/7$, and $b = 0.697$, $c = 0.117$ at $n = 1/8$ mentioned in [76] are, however, erroneous.

For instance, for $b = 0.7$ at $n = 1/7$, one must use a value of $c = 1.2$ (Figs. 2.3 and 2.4). The model [76, 77] is more complicated than the von Karman's approach. Expressions for the Nusselt number that could have been obtained (but actually have not been obtained!) on the base of the model (2.44) would have been again too cumbersome.

For N = const., the following relations were used in [78] and [79–81], respectively:

$$\tan\varphi = \alpha + (N - \alpha)\xi, \tag{2.45}$$

$$\tan\varphi = \alpha(1 - \xi) + \kappa, \tag{2.46}$$

where $\kappa = \dot{m}/[2\pi\rho sr(1-\beta)\omega r]$. Equation (2.46) is least justified, since it does not agree with the condition $\tan\varphi_w = \alpha$ and complicates the solution of Eqs. (2.17) and (2.18).

Thus, Eqs. (2.43)–(2.46) demonstrate lower accuracy than models (2.37)–(2.45).

Integration of Eqs. (2.17) and (2.18) in view of Eqs. (2.37)–(2.45) yields ordinary differential equations with the unknown variables $\alpha(r)$ and $\delta(r)$ for a pre-set function of $\beta(r)$, or $\alpha(r)$ and $\beta(r)$ for a pre-set function $\delta(r)$. In view of an assumption N = const. or β = const., the parameter α becomes constant as well. In this case, δ = const. in laminar flow, or $\delta \sim r^m$ in turbulent flow [1, 2, 4].

Given the velocity profiles in the form of power-law functions, shear stresses τ_{wr} and $\tau_{w\varphi}$ on the right-hand sides of Eqs. (2.17)–(2.19) can be written as [1, 2, 4]

$$\tau_{wr} = -\alpha\tau_{w\varphi}, \quad \tau_{w\varphi} = -\mathrm{sgn}(1-\beta)\tau_w(1+\alpha^2)^{1/2}, \tag{2.47}$$

$$c_f = C_n^{-2/(n+1)} Re_{V_*}^{-2n/(n+1)}, \tag{2.48}$$

$$C_n = 2.28 + 0.924/n. \tag{2.49}$$

Equation (2.49) was proposed in [69]. The constant C_n takes the values 8.74, 9.71, 10.6, and 11.5 for n = 1/7, 1/8, 1/9, and 1/10, accordingly [1, 2, 4, 9, 69].

In frames of logarithmic models of the velocity profiles [82], their near-wall approximations look as

$$v_r = \alpha\omega r + \frac{2.5\alpha V_\tau}{(1+\alpha^2)^{1/2}}\ln(\xi), \quad v_\varphi = -\frac{2.5 V_\tau}{(1+\alpha^2)^{1/2}}\ln(\xi). \tag{2.50}$$

A validation of the logarithmic model has been performed only for a free disk, with the heat transfer problem being not modeled. The moment coefficient C_M is given by a transcendental algebraic equation (see Sect. 3.3) [82]. Inconvenience and complexity prevented further development and use of the logarithmic approach [82].

The integral method described in the work [48] and references includes special arrangements for rotor–stator systems, which fall out of the scope of the present work.

2.3.2 *Thermal Boundary Layer*

Heat transfer modeling in the frames of integral methods performed in the majority of the known works [1, 2, 4, 9, 48, 64, 68, 79–81, 83–89] was based on a "theory of local modelling" (which is a direct translation of the name used in the Russian language literature) that stems from the method of Loytsyanskiy [60] (see also [90]). This theory was for the first time applied to rotating-disk systems by Dorfman [4], who postulated a so-called heat transfer law for the Stanton number:

$$St = M_s Re_T^{**-\sigma} Pr^{-n_s}. \tag{2.51}$$

Universal constants M_s, σ, and n_s do not depend on the disk surface temperature T_w and the Prandtl number. These constants take the values $\sigma = 0.25$, $n_s = 0.5$, and $M_s = 7.246 \times 10^{-3}$ for turbulent flow, and $\sigma = 1.0$, $n_s = 1.0$, and $M_s = 0.07303$ for laminar flow [1–4]. Equation (2.51) is substituted into the thermal boundary layer Eq. (2.23). In doing so, the only remaining unknown parameter is δ_T^{**}.

In the books [1, 2], the Reynolds analogy parameter χ was involved in the integral method instead of the enthalpy thickness δ_T^{**}

$$\frac{q_w}{\tau_{w\phi}} = \chi \frac{c_p(T_\infty - T_w)}{\omega r(1 - \beta)}. \tag{2.52}$$

The unknown parameter χ was found as a result of the solution of Eq. (2.23) by the authors [1, 2] based on the models (2.51) and (2.52).

A power-law temperature profile in turbulent flow regime at $n_T = 1/5$

$$\Theta = \frac{T - T_w}{T_\infty - T_w} = \xi_T^{n_T}, \quad \theta = \frac{T - T_\infty}{T_w - T_\infty} = 1 - \Theta = 1 - \xi_T^{n_T} \tag{2.53}$$

was employed in the work [63], which for a long time had been the only one that used the model (2.53). An additional assumption $\Delta = \delta_T/\delta = 6$ at T_w = const. used in the work [63] is apparently erroneous and must be replaced with a model that enables finding the parameter Δ and its dependence on the other factors (like the model described in Sect. 2.4).

2.4 Improved Integral Method

2.4.1 Structure of the Method

Original results of the studies of fluid flow and heat transfer in rotating-disk configurations outlined here stem from the investigations performed using an improved integral method developed by the author of this work and described in the publications [3, 5, 61, 91–109]. Throughout this work, this methodology is always named as ***the present integral method***.

The basic statements of the present integral method are

(a) the system of Eqs. (2.17)–(2.23);
(b) turbulent velocity and temperature profiles given by improved approximations;
(c) a novel enthalpy thickness model for laminar/turbulent flow;
(d) power-law model for shear stresses and heat fluxes on the wall; and
(e) specified disk temperature distribution, together with the boundary conditions for the temperature and velocity in inviscid (i.e., potential) flow.

The present integral method employs the bedrock assumption that the same mathematical model can be used for modeling laminar and turbulent boundary layers, where the difference is made by numerical values of the certain empirical constants of the model. This model is a mathematical expression of the analogy between the basic characteristics of the laminar and turbulent flow under the same boundary conditions [60]. Authors [1, 2, 4, 9] have already validated this idea with respect to convective heat transfer in rotating-disk systems. However, the imperfect mathematical model used in these works caused noticeable inaccuracy in the simulation of heat transfer under certain thermal boundary conditions (see Sect. 3.2 of Chap. 3).

In the present integral method, we do not attempt to use power-law approximations of the velocity/temperature profiles in laminar flow, which involve polynomials of seventh order or higher and result in cumbersome equations for the friction coefficient and the Nusselt number. We wish to make use of simple and transparent power-law relations for the friction coefficient and the Nusselt number derived using power-law models of the velocity/temperature profiles for turbulent flow. Mathematical expressions for these parameters for turbulent flow can be extended onto laminar flow with particular constants remaining unknowns to be found empirically via validations against the exact solution.

Consequently, the logic of the method is following: firstly, an integral method for turbulent boundary layer is created and validated against experiments; and secondly, the mathematical form of the integral method is elaborated and validated for laminar flow.

2.4.2 Turbulent Flow: Velocity and Temperature Profiles

Velocity profiles are approximated using power-law models, Eq. (2.37) for v_r, as well as the first of Eqs. (2.38) and (2.40) for v_φ. A quadratic polynomial approximates the tangent of the flow swirl angle $\tan\varphi$. The coefficients a, b, and c must comply with the boundary conditions at the wall and at the outer edge of the boundary layer

$$\tan\varphi = a + b\xi + c\xi^2, \tag{2.54}$$

$$\xi = 0, \quad \tan\varphi = \tan\varphi_w = \alpha, \tag{2.55}$$

$$\xi = 1, \quad \tan\varphi = \tan\varphi_\infty = v_{r,\infty}/(\omega r - v_{\varphi,\infty}) = N/(1-\beta) = \kappa, \tag{2.56}$$

$$\xi = 1, \quad \mathrm{d}(\tan\varphi)/\mathrm{d}\xi = 0. \tag{2.57}$$

Conditions (2.54)–(2.57) yield

$$a = \alpha, \quad b = -2(\alpha - \kappa), \quad c = \alpha - \kappa, \quad (\tan\varphi - \kappa)/(\alpha - \kappa) = (1 - \xi)^2. \tag{2.58}$$

Figures 2.2 and 2.3 show profiles of the radial and tangential velocity components for a free rotating disk ($\kappa = 0$) calculated by Eqs. (2.37) and (2.58).

The present integral method enabled finding wall values of the tangent of the flow swirl angle α presented in Table 3.4 of Chap. 3 in comparison with α values obtained by von Karman's method [9]. Power-law profiles for the $\bar{v}_r$ and $\bar{v}_\varphi$ jointly with a quadratic Eq. (2.58) for tan φ agree well with experiments in the outer part of the boundary layer. Here, the profiles at $n = 1/9$ agree with the experiments [70, 71] (Figs. 2.2 and 2.3). The same trend demonstrates Fig. 2.4, where velocity components $\bar{v}_r$ on $\bar{v}_\varphi$ are interconnected via an equation resulting from Eqs. (2.37), (2.38), (2.40), and (2.58) [61]:

$$\bar{v}_r = \alpha\,\bar{v}_\varphi(1 - \bar{v}_\varphi^{1/n})^L. \tag{2.59}$$

Here, $L = 2$ in the present method and $L = 1$ in the method [9]. In the vicinity of the wall, the value of the exponent n = 1/7–1/8 yields, however, the best agreement with experiments. Based on Eq. (2.59) [61], a maximum in the dependence of $\bar{v}_r$ on $\bar{v}_\varphi$ is observed at

$$\bar{v}_{\varphi,\max} = \xi_{\max}^n, \quad \xi_{\max} = n/(n + L). \tag{2.60}$$

In frames of the present integral method, temperature distributions in the boundary layer are approximated with Eq. (2.53). This appears to be in a good agreement with the experimental data of different authors depicted in Fig. 2.5.

2.4.3 *Surface Friction and Heat Transfer*

Shear stresses $\tau_{w\varphi}$, τ_{wr} and wall heat flux q_w can be expressed with the help of a two-layer model of the velocity and temperature profiles non-dimensionalized using the law of the wall. Power-law profiles (2.40) and (2.53) can be re-written in wall coordinates as

$$V^+ = \xi^n/\sqrt{c_f/2}, \quad T^+ = \xi_T^{n_T}\sqrt{c_f/2}/St. \tag{2.61}$$

These relations are not valid in the viscous sub-layer; therefore, their place is taken here by the linear equations

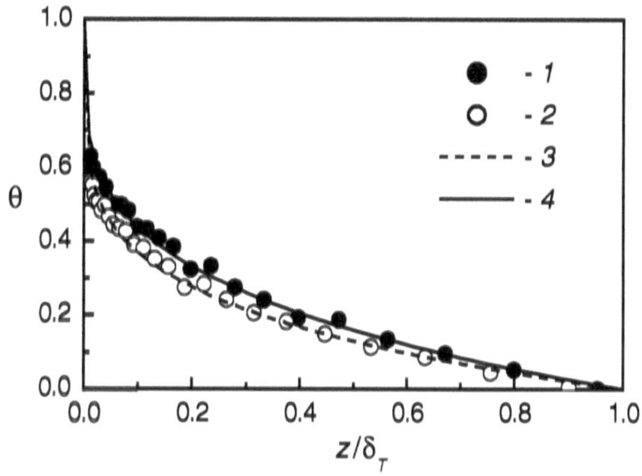

Fig. 2.5 Profiles of the non-dimensional temperature θ in the turbulent boundary layer over a free rotating disk [3]. Experiments [72], q_w = const., $Re_\omega = 1.0 \times 10^6$: *1*—inner heater on, *2*—inner heater off. Calculations by Eq. (2.53): *3*—$n_T = 1/5$, *4*—1/4

$$V^+ = z^+, \quad T^+ = Pr\, z^+. \tag{2.62}$$

Equations (2.61) and (2.62) must be spliced at the boundary z_1^+ of the viscous sub-layer and at the boundary z_{1T}^+ of the heat conduction sub-layer, accordingly. In doing so, one can come to relations for the friction coefficient and the Stanton number:

$$c_f/2 = (z_1^+)^{2(n-1)/(n+1)}\, Re_{V_*}^{-2n/(n+1)}, \tag{2.63}$$

$$St = (z_1^+)^{n_T-1} Re_{V*}^{-n_T} (c_f/2)^{(1-n_T)/2} \Delta^{-n_T} (z_{1T}^+/z_1^+)^{n_T-1} Pr^{-n_T}. \tag{2.64}$$

Instead of the coordinate z_1^+, its modification $C_n = (z_1^+)^{1-n}$ is often used. C_n is a constant whose dependence on the exponent n is clarified in the comments to Eq. (2.49). The constant z_1^+ takes the values 12.54, 13.44, 14.23, and 15.09 for n = 1/7, 1/8, 1/9, and 1/10, respectively. Based on Eq. (2.47), shear stresses τ_w, $\tau_{w\varphi}$, and τ_{wr} are mutually related as

$$\tau_{wr}/\rho = C_n^{-2/(n+1)} \mathrm{sgn}(1-\beta)(\nu/\delta)^{2n/(n-1)} (\omega r|1-\beta|)^{2/(n-1)} \alpha (1+\alpha^2)^{0.5(1-n)/(1+n)},$$

$$\tau_{w\phi}/\rho = -C_n^{-2/(n+1)} \mathrm{sgn}(1-\beta)(\nu/\delta)^{2n/(n-1)} (\omega r|1-\beta|)^{2/(n-1)} \times (1+\alpha^2)^{0.5(1-n)/(1+n)}. \tag{2.65}$$

In Eq. (2.64), the unknown Δ to be found is a function of the Prandtl number Pr, as well as the distribution of $T_w(r)$. The ratio (z_{1T}^+/z_1^+) depends on the Pr number only. One can denote $(z_{1T}^{+}/z_{1}^{+})^{n_T-1}Pr^{-n_T} = Pr^{-n_p}$, with the exponent n_p remaining so far unknown.

A condition $n_T = n$ will be often employed below, which leads to a simplification of the expressions for the Stanton number and the Nusselt number:

$$St = (c_f/2)\Delta^{-n}Pr^{-n_p}, \tag{2.66}$$

$$Nu = St\frac{V_* r}{\nu}Pr = St\,Re_\omega Pr|\beta - 1|(1+\alpha^2)^{1/2}. \tag{2.67}$$

2.4.3.1 Integral Equations

Having integrated Eqs. (2.17) and (2.18) with respect to the z-coordinate in view of Eqs. (2.37)–(2.40), (2.58), one can derive the following ordinary differential equations [61]:

$$\begin{aligned}&\frac{\mathrm{d}}{\mathrm{d}r}\left\{\delta r(\omega r)^2(1-\beta)^2\left[\kappa(A_1\alpha + A_2\kappa) - \left(B_1\alpha^2 + B_2\alpha\kappa + B_3\kappa^2\right)\right]\right\}\\&\quad + \delta\omega r^2(1-\beta)\frac{\mathrm{d}(N\omega r)}{\mathrm{d}r}[\kappa - (A_1\alpha + A_2\kappa)]\\&\quad + \rho\delta(\omega r)^2\left(C_1 + C_2\beta + C_3\beta^2\right) = r\tau_{wr}/\rho,\end{aligned} \tag{2.68}$$

$$\begin{aligned}&\frac{\mathrm{d}}{\mathrm{d}r}\{\delta\omega^2 r^4(1-\beta)[\alpha(D_1 + \beta D_2) + \kappa(D_3 + \beta D_4)]\}\\&\quad - (\omega r)^2\beta\frac{\mathrm{d}}{\mathrm{d}r}\left[\delta\omega r^2(1-\beta)(A_1\alpha + A_2\kappa)\right] = -r^2\tau_{w\varphi}/\rho,\end{aligned} \tag{2.69}$$

where

$$\begin{aligned}
&A_1 = 1/(n+1) - A_2; \quad A_2 = 2/(n+2) - 1/(n+3);\\
&B_1 = 1/(2n+1) - 2/(n+1) + 6/(2n+3) - 2/(n+2) + 1/(2n+5);\\
&B_2 = 2/(n+1) - 10/(2n+3) + 4/(n+2) - 2/(2n+5);\\
&B_3 = 4/(2n+3) - 2/(n+2) + 1/(2n+5);\\
&C_1 = 1 - 2/(n+1) + 1/(2n+1),\\
&C_2 = -2(1/(2n+1) - 1/(n+1));\\
&C_3 = -1 + 1/(2n+1);\\
&D_1 = A_1 - D_2;\\
&D_2 = 1/(2n+1) - D_4;\\
&D_3 = A_2 - D_4;\\
&D_4 = 1/(n+1) - 1/(2n+3).
\end{aligned}$$

Equation (2.20) for the thermal boundary layer integrated with respect to z and account for Eqs. (2.37), (2.38), (2.40), (2.53), and (2.58) can be written as

$$\frac{\mathrm{d}}{\mathrm{d}r}\left[\delta\omega r^2(1-\beta)F_1(T_\infty - T_w)\right] + \frac{\mathrm{d}T_\infty}{\mathrm{d}r}\delta\omega r^2(1-\beta)F_2$$
$$= -St_{\,*}r\Delta^{-n_T}Pr^{-n_P}(T_\infty - T_w), \tag{2.70}$$

where

$$\begin{aligned}
F_1 &= E_1, \quad F_2 = E_2 \text{ at } \Delta \le 1;\ F_1 = E_3,\ F_2 = E_4 \text{ at } \Delta \ge 1;\\
E_1 &= \Delta^{n+1}(aa_{*T} + bb_{*T}\Delta + cc_{*T}\Delta^2),\\
a_{*T} &= 1/(1+n+n_T) - 1/(1+n),\\
b_{*T} &= 1/(2+n+n_T) - 1/(2+n),\\
c_{*T} &= 1/(3+n+n_T) - 1/(3+n),\\
E_2 &= \Delta^{n+1}\left[a/(n+1) + b\Delta/(n+2) + c\Delta^2/(n+3)\right],\\
E_3 &= E_5 + \kappa E_6,\\
E_4 &= \alpha A_1 + \kappa(\Delta - 1) + \kappa A_2,\\
E_5 &= \alpha(-A_1 + \Delta^{-n_T}D_{2T}),\\
D_{2T} &= 1/(1+n+n_T) - D_{4T},\\
E_6 &= (\Delta - \Delta^{-n_T})/(n_T+1) - \Delta + 1 - A_2 + \Delta^{-n_T}D_{4T},\\
D_{4T} &= 2/(2+n+n_T) - 1/(3+n+n_T).
\end{aligned}$$

The mass flow rate through the boundary layer can be expressed as

$$\dot{m}_d/(\rho\omega r^3) = 2\pi(1-\beta)(A_1\alpha + A_2\kappa)\delta/r. \tag{2.71}$$

Equations (2.68)–(2.70) involve three unknowns:

(a) *in the entraining boundary layers*: α, δ, and Δ for specified β, as well as T_∞; and
(b) *in the Ekman-type boundary layers*: α, β for a specified mass flowrate $\dot{m}_d =$ const. (i.e., specified distribution of δ), as well as unknown T_∞ for a specified $\Delta =$ const.

In case (a), Eqs. (2.68)–(2.70) can be solved analytically at the boundary conditions (2.27)–(2.31) (and $N =$ const.), assumptions $\alpha =$ const. and $\Delta =$ const. and a power law for the radial distribution of the boundary layer thickness $\delta \sim r^m$. If the boundary conditions are approximated with arbitrary functions, Eqs. (2.68)–(2.70) are to be solved numerically being re-written to a notation that enables using the Runge–Kutta method [92, 96]:

$$\begin{cases} \alpha' = (\Phi_1\Phi_4 + \Phi_2)/(1 - \Phi_1\Phi_3), \\ \bar{\delta}' = (\Phi_2\Phi_3 + \Phi_4)/(1 - \Phi_1\Phi_3), \end{cases} \tag{2.72}$$

$$\Delta' = (S_1 - S_2 - S_3)/S_4. \tag{2.73}$$

Here,

$$\begin{aligned}
\Phi_2 &= \{[\mathrm{sgn}(1-\beta)|c_{\mathrm{fr}}/2|\bar{r}^3 Re^2_{V_*}/\bar{\delta}^2 - Z_1\bar{\delta} \\
&\quad - G_1\bar{\delta} - G_2]/(\bar{\delta}\bar{r}) - Q_2\bar{r}^2\}/Q_1; \\
\Phi_4 &= \{-\mathrm{sgn}(1-\beta)|c_{f\varphi}/2|\bar{r}^2 Re^2_{V_*}/\bar{\delta}^2 - \bar{\delta}[\alpha Q_3' + \bar{Q}_4' \\
&\quad + (\beta Re_\omega)'(\alpha Q_5 + Q_6)]/Q_7\}; \\
\Phi_1 &= -Z_1/(\bar{\delta}Q_1);\ \Phi_3 = -\bar{\delta}Q_3/Q_7; \\
Z_1 &= Re^2_\omega(1-\beta)^2[-B_1\alpha^2 + \alpha\kappa(A_1 - B_2) + \kappa^2(A_2 - B_3)]; \\
G_1 &= Re^2_\omega(C_1 + C_2\beta + C_3\beta^2); \\
G_2 &= Re^2_\omega(1-\beta)\bar{\delta}[-A_1\alpha + \kappa(1 - A_2)]\bar{v}'_{r,\infty}; \\
Q_1 &= Re^2_\omega(1-\beta)^2[-2\alpha B_1 + \kappa(A_1 - B_2)];\ Q_3 = -Re^2_\omega(1-\beta)^2 D_1 \\
Q_2 &= Re^2_{\omega i}\{-\alpha^2 B_1[\bar{r}^2(1-\beta)^2]' \\
&\quad + \alpha(A_1 - B_2)[\bar{r}(1-\beta)\bar{v}_{r,\infty}]' + (A_2 - B_3)(\bar{v}^2_{r,\infty})'\}; \\
Q_4 &= -Re^2_\omega(1-\beta)\bar{v}_{r,\infty}D_3/\bar{r};\ Q_5 = -Re_\omega(1-\beta)A_1; \\
Q_6 &= -Re_{\omega i}\bar{r}\bar{v}_{r,\infty}A_2;\ Q_7 = \alpha Q_3 + Q_4; \\
\bar{v}_{r,\infty} &= v_{r,\infty}/v_{r,\infty}(\omega a);\ Re_{\omega i} = \omega r_i^2/\nu;\ |c_{fr}/2| = (c_f/2)\alpha/(1+\alpha^2)^{1/2}; \\
|c_{f\varphi}/2| &= (c_f/2)/(1+\alpha^2)^{1/2};\ \bar{\delta} = \delta/r_i;\ \bar{r} = r/r_i.
\end{aligned}$$

Given $\Delta \leq 1$ in Eq. (2.73), we have

$$\begin{aligned}
S_1 &= -Re_\omega|1-\beta|(1+\alpha^2)^{1/2} St(\bar{T}_\infty - \bar{T}_w); \\
S_2 &= \bar{T}'_\infty\bar{\delta}Re_\omega\Delta^{n+1}[\alpha(1-\beta)/(n+1) - 2\Delta(\alpha(1-\beta) - N)/(n+2) \\
&\quad + \Delta^2(\alpha(1-\beta) - N)/(n+3)]; \\
S_3 &= \Delta^{n+1}L_1' + \Delta^{n+2}L_2' + \Delta^{n+3}L_3';\ S_4 = L_1(n+1)\Delta^n + L_2(n+2)\Delta^{n+1} + L_3(n+3)\Delta^{n+2}; \\
L_1 &= L_0 a_{*T}\alpha(1-\beta);\ L_2 = L_0 b_{*T}(-2)[\alpha(1-\beta) - N];\ L_3 = L_0 c_{*T}[\alpha(1-\beta) - N]; \\
L_0 &= \bar{\delta}Re_\omega(\bar{T}_\infty - \bar{T}_w),\ \bar{T} = T/T_{\mathrm{ref}}.
\end{aligned}$$

The function S_1 has the identical form for $\Delta \geq 1$ and $\Delta \leq 1$.

$$\begin{aligned}
\text{For}\,\Delta \geq 1: S_2 &= \bar{T}'_{\infty}\bar{\delta}Re_{\omega}[\alpha(1-\beta)A_1 + NA_2 + N(\Delta - 1)];\\
S_3 &= L'_{1*}C^*_{6T} + L'_{2*}C^*_{7T};\\
S_4 &= -n_T\Delta^{-n_T-1}D_{2T}L_{1*} + [(1 + n_T\Delta^{-n_T-1})/(n_T+1)\\
&\quad - D_{4T}n_T\Delta^{-n_T-1} - 1]L_{2*};\\
L_{1*} &= L_0(1-\beta)\alpha;\\
L_{2*} &= L_0N;\\
C^*_{6T} &= -A_1 + \Delta^{-n_T}D_{2T};\\
C^*_{7T} &= (\Delta - \Delta^{-n_T})/(n_T+1) - \Delta + 1 - A_2 + D_{4T}\Delta^{-n_T}.
\end{aligned}$$

Derivatives with respect to the radial coordinate $\mathrm{d}/\mathrm{d}\bar{r}$ are denoted here with primes; r_l is a characteristic radius (for instance, the inlet radius that is used here).

In case (b), i.e., in the Ekman-type layers

$$\begin{cases}
\alpha' = \frac{c_f}{2}\alpha(\beta-1)Re_{\omega}(1+\alpha^2)^{1/2}\frac{4\pi A_1 r_l}{B_1 C_w b} + \frac{\mathrm{d}\beta}{\mathrm{d}\bar{r}}\frac{\alpha}{\beta-1} - \frac{C_3[\beta + n/(n+1)]}{\bar{r}(\beta-1)\alpha B_1} - \frac{\alpha}{\bar{r}},\\
\beta' = \left\{-\frac{c_f}{2}(1-\beta)^2 Re_{\omega}(1+\alpha^2)^{1/2}\frac{4\pi A_1 r_l}{D_1 C_w b} - \frac{2}{\bar{r}}\left[\beta\left(1-\frac{A_1}{D_1}\right) - 1\right]\right\}\Big/\left(1-\frac{A_1}{D_1}\right),
\end{cases} \tag{2.74}$$

$$\frac{\mathrm{d}\bar{T}_{\infty}}{\mathrm{d}\bar{r}} = \left[St\frac{V_* r}{\nu}\frac{2\pi}{0.5C_w}\frac{r_l}{b}\frac{1}{K_H}(\bar{T}_{\infty} - \bar{T}_w) + \frac{\mathrm{d}\bar{T}_w}{\mathrm{d}\bar{r}}\right]\frac{K_H}{K_H - 1}. \tag{2.75}$$

In the Ekman-type layers, authors [1, 2] recommended to assign the parameter K_H to be constant [92, 95, 97]:

$$K_H = 1 - (D_{2T}/A_1)\Delta^{-n_T} = \text{const.} \quad \text{or} \quad \Delta = \text{const.} \tag{2.76}$$

2.5 Disk Rotation in a Fluid Rotating as a Solid Body and Simultaneous Accelerating Imposed Radial Flow

We will consider here flows where β = const., N = const., and $\kappa > 0$. The assumption β = const. outlines the solid-body rotation case that occurs in rotor–stator geometries. The assumption N = const. describes the case of accelerating radial flow, which occurs around the stagnation point of flow impinging onto a perpendicular plate. If $\kappa > 0$, fluid flow over a rotating disk never exhibits recirculation [1, 2, 68]. Given these assumptions, one can solve Eqs. (2.68) and (2.69) analytically. This solution can be written as [3]

$$\delta = C_\delta r^m, \quad C_\delta = \gamma(\omega/\nu)^{-2n/(3n+1)}, \quad \delta/r = \gamma Re_\omega^{-2n/(3n+1)}, \tag{2.77}$$

$$\alpha = \text{const.}, \quad m = (1-n)/(3n+1), \tag{2.78}$$

$$\gamma = \gamma_* |1-\beta|^{(1-n)/(3n+1)}, \tag{2.79}$$

$$C_\mathrm{M} = \varepsilon_M Re_\varphi^{-2n/(3n+1)}, \tag{2.80}$$

$$\dot{m}_\mathrm{d}/(\mu r) = \varepsilon_m Re_\omega^{(n+1)/(3n+1)}, \tag{2.81}$$

$$c_f/2 = A_c Re_\omega^{-2n/(3n+1)}, \tag{2.82}$$

$$\alpha = -H_2/2H_3 + \left[(H_2/2H_3)^2 - H_1/H_3\right]^{1/2}, \tag{2.83}$$

$$\gamma_* = C_n^{-2/(3n+1)}(1+\alpha^2)^{0.5(1-n)/(3n+1)} H_9^{-(n+1)/(3n+1)}, \tag{2.84}$$

$$\varepsilon_m = \varepsilon_m^* |1-\beta|^{2(n+1)/(3n+1)}, \quad \varepsilon_m^* = 2\pi\gamma(A_1\alpha + A_2\kappa)\mathrm{sgn}(1-\beta), \tag{2.85}$$

$$\varepsilon_\mathrm{M} = \frac{8\pi}{5-4n/(3n+1)} C_n^{-\frac{2}{n+1}} \gamma_*^{\frac{2n}{n+1}} |1-\beta|^{\frac{2(n-1)}{3n+1}} (1+\alpha^2)^{\frac{1-n}{2(n+1)}} \mathrm{sgn}(1-\beta), \tag{2.86}$$

$$A_c = C_n^{-2/(n+1)} \gamma^{-2n/(n+1)} (1+\alpha^2)^{-n/(n+1)} |\beta - 1|^{-2n/(n+1)}, \tag{2.87}$$

where

$$\begin{aligned}
&H_1 = C_3(\beta - C_5) + (\beta - 1)\kappa^2 H_4;\ H_2 = \kappa(\beta H_5 + H_6);\\
&\quad H_3 = \beta H_7 + H_8;\\
&H_4 = 1 + (2+m)A_2 - (3+m)B_3;\ H_5 = A_1(2+m)\\
&\quad - B_2(3+m) + D_4(m+4) - A_2(2+m);\\
&H_6 = -A_1(2+m) + B_2(3+m) + D_3(4+m);\\
&H_7 = -(3+m)B_1 + (4+m)D_2 - (2+m)A_1;\\
&H_8 = (3+m)B_1 + (4+m)D_1;\ C_5 = C_1/C_3;\\
&H_9 = \alpha[(D_1 + \beta D_2)(4+m) - \beta(2+m)A_1]\\
&\quad + \kappa[(D_3 + \beta D_4)(4+m) - \beta A_2(2+m)].
\end{aligned} \tag{2.88}$$

Equation (2.70) can be solved analytically at the boundary conditions (2.29)–(2.31) provided that Δ = const., Pr = const., and $n = n_T$. An additional condition is $D_{2T} = D_2$ and $D_{4T} = D_4$.

Equation (2.70) is to be solved jointly with Eq. (2.69), in view of Eqs. (2.31), (2.67), (2.77) and (2.78). As a result, one can derive [91]

$$\left[F_1(2+m+n_*)+\frac{\beta n_*}{\beta-1}F_2\right]\Delta^n Pr^{n_p}=(4+m)C_4+\frac{2\beta}{\beta-1}C_5. \tag{2.89}$$

Functions F_1 and F_2 are clarified in explanations to Eq. (2.70); $C_4=-(\alpha D_1+\kappa D_2)$, $C_5=1/(n+1)+1/(n+2)+1/(n+3)$. Solutions of Eq. (2.89) for the cases $\Delta \geq 1$ and $\Delta \leq 1$ are different (which is manifested via different mathematical expressions for F_1 and F_2 at $\Delta \geq 1$ and $\Delta \leq 1$). Heat transfer conditions at $\Delta \geq 1$ can be observed for gases at Pr ≤ 1. Conditions with $\Delta \leq 1$ take place at heat transfer in liquids for Pr ≥ 1 (see Chap. 6).

Given simultaneously non-zero values of β and N, the algebraic Eq. (2.89) is transcendental. At N = 0 and $\Delta \geq 1$, there exists an explicit solution for the parameter Δ. The exponent n_p for flow over the free rotating disk is specified below.

Nusselt and Stanton numbers are given by the following equations:

$$St=A_c Re_\omega^{-2n/(3n+1)}\Delta^{-n}Pr^{-n_p}, \tag{2.90}$$

$$Nu=A_c(1+\alpha^2)^{1/2}|1-\beta|Re_\omega^{(n+1)/(3n+1)}\Delta^{-n}Pr^{1-n_p}. \tag{2.91}$$

The *present integral method* is thoroughly validated for turbulent air flow and extended to laminar flow in Chaps. 3 and 4. Simulations for a free rotating disk ($\beta=0$, $N=0$) are described in detail in Chap. 3. Cases of a rotating disk in a fluid that (a) co-rotates as a solid body (β = const., N = 0), and (b) is uniformly accelerating and non-rotating (β = 0, N = const.), as well as for the case of turbulent through flow between parallel co-rotating disks are analyzed in Chap. 4. A description of an extension of the integral method for gases and liquids at Prandtl or Schmidt numbers larger than unity is documented in Chap. 6.

References

1. Owen JM, Rogers RH (1989) Flow and heat transfer in rotating-disc systems. In: Rotor-stator systems, vol 1. Research Studies Press Ltd., Taunton
2. Owen JM, Rogers RH (1995) Flow and heat transfer in rotating-disc systems. In: Rotating cavities, vol 2. Research Studies Press Ltd., Taunton
3. Shevchuk IV (2009) Convective heat and mass transfer in rotating disk systems. Springer, Berlin
4. Dorfman LA (1963) Hydrodynamic resistance and the heat loss of rotating solids. Oliver and Boyd, Edinburgh
5. Shevchuk IV (2002) Laminar heat transfer in a rotating disk under conditions of forced air impingement cooling: approximate analytical solution. High Temp 40(5):684–692
6. Cochran WG (1934) The flow due to a rotating disk. Proc Cambridge Phil Soc 30:365–375
7. Ginzburg IP (1970) Theory of resistance and heat transfer. Izd. Leningrad. University, Leningrad, USSR (in Russian)
8. Hartnett JP, Deland EC (1961) The influence of Prandtl number on the heat transfer from rotating nonisothermal disks and cones. Trans ASME J Heat Transfer 83(1):95–96

9. Karman Th von (1921) Über laminare und turbulente Reibung. Z Angew Math Mech 1(4):233–252
10. Lin H-T, Lin L-K (1987) Heat transfer from a rotating cone or disk to fluids at any Prandtl number. Int Commun Heat Mass Transfer 14(3):323–332
11. Schlichting G (1968) Boundary-layer theory. McGraw-Hill Book Company, New York
12. Sparrow EM, Gregg JL (1959) Heat transfer from a rotating disc to fluids at any Prandtl number. Trans ASME J Heat Transfer 81:249–251
13. Turkyilmazoglu M (2010) Heat and mass transfer on the unsteady magnetohydrodynamic flow due to a porous rotating disk subject to a uniform outer radial flow. Trans ASME J Heat Transfer 132(6). Paper 061703
14. Turkyilmazoglu M (2011) Wall stretching in magnetohydrodynamics rotating flows in inertial and rotating frames. AIAA J Thermophys Heat Transfer 25(4):606–613
15. Turkyilmazoglu M (2011) Thermal radiation effects on the time-dependent MHD permeable flow having variable viscosity. Int J Thermal Sci 50(1):88–96
16. Turkyilmazoglu M (2012) Effects of uniform radial electric field on the MHD heat and fluid flow due to a rotating disk. Int J Eng Sci 51:233–240
17. Turkyilmazoglu M (2012) Three dimensional MHD stagnation flow due to a stretchable rotating disk. Int J Heat Mass Transfer 55(23–24):6959–6965
18. Turkyilmazoglu M (2012) MHD fluid flow and heat transfer with varying Prandtl numbers due to a rotating disk subject to a uniform radial electric field. Appl Thermal Eng 35:127–133
19. Asghar S, Jalil M, Hussan M, Turkyilmazoglu M (2014) Lie group analysis of flow and heat transfer over a stretching rotating disk. Int J Heat Mass Transfer 69:140–146
20. Makukula ZG, Sibanda P, Motsa SS (2012) On a quasilinearisation method for the von Karman flow problem with heat transfer. Latin American Appl Res 42(1):97–102
21. Arikoglu A, Komurgoz G, Ozkol I, Gunes AY (2010) Combined effects of temperature and velocity jump on the heat transfer, fluid flow, and entropy generation over a single rotating disk. Trans ASME J Heat Transfer 132(7). Paper 111703
22. Aly EH, Soliman HA (2011) Solution for steady flow over a rotating disk in porous media with heat transfer. J Appl Sci Res 7(11):1485–1491
23. Devi SPA, Devi RU (2011) On hydromagnetic flow due to a rotating disk with radiation effects. Nonlin Anal Model Control 16(1):17–29
24. Ariel PD (1996) The flow near a rotating disk: an approximate solution. Trans ASME J Appl Mech 63(2):436–438
25. Chen JX, Gan X, Owen JM (2001) Heat transfer from air-cooled contra-rotating disks. Trans ASME J Turbomach 119(1):61–67
26. Karabay H, Chen J-X, Pilbrow R, Wilson M, Owen JM (1999) Flow in a "cover-plate" preswirl rotor-stator system. Trans ASME J Turbomach 121(1):161–166
27. Karabay H, Wilson M, Owen JM (2001) Predictions of effect of swirl on flow and heat transfer in a rotating cavity. Int J Heat Fluid Flow 22:143–155
28. Kilic M, Owen JM (2003) Computation of flow between two disks rotating at different speeds. Trans ASME J Turbomach 125(2):394–400
29. Ong CL, Owen JM (1991) Computation of the flow and heat transfer due to a rotating disc. Int J Heat Fluid Flow 12(2):106–115
30. Oehlbeck DL, Erian FF (1979) Heat transfer from axisymmetric sources at the surface of a rotating disk. Int J Heat Mass Transfer 22(6):601–610
31. Pilbrow R, Karabay H, Wilson M, Owen JM (1999) Heat transfer in a 'cover-plate' pre-swirl rotating-disc system. Trans ASME J Turbomach 121(2):249–256
32. Prata AT, Pilichi CDM, Ferreira RTS (1995) Local heat transfer in axially feeding radial flow between parallel disks. Trans ASME J Heat Transfer 117(1):47–53
33. Roy RP, Xu G, Feng J (2001) A study of convective heat transfer in a model rotor-stator disk cavity. Trans ASME J Turbomach 123(3):621–632
34. Tadros SE, Erian FF (1983) Heat and momentum transfer in the turbulent boundary layer near a rotating disk. Paper ASME 83–WA/HT–6

35. Tadros SE, Erian FF (1982) Generalized laminar heat transfer from the surface of a rotating disk. Int J Heat Mass Transfer 25(11):1651–1660
36. Volkov KN, Hills NJ, Chew JW (2008) Simulation of turbulent flows in turbine blade passages and disc cavities. In: Proceedings of ASME Turbo Expo 2008, 4(Pts. A & B). Berlin, Germany, pp 1543–1554
37. Wu CM, Li YR, Ruan DF (2013) Aspect ratio and radius ratio dependence of flow pattern driven by differential rotation of a cylindrical pool and a disk on the free surface. Phys Fluids 25(8). Paper 084101
38. Yan Y, Gord MF, Lock GD, Owen JM (2003) Fluid dynamics of a pre-swirl rotor-stator system. Trans ASME J Turbomach 125(4):641–647
39. Cebeci T, Bradshaw P (1984) Physical and computational aspects of convective heat transfer. Springer, Berlin
40. Launder BE, Sharma BI (1974) Application of the energy dissipation model of turbulence to the calculation of the flow near a spinning disc. Lett Heat Mass Transfer 1:131–138
41. Morse AP (1988) Numerical prediction of turbulent flow in rotating cavities. Trans ASME J Turbomach 110(2):202–215
42. Morse AP (1991) Assessment of laminar-turbulent transition in closed disc geometries. Trans ASME J Turbomach 113(2):131–138
43. Derevich IV (2005) Modeling of the motion of particles in a rotary crusher. Theor Foundations Chem Eng 39(2):213–219
44. Lam CKG, Bremhorst KA (1981) Modified form of the k–ε model for predicting wall turbulence. Trans ASME J Fluid Eng 103(3):456–460
45. Arani AAA, Shahmohamadia P, Sheikhzadeha G, Mehrabianb MA (2013) Convective heat transfer from a heated rotating disk at arbitrary inclination angle in laminar flow. IJE Trans B Appl (Int J Eng) 26(8):865–874
46. Harmand S, Pellé J, Poncet S, Shevchuk IV (2013) Review of fluid flow and convective heat transfer within rotating disk cavities with impinging jet. Int J Thermal Sci 67:1–30
47. Indinger T, Shevchuk IV (2004) Transient laminar conjugate heat transfer of a rotating disk: theory and numerical simulations. Int J Heat Mass Transfer 47(14–16):3577–3581
48. Kumar BGV, Chew JW, Hills NJ (2013) Rotating flow and heat transfer in cylindrical cavities with radial Inflow. ASME J Eng Gas Turb Power 135(3). Paper 032502
49. Poncet S, Schiestel R (2007) Numerical modeling of heat transfer and fluid flow in rotor-stator cavities with throughflow. Int J Heat Mass Transfer 50:1528–1544
50. Poncet S, Nguyen TD, Harmand S, Pellé J, Da Soghe R, Bianchini C, Viazzo S (2013) Turbulent impinging jet flow into an unshrouded rotor–stator system: hydrodynamics and heat transfer. Int J Heat Fluid Flow 44:719–734
51. Roy KER, Prasad BVSSS, Murthy SS, Gupta NK (2005) Conjugate heat transfer in an arbitrary shaped cavity with a rotating disk. Heat Transfer Eng 25(8):69–79
52. Shevchuk IV, Delas NI (2005) Aerodynamics and turbulent flow heat exchange in the rotary disk air cleaner. Heat Transfer Res 36(1–2):104–113
53. Wu X, Squires KD (2000) Prediction and investigation of the turbulent flow over a rotating disk. J Fluid Mech 418:231–264
54. Helcig C, aus der Wiesche S (2013) The effect of the incidence angle on the flow over a rotating disk subjected to forced air streams. In: Proceedings of FEDSM 2013, Incline Village, Nevada, USA. Paper FEDSM2013-16360
55. Poncet S, Serre E (2009) High-order LES of turbulent heat transfer in a rotor-stator cavity. Int J Heat Fluid Flow 30(4):590–601
56. Tuliszka-Sznitko E, Zielinski A, Majchrowski W (2009) LES of the non-isothermal transitional flow in rotating cavity. Int J Heat Fluid Flow 30(3):543–548
57. aus der Wiesche S (2004) LES study of heat transfer augmentation and wake instabilities of a rotating disk in a planar stream of air. Heat Mass Transfer 40(3–4):271–284
58. aus der Wiesche S (2007) Heat transfer from a rotating disk in a parallel air crossflow. Int J Thermal Sci 46(8):745–754
59. Childs PRN (2011) Rotating flow. Elsevier Inc., Amsterdam

60. Loitsyanskii LG (1966) Mechanics of liquids and gases. Pergamon, Oxford
61. Shevchuk IV, Khalatov AA (1997) Integral method for calculating the characteristics of a turbulent boundary layer on a rotating disk: quadratic approximation of the tangent of the flow swirl angle. Heat Transfer Res 28(4–6):402–413
62. Baibikov AS (1998) Method of calculation of a turbulent flow in an axial gap with a variable radius between a rotating disk and an axisymmetric casing. J Eng Phys Thermophys 71(6):1072–1081
63. McComas ST, Hartnett JP (1970) Temperature profiles and heat transfer associated with a single disc rotating in still air. In: Proceedings of IV IHTC, Paris-Versailles, France, vol 3, Paper FC 7.7
64. Newman BG (1983) Flow and heat transfer on a disk rotating beneath a forced vortex. AIAA J 21(8):1066–1070
65. Cobb EC, Saunders OA (1956) Heat transfer from a rotating disk. Proc Roy Soc A 236:343–349
66. Kabkov VIa (1978) Heat transfer at air flow in a gap between a disk and a deflector rotating together with it. In: Teploobmen v Energeticheskikh Ustanovkakh, Naukova Dumka, Kiev, pp 121–124 (in Russian)
67. Mager A (1952) Generalization of boundary-layer momentum-integral equations to three-dimensional flows including those of rotating system. NACA Rep 1067:1–37
68. Shvets IT, Dyban EP (1974) Air cooling of gas turbine parts. Naukova Dumka, Kiev (in Russian)
69. Kabkov VIa (1974) Characteristics of turbulent boundary-layer on a smooth disk rotating in a large volume. Teplofizika i Teplotekhnika, Naukova Dumka, Kiev 28:119–124 (in Russian)
70. Littel HS, Eaton JK (1994) Turbulence characteristics of the boundary layer on a rotating disk. J Fluid Mech 266:175–207
71. Itoh M, Hasegawa I (1994) Turbulent boundary layer on a rotating disk in infinite quiescent fluid. JSME Int J Ser B 37(3):449–456
72. Elkins CJ, Eaton JK (1997) Heat transfer in the rotating disk boundary layer. Stanford University, Department of Mechanical Engineering, Thermosciences Division Report TSD–103. Stanford University (USA)
73. Cheng W-T, Lin H-T (1994) Unsteady and steady mass transfer by laminar forced flow against a rotating disk. Wärme und Stoffübertragung 30(2):101–108
74. Elkins CJ, Eaton JK (2000) Turbulent heat and momentum transfer on a rotating disk. J Fluid Mech 402:225–253
75. Case P (1966) Measurements of entrainment by a free rotating disk. J Roy Aero Soc 71:124–129
76. Abrahamson S, Lonnes S (1993) An integral method for turbulent boundary layers on rotating disks. Trans ASME J Fluids Eng 115(4):614–619
77. Abrahamson S, Lonnes S (1994) An improved model for radial injection between corotating disks. Trans ASME J Fluids Eng 116(2):255–257
78. Truckenbrodt E (1954) Die turbulente Strömung an einer angeblasenen rotierenden Scheibe. Z Angew Math Mech 34:150–162
79. Kapinos VM (1964) Heat transfer of a turbine rotor with a radial coolant flow. Inzhenerno-Fizicheskiy Zhurnal 7(1):3–11 (in Russian)
80. Kapinos VM (1965) Heat transfer from a disc rotating in a housing with a radial flow of coolant. J Eng Phys Thermophys 8(1):35–38
81. Kapinos VM (1965) Heat transfer of a disk rotating in a housing. Izvestiya vuzov. Aviatsyonnaya Tekhnika 2:76–86 (in Russian)
82. Goldstein S (1935) On the resistance to the rotation of a disc immersed in a fluid. Proc Cambridge Phil Soc 31:232–241
83. Johnson MC (1980) Turbulent heat transfer to a rotating disk: a review and extension of Dorfman. Trans ASME J Heat Transfer 102(4):780–781
84. Kapinos VM (1964) Hydraulic resistance and heat transfer of a free disc with a cob. Izvestiya vuzov Energetika 11:85–92 (in Russian)

85. Kapinos VM (1965) Effect of a radial gradient of a relative circumferential velocity component on heat transfer in forced flow between two rotating disks. Izvestiya vuzov Energetika 3:111–120 (in Russian)
86. Kapinos VM, Pustovalov VN, Rud'ko AP (1971) Heat transfer in a coolant flow from a center to a periphery between two rotating disks. Izvestiya vuzov Energetika 6:116–124 (in Russian)
87. Lokai VI, Bodunov MN, Zhuikov VV, Shchukin AV (1985) Heat transfer in cooled parts of aircraft gas-turbine engines. Izdatel'stvo Mashinostroenie, Moscow, USSR (in Russian)
88. Northrop A, Owen JM (1988) Heat transfer measurements in rotating-disc systems. Part 1: The free disc. Int J Heat Fluid Flow 9(1): 19–26
89. Northrop A (1984) Heat transfer in a cylindrical rotating cavity. D. Phil. thesis. University of Sussex, Brighton, UK
90. Kays WM, Crawford ME, Weigand B (2005) Convective heat and mass transfer, 4th edn. Mc-Graw-Hill, New York. ISBN:0-07-246876-9
91. Shevchuk IV, Khalatov AA (1997) The integral method of calculation of heat transfer in the turbulent boundary layer on a rotating disk: quadratic approximation of the tangent of the flow-swirl angle. Promyshlennaya Teplotekhnika 19(4–5):145–150 (in Russian)
92. Shevchuk IV (1998) Heat transfer in underswirled radial turbulent outflow in the gap between co-rotating discs of gas turbine rotors. Promyshlennaya Teplotekhnika 20(1):54–58 (in Russian)
93. Shevchuk IV (1998) Simulation of heat transfer in a rotating disk: the effect of approximation of the tangent of the angle of flow swirling. High Temp 36(3):522–524
94. Shevchuk IV, Khalatov AA, Karabay H, Owen JM (1998) Heat transfer in turbulent centrifugal flow between rotating discs with flow swirling at the inlet. Heat Transfer Res 29(6–8):383–390
95. Shevchuk IV (1998) The effect of distribution of radial velocity in the flow core on heat transfer under conditions of centrifugal flow in a gap between parallel rotating disks. High Temp 36(6):972–974
96. Shevchuk IV (1999) Integral method of calculation of a turbulent centrifugal underswirl flow in a gap between parallel rotating disks. Heat Transfer Res 30(4–6):238–248
97. Shevchuk IV (1999) Effect of wall-temperature distribution on heat transfer in centrifugal flow in the gap between parallel rotating disks. J Eng Phys Thermophys 72(5):896–899
98. Shevchuk IV (1999) Simulation of heat transfer and hydrodynamics over a free rotating disk using an improved radial velocity profile. J. Thermal Sci 8(4):243–249
99. Shevchuk IV (2000) Turbulent heat transfer of rotating disk at constant temperature or density of heat flux to the wall. High Temp 38(3):499–501
100. Shevchuk IV (2001) Effect of the wall temperature on laminar heat transfer in a rotating disk: an approximate analytical solution. High Temp 39(4):637–640
101. Shevchuk IV (2001) An improved enthalpy-thickness model for predicting heat transfer of a free rotating disk using an integral method. In: Proceedings of 35th ASME NHTC'01, Anaheim, California, USA. Paper NHTC2001–20195
102. Shevchuk IV, Saniei N, Yan XT (2001) Effect of the wall temperature distribution in direct and inverse problems of laminar heat transfer over a free rotating disk. In: Proceedings of 35th ASME NHTC'01, Anaheim, California, USA. Paper NHTC2001–20196
103. Shevchuk IV (2001) Turbulent heat transfer over a free rotating disk: analytical and numerical predictions using solutions of direct and inverse problems. In: Proceedings of IMECE'01, New York, USA, pp 1–8
104. Shevchuk IV (2002) Numerical modeling of turbulent heat transfer in a rotating disk at arbitrary distribution of the wall temperature. J Eng Phys Thermophys 75(4):885–888
105. Shevchuk IV, Saniei N, Yan XT (2002) Impinging jet heat transfer over a rotating disk: exact solution and experiments. In: Proceedings of 8th AIAA/ASME Joint Thermophys Heat Transfer Conference, St. Louis, Missouri, USA, Paper AIAA 2002–3015
106. Shevchuk IV, Saniei N, Yan XT (2003) Impingement heat transfer over a rotating disk: integral method. AIAA J Thermophys Heat Transfer 17(2):291–293

107. Shevchuk IV (2005) A new type of the boundary condition allowing analytical solution of the thermal boundary layer equation. Int J Thermal Sci 44(4):374–381
108. Shevchuk IV, Buschmann MH (2005) Rotating disk heat transfer in a fluid swirling as a forced vortex. Heat Mass Transfer 41(12):1112–1121
109. Shevchuk IV (2009) An integral method for turbulent heat and mass transfer over a rotating disk for the Prandtl and Schmidt numbers much larger than unity. Heat Mass Transfer 45 (10):1313–1321

Chapter 3
Free Rotating Disk

3.1 Laminar Flow

A flow and heat transfer pattern over a single rotating disk is schematically depicted in Fig. 2.1. Self-similar velocity and temperature profiles, F_0, G_0, H_0 and θ, computed numerically using the Mathcad software as a solution of Eqs. (2.32)–(2.36) for $\beta = 0$, $N = 0$ [1] agree well with computations [2–5] and experiments [6–8] (Fig. 3.1). Derivatives of the F_0 and G_0profiles at the disk wall, as well as the mass flow rate in the boundary layer are given by the following relations [3, 4]

$$G'_{0w} = (\mathrm{d}G_0/\mathrm{d}\zeta_0)_{\zeta=0} = -0.6159, \quad F'_{0w} = (\mathrm{d}F_0/\mathrm{d}\zeta_0)_{\zeta=0} = 0.5102, \tag{3.1}$$

$$\alpha_0 = -F'_{0w}/G'_{0w} = 0.8284, \quad \dot{m}_{\mathrm{d}}/(\mu r) = 0.8845 Re_{\omega}^{1/2}. \tag{3.2}$$

In laminar flow, the boundary layer over a rotating disk has a constant thickness δ_0. Assuming $G_0 = 0.01$ at the outer edge of the boundary layer, one can obtain $\delta_0 = 5.5(\omega/v)^{1/2}$ [3]. The numerical coefficient is higher for smaller values of G_0 defining the location of δ_0, which, however, is unimportant in the integral method approach [1, 3].

For a single rotating disk, the moment coefficient of C_{M} is given by [2]

$$C_{\mathrm{M}} = 3.87 Re_{\varphi}^{-1/2}. \tag{3.3}$$

Equation (3.3) agrees with experiments [9] and differs modestly from the data [10, 11] (Fig. 3.2), which might be attributed to a poorer accuracy of the measurements.

Generally, the heat transfer rate over rotating disks follows the relations

$$Nu = K_1 Re_{\omega}^{n_R}, \quad Nu_{\mathrm{av}} = K_2 Re_{\varphi}^{n_R}. \tag{3.4}$$

I.V. Shevchuk, *Modelling of Convective Heat and Mass Transfer in Rotating Flows*, Mathematical Engineering, DOI 10.1007/978-3-319-20961-6_3

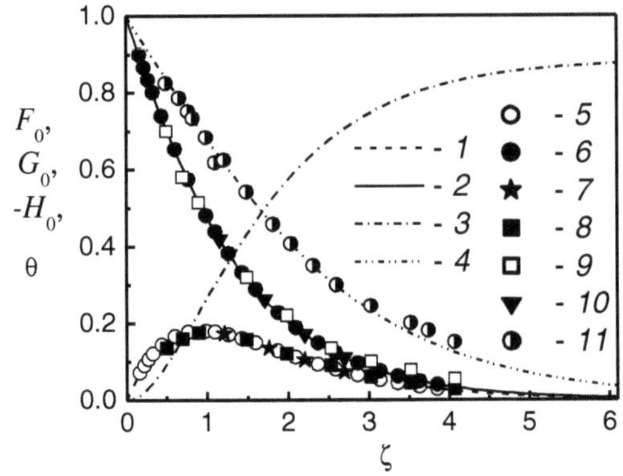

Fig. 3.1 Velocity and temperature profiles in laminar flow over a free rotating disk [1]. Computations: *1*—F_0, *2*—G_0, *3*—$(-H_0)$, *4*—θ for $n_* = 0$, $Pr = 0.71$. Experiments: *5*—F_0 [7], *6*—F_0 [6], *7*—F_0 [8], *8*—G_0 [7], *9*—G_0 [6], *10*—G_0 [8], *11*—θ for $n_* = 0$, $Pr = 0.71$ [6]. Subscript "0": a free disk

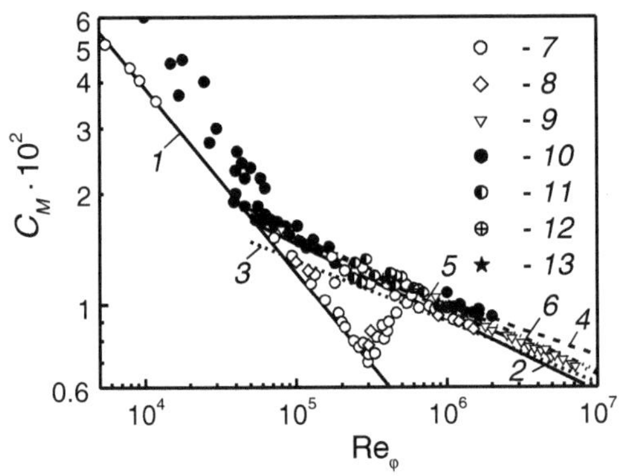

Fig. 3.2 Moment coefficient of a free rotating disk [1]. Laminar flow: *1*—Eq. (3.3). Turbulent flow: *2*—Eq. (2.80), $n = 1/7$; *3*—(2.80), 1/9: *4*—(3.18); *5*—(3.19), *6*—(3.20). Experiments: *7, 8, 9*—[9], *10*—[10], *11*—[11], *12*—[12], *13*—[13]. Data *7–11* cited according to [2, 5]; data [12, 13] processed in [14]. Subscript "0": a free disk

Flow regime, Prandtl number and function of the wall temperature determine coefficients K_1 and K_2 in Eq. (3.4). The constant n_R is determined by the flow regime. For instance, $K_1 = K_2$, $n_R = 1/2$ in laminar flow [2–4, 6]. For an isothermal rotating disk at $Pr = 0.71$–0.72 (air), the most reliable experiments yield $K_1 = 0.32$–0.34 [3, 4, 6, 15–25] (Fig. 3.3).

Values of the constant K_1 obtained by the author [1] are presented in Table 3.1.

They result from a solution of Eqs. (2.32)–(2.36) for $\beta = 0$, $N = 0$ under condition (2.30)

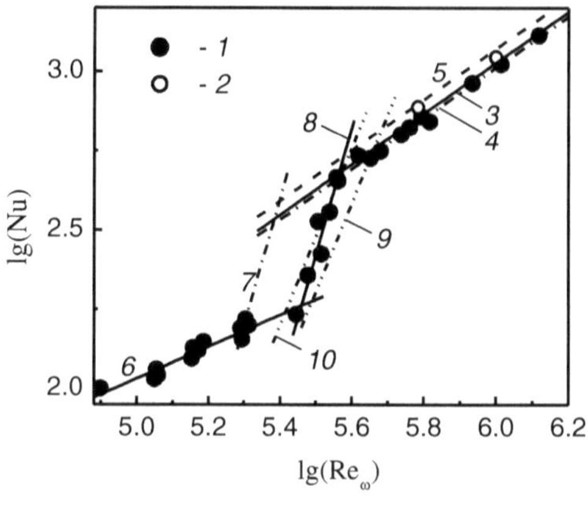

Fig. 3.3 Local Nusselt numbers on a rotating disk [1]. Experiments [6]: *1*—q_w = const.; *2*—$T_w \approx$ const. *3–10*—calculations, Eq. (3.4). Turbulent flow, $n_R = 0.8$: *3*—$K_1 = 0.0169$ [26]; *4*—$K_1 = 0.0163$ [6]; *5*—$K_1 = 0.0187$ [26]. Laminar flow, $n_R = 0.5$: *6*—$K_1 = 0.34$ [6]. Transitional flow: *7*—$n_R = 4$, $K_1 = 10.0 \times 10^{-20}$ [23]; *8*—$n_R = 4$, $K_1 = 2.65 \times 10^{-20}$ [6]; *9*—$n_R = 2.8$, $K_1 = 8.01 \times 10^{-14}$ [27, 28]; *10*—$n_R = 2.8$, $K_1 = 1.2 \times 10^{-13}$ [1]

Table 3.1 Constant K_1 according to the exact solution of Eqs. (2.32)–(2.36) [1, 29, 30]

Pr	$n_* = -2$	$n_* = -1.5$	$n_* = -1$	$n_* = -0.5$	$n_* = 0$	$n_* = 1$	$n_* = 2$	$n_* = 3$	$n_* = 4$
1.0	0.0	0.1305	0.2352	0.3221	0.3963	0.5180	0.6159	0.6982	0.7693
0.9	0.0	0.1217	0.2204	0.3029	0.3737	0.4905	0.5849	0.6643	0.7331
0.8	0.0	0.1124	0.2046	0.2824	0.3495	0.4608	0.5513	0.6276	0.6939
0.72	0.0	0.1045	0.1911	0.2647	0.3286	0.4352	0.5223	0.5959	0.6599
0.71	0.0	0.1035	0.1893	0.2624	0.3259	0.4319	0.5185	0.5918	0.6555
0.6	0.0	0.0917	0.1691	0.2358	0.2943	0.3929	0.4742	0.5433	0.6036
0.5	0.0	0.0802	0.1490	0.2091	0.2623	0.3531	0.4287	0.4935	0.5502
0.4	0.0	0.0675	0.1267	0.1792	0.2263	0.3078	0.3767	0.4362	0.4887
0.3	0.0	0.0536	0.1017	0.1452	0.1849	0.2550	0.3153	0.3682	0.4153
0.2	0.0	0.0381	0.0732	0.1058	0.1362	0.1912	0.2400	0.2838	0.3234
0.1	0.0	0.0204	0.0399	0.0586	0.0766	0.1104	0.1417	0.1709	0.1981

$$K_1 = -\theta'_{\zeta=0}. \tag{3.5}$$

For T_w = const. and q_w = const., the exponent in Eq. (2.30) for laminar flow is the same: $n_* = 0$. The constant K_1 in an increasing function of the parameter n_*: e.g. for air at $Pr = 0.71$, the value of K_1 grows up by 313 % with n_* increasing from −1 to 3 (see Table 3.1). For non-isothermal disks and smaller values of the Pr number, there are no experimental data in literature that would enable validation of the constant K_1.

The first known approximate solution valid for gases was derived by Dorfman [2]

$$K_1 = K_2 = 0.308(n_* + 2)^{1/2} Pr^{1/2}. \tag{3.6}$$

Values of K_1 predicted by Eq. (3.6) by 34–238 % surpass the data from Table 3.1 for $n_* = -1.5$–0. These deviations are magnified with decreasing Pr number.

For the conditions of T_w = const. and Pr = 0–∞, the following theoretical solutions for the constant K_1 were derived in the works [31, 32], respectively

$$K_1 = 0.6109Pr/(0.5301 + 0.3996Pr^{1/2} + Pr)^{2/3}, \tag{3.7}$$

$$K_1 = 0.6Pr/(0.56 + 0.26Pr^{1/2} + Pr)^{2/3}. \tag{3.8}$$

Predictions by Eqs. (3.7) and (3.8) divert from the data in Table 3.1 by maximum 4 and 5 %, accordingly. In the limit at $Pr \to 0$, Eqs. (3.7) and (3.8) tend to a relation $K_1 \sim Pr$ that conforms to the asymptotic equation $K_1/Pr = 0.885$ derived in [33]. For $Pr \to \infty$, Eqs. (3.7) and (3.8) reduce to a relation $K_1 \approx 0.62Pr^{1/3}$, which coincides with that obtained in the work [33]. Validity of Eqs. (3.7) and (3.8) is restricted by the case of $n_* = 0$.

3.2 Transition to Turbulent Flow

Increasing the angular velocity of the disk rotation causes instability of laminar flow and set up of spiral vortices followed by a development of turbulence. Authors [34] detected 32 spiral waves over a rotating disk, with the angle ε between a perpendicular to the spiral and the radial direction being about 14°. Visualization [35] by means of a thin layer of Kaolin on the surface of a disk revealed simultaneous co-existence of the regions of laminar flow, laminar flow with 28–31 spiral vortices and turbulent flow. The number of spiral vortices is a function of the local Reynolds number Re_ω [36–40]. For instance, 14–16 vortices with the angle ε = 20° were detected in [37]. The number of vortices n_v can be predicted by an empirical equation [40]

$$n_v = 0.0698 \cdot Re_\omega^{1/2}. \tag{3.9}$$

Equation (3.9) is confirmed by experimental data [37, 40, 41].

Detailed theoretical investigations of instabilities emerging in flow over a rotating disk were performed in [8, 42–54]. At the onset of instability, the predicted value $Re_\omega^{1/2} = 286$ conforms to the experimental values [40]. Averaging of the Reynolds number at the onset of transition to turbulent regime yields the value $Re_\omega^{1/2} = 513$ [42].

Experimental data of different authors summarizing the Reynolds numbers at the onset of instability, beginning and end of transition to turbulent flow are documented in Table 3.2. Table 3.3 lists results of averaging within every sub-group of techniques (heat and mass transfer techniques; thermo-anemometry; visualization and acoustics).

Table 3.2 Values of the Reynolds number at the onset of instability, beginning and end of transition to turbulent flow [1]

No.	Experimental technique	Values of the Reynolds number Re_ω		
		Onset of instability	Beginning of transition	End of transition
1	Thermoanemometer [34]	2.1×10^5		3.1×10^5
2	Thermoanemometer [56]	8.8×10^4	2.5×10^5	3.2×10^5
3	Thermoanemometer [40]	8.6×10^4	$(2.6–2.8) \times 10^5$	
4	Thermoanemometer [41]	9×10^4		$(2.95–3.1) \times 10^5$
5	Thermoanemometer/visualization [57]		$(2.4–2.63) \times 10^5$	
6	Thermoanemometer [38, 39]		$(2.4–2.6) \times 10^5$	
7	Thermoanemometer [8]		$(2.5–2.64) \times 10^5$	
8	Thermoanemometer [7]		2.5×10^5	
9	Thermoanemometer [54, 58]	9.5×10^4		3.4×10^5
10	Acoustic measurements [59]	1.35×10^5	2.55×10^5	2.75×10^5
11	Visualization (Kaolin) [35]	1.8×10^5		3.0×10^5
12	Visualization (naphthalene) and acoustic measurements [37]	1.5×10^5 $(0.33–1.9) \times 10^5$		2.65×10^5
13	Visualization in water [60]		$(2.8–2.86) \times 10^5$	$(3.2–4.6) \times 10^5$
14	Heat transfer coefficient [17]	2.0×10^5	2.4×10^5	
15	Heat transfer coefficient [6]		2.9×10^5	3.6×10^5
16	Heat transfer coefficient [23]		1.95×10^5	2.5×10^5
17	Heat transfer coefficient [27, 28]		2.5×10^5	3.2×10^5
18	Heat transfer coefficient [61]		2.4×10^5	
19	Heat transfer coefficient [62]		2.4×10^5	
20	Heat transfer coefficient [63]		2.2×10^5	3.4×10^5
21	Mass transfer coefficient (naphthalene sublimation) [18]		2.7×10^5	
22	Mass transfer coefficient (naphthalene sublimation) [16]		1.9×10^5	2.75×10^5
23	Mass transfer coefficient (naphthalene sublimation) [64]		1.8×10^5	
24	Mass transfer coefficient (naphthalene sublimation) [20]		2.0×10^5	
25	Mass transfer coefficient (electrochemistry) [42]	1.7×10^5	2.6×10^5	3.5×10^5
26	Mass transfer coefficient (electro-chemistry) [65–67]		2.3×10^5	2.9×10^5
27	Mass transfer coefficient (electrochemistry) [68]		2.2×10^5	3.0×10^5

Table 3.3 Averaged Re_ω numbers for boundaries of flow regimes over a rotating disk [1]

No.	Experimental technique	Values of the Reynolds number Re_ω		
		Onset of instability	Beginning of transition	End of transition
1	Thermoanemometer	1.14×10^5	2.54×10^5	3.15×10^5
2	Visualization and acoustic measurements	1.4×10^5	3.08×10^5	3.3×10^5
3	Methods of heat and mass transfer	1.85×10^5	2.3×10^5	3.1×10^5
4	Average value	1.46×10^5	2.64×10^5	3.18×10^5

Different criteria for determining the critical values of Re_ω apparently entailed inconsistency of experimental data. In spite of the setup of spiral vortices already for $Re_\omega = (1.14–1.4) \times 10^5$, surface heat and mass transfer rates deviate from the laminar flow data at larger Reynolds numbers $Re_\omega = 1.85 \times 10^5$. Also, disagreement between experiments may have been resulted from possible vibrations, different roughness etc. The ratio of the Reynolds numbers at the setup and end of the transition to turbulent flow yields the value 1.21; this agrees with the respective ratio 1.3 for a flat plate flow [55].

The heat/mass transfer rate measured in the transitional flow regime at the same local Reynolds number Re_ω was different in various experiments

$$Nu = 10.0 \times 10^{-20} \cdot Re_\omega^4 \quad \text{for} \quad Re_\omega = (1.95–2.5) \times 10^5 \quad \text{Ref. [23]}, \tag{3.10}$$

$$Nu = 2.65 \times 10^{-20} \cdot Re_\omega^4 \quad \text{for} \quad Re_\omega = (2.9–3.6) \times 10^5 \quad \text{Ref. [6]}, \tag{3.11}$$

$$Nu = 8.01 \times 10^{-14} \cdot Re_\omega^{2.8} \quad \text{for} \quad Re_\omega = (2.6–3.2) \times 10^5 \quad \text{Ref. [27, 28]}, \tag{3.12}$$

$$Sh = 20.0 \times 10^{-20} \cdot Re_\omega^4 \quad \text{for} \quad Re_\omega = (2.0–2.5) \times 10^5 \quad \text{Ref. [16]}, \tag{3.13}$$

$$Sh = 3.4 \times 10^{-14} \cdot Re_\omega^3 Sc^{1/3} \quad \text{for} \quad Re_\omega = (2.0–3.0) \times 10^5 \quad \text{Ref. [67]}. \tag{3.14}$$

In fact, Eqs. (3.11) and (3.12) from one side, and Eqs. (3.10), (3.13) and (3.14) from the other side form two different groups that suggest the different ranges of the Reynolds numbers Re_ω for transitional flow, which is in line with the data from Table 3.2.

Equation (3.12) does not follow the original experiments [27, 28]. Corrected coefficient $K_1 = 1.2 \times 10^{-13}$ (with the end of transition to turbulent flow at $Re_\omega = 3.7 \times 10^5$) [1] is consistent with experiments [6, 27, 28] and Eq. (3.11) (Fig. 3.3). Equation (3.10) is in disagreement with Eqs. (3.12) and (3.13) for $Re_\omega = (2.5–2.9) \times 10^5$ (Fig. 3.3).

Equation (3.13) conforms to the experiments [16], though its validity should be modified to the range $Re_\omega = (1.9–2.75) \times 10^5$ [1]. Equation (3.14) deduced for large Schmidt numbers $Sc = 1192–2465$ is in a good consistency with Eq. (3.13) for $Sc = 2.28$.

Roughness of the disk surface may provoke earlier transition to turbulent flow in comparison with a smooth disk. In laminar flow, surface roughness does not influence heat transfer. In experiments for $Re_\omega > 5.6 \times 10^4$ [19, 69, 70], heat transfer increased very moderately at the expense of the enlarged surface area of a rough disk.

Surface roughness noticeably affects the boundaries of the flow regimes. The Reynolds number at onset of the spiral vortices (whose number diminished from 32 to 25) decreased from $Re_\omega = 0.95 \times 10^5$ on a smooth disk to $Re_\omega = 0.5 \times 10^5$ on a rough disk [58].

In experiments [57] on a rough disk, the onset of transition to turbulence began at $Re_\omega = (0.23–1.23) \times 10^5$ (on a smooth disk, at $Re_\omega = (2.4–2.63 \times 10^5)$. The boundary of the end of transition also shifted down to $Re_\omega = 2.55 \times 10^5$ on a rough disk in comparison with $Re_\omega = 3.4 \times 10^5$ on the smooth disk in [58]. In the paper [18], the end of transition shifted down to $Re_\omega = 2.0 \times 10^5$ on a rough disk as compared to $Re_\omega = 2.7 \times 10^5$ on a smooth disk.

At the beginning of the transition to turbulent flow, the heat and mass transfer rate in the experiments [18] on a rough disk was by 34 % higher in comparison with a smooth disk over the range of the Reynolds numbers studied in that work.

3.3 Turbulent Flow

3.3.1 Parameters of the Boundary Layer

Tangential v_φ and radial v_r velocity components described by the power-law function (2.40) and the quadratic Eq. (2.58), respectively, are in a good agreement with the experiments [7, 12] (see Figs. 2.2, 2.3 and 2.4). In addition, Eq. (2.58) conforms to the experimental data in the outer part of the boundary layer better than Eq. (2.41) (Fig. 2.4).

Temperature profiles θ approximated by the power-law Eq. (2.53) at $n_T = 1/4–1/5$ match well to the profiles measured in the work [6] for $q_w = \text{const.}$ (see Fig. 2.5).

This chapter represents a validation of the present integral method (described in Chap. 2) in comparison with the von Karman's method, Eq. (2.41), often incorporated in many integral methods. On the basis of Eqs. (2.40) and (2.41) [71], the rest of boundary layer parameters can be written as Eqs. (2.77)–(2.81) with their constants expressed as [3, 4]

$$\alpha^2 = \frac{4(2+3/n)(1+2/n)(3+1/n)}{(16n^{-3}+85n^{-2}+145n^{-1}+66)/n^2}, \tag{3.15}$$

$$\gamma^{(3n+1)/(n+1)} = C_n^{-2/(n+1)} \frac{2(n+1)(2n+2)(3n+1)(n+2)(1+\alpha^2)^{(1-n)/[2(n+1)]}}{3(11n+5)\alpha n}, \tag{3.16}$$

$$\varepsilon_m = \frac{2\pi\alpha\gamma}{(1+n)(2+n)}, \quad \varepsilon_M = \frac{6\pi\alpha\gamma/n^2}{(1+1/n)(2+1/n)(1+2/n)}. \tag{3.17}$$

Constants α, γ, ε_m and ε_M computed by Eqs. (2.83)–(2.86) (present integral method) and (3.15)–(3.17) (von Karman's method) are listed in Table 3.4. Computed values for γ, and ε_M are practically identical for both methods at the same values of n, while the constants α and ε_m representing the approximation of v_r are noticeably different.

Values of the flow swirl angle $\alpha = \tan\varphi_w$ from Table 3.4 are plotted in Fig. 3.4 together with experiments [6, 7] and predictions by other models [72, 76]. Experimental data for the parameter α behave as a decreasing function of the Reynolds number Re_ω within the limits suggested by Eq. (2.83) at $n = 1/7$ (upper) and $n = 1/9$ (lower). The exponent n diminishes for larger Re_ω [2, 3], which is followed with a decrease in the α values (like in the experiments). The highest magnitude of $\alpha = 0.162$ (at $n = 1/7$) by von Karman's Eq. (3.15) matches to the lowest level of α found experimentally [6, 7, 72].

Table 3.4 Constants of the solution for a free disk [1]

Coefficient	Equation, source	$n = 1/7$	$n = 1/8$	$n = 1/9$	$n = 1/10$
C_n	(2.49), [71]	8.74	9.71	10.6	11.5
α	(2.83), [73]	0.2087	0.1842	0.1649	0.1493
α	(3.15), [3]	0.162	0.143	0.128	0.116
γ	(2.84), [73]	0.5299	0.4977	0.4773	0.4597
γ	(3.16), [3]	0.526	0.497	0.479	0.463
ε_m	(2.85), [73]	0.1806	0.1542	0.1355	0.1204
ε_m	(3.17), [3]	0.219	0.187	0.164	0.146
ε_M	(2.86), [73]	0.1466	0.1127	0.0901	0.0734
ε_M	(3.17), [3]	0.1458	0.1122	0.0896	0.073

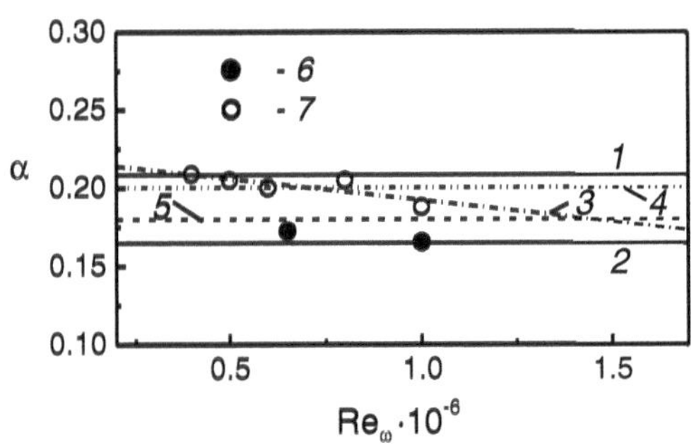

Fig. 3.4 Tangent α of the flow swirl angle at the wall [1]. Present integral method [1, 74, 75], Eq. (3.15): *1*—$n = 1/7$; *2*—1/9 (or 1/7, von Karman [71]). *3*—model [72]; *4*—Eq. (2.44) [76], $\alpha = 0.2003$ for $n = 1/7$, $b = 0.7$, $c = 1.2$; *5*—model [77]. Experiments: *6*—[7]; *7*—[6]

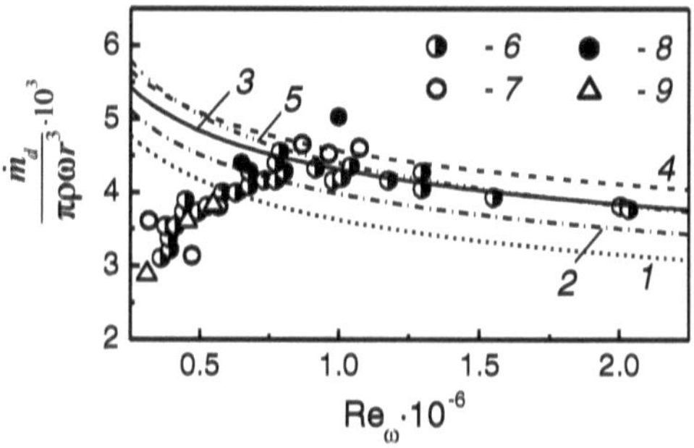

Fig. 3.5 Dimensionless mass flow rate through the boundary layer [1]. Equation (2.85): *1*—n = 1/7; *2*—1/8; *3*—1/9; *4*—1/10; *5*—1/7, von Karman's method (3.17) [71]. Experiments: *6*—[72]; *7*—[78]; *8*—Shevchuk based on experimental data of [6]; *9*—[79]

As it can be seen from Fig. 3.5, for $Re_\omega \geq 0.75 \times 10^6$, the dimensionless mass flow rate predicted by Eq. (2.85) for $n = 1/9$ (the present integral method) is in good agreement with experiments [6, 72, 78] and predictions by the von Karman's method for $n = 1/7$. For $n = 1/7–1/8$ and $Re_\omega = (0.5–0.75) \times 10^6$, Eq. (2.85) also reasonably estimates the mass flow rate, although it incorrectly predicts the functional dependence of $\dot m_d/(\rho_\infty \omega r^3)$ on Re_ω. At the same time, approach [71] yields inaccurate predictions for $Re_\omega = (0.5–0.75) \times 10^6$.

Equation (2.42) for $\alpha \equiv 0.18$ for $n = 1/7$ was employed in [77]. The value $\alpha = 0.$ 18 insignificantly differs from the values by the present integral method given in Table 3.4.

In the work [79], velocity profiles did not exhibit self-similarity in turbulent flow over the range $Re_\omega \approx (3.1–6.64) \times 10^5$, whereas the wall value α increased together with Re_ω (or r/b). The rather complicated integral method based on Eq. (2.43) and original experiments [79] tried to model this phenomenon. However, because of its excessive complexity, the method [79] was not developed further to include a model for heat transfer.

The work [72] also employed Eq. (2.42) together with an empirical equation for the mass flow rate in the boundary layer. However, the accuracy of this integral method did not exceed that of the von Karman's method [71].

To conclude, the present integral method, Chap. 2, enables a more accurate prediction of the radial velocity distributions, values of α and mass flow rate in the boundary layer than other integral methods. The values of the parameter n must be selected based on Figs. 2.2, 2.3, 2.4, 3.5 and 3.6, as well as the data for the Nusselt number presented below.

Predictions of the moment coefficient C_M by Eq. (2.80) for $n = 1/7$ are in good agreement with the experiments in Fig. 3.2 for $Re_\varphi < 2.0 \times 10^6$. For larger values of Re_φ, predictions by Eq. (2.80) lie below the experiments. Values $n = 1/8–1/10$ used in Eq. (2.80) shift the predictions close to experiments at $Re_\varphi \geq 3.0 \times 10^6$ (see Fig. 3.2).

Logarithmic velocity profiles yield in the end the following equations for C_M

$$C_M^{-1/2} = 1.97 \lg\left(Re_\varphi \sqrt{c_M}\right) + 0.03, \tag{3.18}$$

$$C_M = 0.982\left(\lg Re_\varphi\right)^{-2.58}. \tag{3.19}$$

Equation (3.19) of Dorfman [2] yields the best overall match with the experiments, while Eq. (3.18) [80] sets the upper level restriction for the experimental data in Fig. 3.2.

In the paper [14] also employing a logarithmic model, the following approximation for the coefficient C_M was derived, which is valid over the range $Re_\varphi \approx 4.0 \times 10^5$–$2.0 \times 10^6$

$$C_M = 0.13 Re_\varphi^{-0.185}. \tag{3.20}$$

Equation (3.20) agrees well with the experiments [12, 13] re-evaluated in [14], though it is by 5–10 % higher than the data of other authors for $Re_\varphi \geq 2.0 \times 10^6$ (see Fig. 3.2).

Logarithmic velocity profiles entail noticeable complications of mathematical models; therefore they were practically rarely used in integral methods.

3.3.2 Surface Heat Transfer: Different Experiments and Solutions

The integral method of Dorfman [2], applied together with the boundary condition (2.30) and $n_R = 0.8$, yielded known solutions for the coefficients K_1 and K_2 in Eq. (3.4)

$$K_1 = 0.0197(n_* + 2.6)^{0.2} Pr^{0.6}, \tag{3.21}$$

$$K_2 = K_1(n_* + 2)/(n_* + 2.6). \tag{3.22}$$

The multiplier $Pr^{0.6}$ was obtained by the authors [3, 4].

Table 3.5 represents results for the coefficients K_1 and K_2 depending on n_* and computed by Eqs. (3.21) and (3.22) together with experimental data from different sources. Measurements [17, 61, 81–83] for T_w = const. (or $n_* = 0$) made mainly in 1950th and 1960th are in good agreement with Eqs. (3.21) and (3.22). However, the modern accurate measurements for K_1 are lower than the Dorfman's predictions in Table 3.5: by 4.4 % [6, 15, 23, 62, 63] and 9.5 % [84]. Experimental values [62, 63] are estimations made by the author of the present work based on the published data.

For the thermal boundary condition q_w = const. (or $n_* = -0.6$), the coefficient K_1 in measurements [6, 27, 28] was by 14.1 % smaller than the estimation by

Table 3.5 Values of the constants K_1 and K_2 for $n_R = 0.8$ and $Pr = 0.72$ [1]

Coefficient	Source or equation	$n_* = -0.6$	$n_* = 0$	$n_* = 2$	$n_* = 6$
K_2	[17, 61, 81, 82]		0.015		
K_1	[83]		0.0194		
K_1	[84]		0.0179		
K_2	[84]		0.0138		
K_1	[6, 15, 23, 62, 63]		0.0188		
K_2	[23, 62]		0.0145		
K_1	[6, 27, 28]	0.0163			
K_1	Dorfman, Eq. (3.21)	0.0186	0.0196	0.022	0.0249
K_2	Dorfman, Eq. (3.22)	0.013	0.0151	0.0191	0.0231

Eq. (3.21) (Table 3.5). CFD simulations [85] for q_w = const. are approximated by Eq. (3.4) at $n_R = 0.83$ and $K_1 = 0.0111$, $K_2 = 0.0086$. Authors [81] obtained the same value $K_2 = 0.015$ for the cases q_w = const. and T_w = const., which does not look trustworthy.

For $n_* = -0.2$, predictions by Eqs. (3.21) and (3.22) by up to 10 % surpass the Nusselt numbers measured in [86, 87]. Empirical values of the coefficient K_1 were not estimated in [86, 87], apparently because the T_w distribution did not comply with Eq. (2.30).

The so called theory of local modelling [2–4, 88] (see Sect. 2.3.2) yields a solution for the Nusselt number for the situation, where Eq. (2.30) does not hold. This solution proved to be insufficiently accurate and has therefore not been further developed.

Average Nusselt number for an entire disk. Average Nusselt numbers Nu_{av} for an *entire disk*, where laminar, transitional and turbulent flows co-exist simultaneously, are often of interest in technical applications.

It was assumed in the model [61] that transition to turbulence takes place abruptly at the Reynolds number $Re_{\omega,tr}$ calculated at a coordinate r_{tr}. Following this assumption, one can present Nu_{av} (see its definition in Nomenclature) as follows

$$Nu_{av} = \frac{b\left[\int_0^{tr} Nu_{lam}(T_w - T_\infty)\mathrm{d}r + \int_{tr}^{b} Nu_{turb}(T_w - T_\infty)\mathrm{d}r\right]}{\int_{tr}^{b}(T_w - T_\infty)r\mathrm{d}r}. \tag{3.23}$$

The Nusselt numbers to be substituted in Eq. (3.23) are defined by Eq. (3.4) for Nu, where the constants are $K_{1,lam}$, $n_R = 1/2$ for laminar flow, and $K_{1,turb}$, n_R for turbulent flow.

If the disk temperature is described by Eq. (2.30), this yields [1]

$$Nu_{av} = K_{1,lam}Re_{\omega,tr}^{1/2}\left(\frac{Re_{\omega,tr}}{Re_\varphi}\right)^{n_*/2+1/2} + \frac{2+n_*}{2n_R+1+n_*}K_{1,turb}Re_\varphi^{n_R}\left[1-\left(\frac{Re_{\omega,tr}}{Re_\varphi}\right)^{n_*/2+n_R+1/2}\right]. \tag{3.24}$$

For T_w = const. (or $n_* = 0$), Eq. (3.24) simplifies to [1]

$$Nu_{av} = K_{1,lam} Re_{\omega,tr}^{1/2} \left(\frac{Re_{\omega,tr}}{Re_\varphi}\right)^{1/2} + \frac{2}{2n_R+1} K_{1,turb} Re_\varphi^{n_R} \left[1 - \left(\frac{Re_{\omega,tr}}{Re_\varphi}\right)^{n_R+1/2}\right]. \tag{3.25}$$

Equation (3.24) holds, if $Re_\varphi \geq Re_{\omega,tr}$. Given $Re_\varphi < Re_{\omega,tr}$, the second term in Eq. (3.24) vanishes. Asymptotically at $Re_\varphi \gg Re_{\omega,tr}$, the turbulent flow fully occupies a disk, and Eq. (3.24) turns into Eq. (3.4) for Nu_{av}, where [1]

$$K_{2,turb} = \frac{2+n_*}{2n_R+1+n_*} K_{1,turb}. \tag{3.26}$$

Once $n_R = 0.8$, Eqs. (3.22) and (3.26) become identical.

In Fig. 3.6, curve 4 is based on Eq. (3.25) for the case T_w = const. and the values $n_R = 0.8$, $2K_{1,turb}/(2n_R+1) = 0.015$, $K_{1,lam} = 0.4$, $Re_{\omega,tr} = 2.4 \times 10^5$ [61]. Curve 4 lies by 15 % lower than the original experimental data [61] for $Re_\omega \leq 6.5 \times 10^5$. A smaller value of $Re_{\omega,tr} = 2.0 \times 10^5$ (curve 5) provides a better match of Eq. (3.25) with experiments [1].

3.3.3 Effect of Approximation of the Radial Velocity Profile

Effect of the tangent of the flow swirl angle was taken into account via a model

$$\tan\varphi = \alpha(1-\xi)^\sigma. \tag{3.27}$$

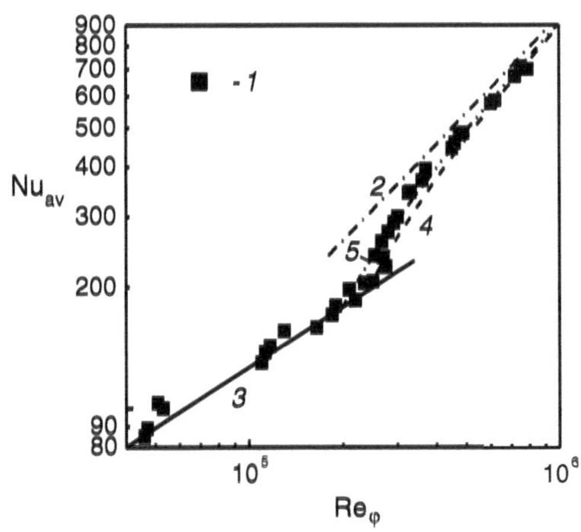

Fig. 3.6 Average Nusselt numbers of an *entire* disk rotating in still air [1]. Experiments: *1*—[61]. Calculations by Eq. (8.4): *2*—developed turbulent flow, $n_R = 0.8$, $K_2 = 0.015$ (Table 3.5) [61]; *3*—laminar flow, $n_R = 1/2$, $K_1 = 0.4$ [61]. Calculation of Nu_{av} for an *entire* disk: *4*—Eq. (3.25) at $Re_{\omega,tr} = 2.4 \times 10^5$ [61]; *5*—Eq. (3.25) at $Re_{\omega,tr} = 2.0 \times 10^5$ [1]

For the constant σ, the values $\sigma = 2$, 1 and 0 were selected, which enabled undertaking parametric studies, whereas the value $\sigma = 2$ remained the major one in the present integral method. In the end, solutions for the parameters in Eqs. (2.77)–(2.87) look as

$$\alpha = \left[\frac{C_1}{(3+m)B_0 + (4+m)D_0}\right]^{1/2}, \tag{3.28}$$

$$H_9 = \alpha D_0(4+m), \tag{3.29}$$

$$\varepsilon_M = 8\pi\alpha\gamma D_0. \tag{3.30}$$

Here $B_0 = B_1$ and $D_0 = D_1$ for $\sigma = 2$; $D_0 = 1/(n+1) - 1/(n+2) - 1/(2n+1) + 1/(2n+2)$ and $B_0 = D_2$ for $\sigma = 1$; $B_0 = 1/(2n+1)$ and $D_0 = 1/(n+1) - 1/(2n+1)$ for $\sigma = 0$ [1]. The case with $\sigma = 2$ coincides with Eqs. (2.83)–(2.87) at $\kappa = 0$, $\beta = 0$.

The effect of the exponent σ on the radial velocity profiles is elucidated in Fig. 3.7.

Table 3.6 lists numerical data for the constants α, γ and ε_M computed by Eqs. (3.28)–(3.30) (data for $\sigma = 1$ and 2 partially repeat those from Table 3.4).

In Fig. 3.7, the radial velocity profile for $\sigma = 0$ is qualitatively and quantitatively different from the other v_r profiles. Anyway, in spite of this, the deviation of the calculated C_M values for $\sigma = 0$ from the basic case of $\sigma = 2$ is 5 % at maximum (Table 3.6).

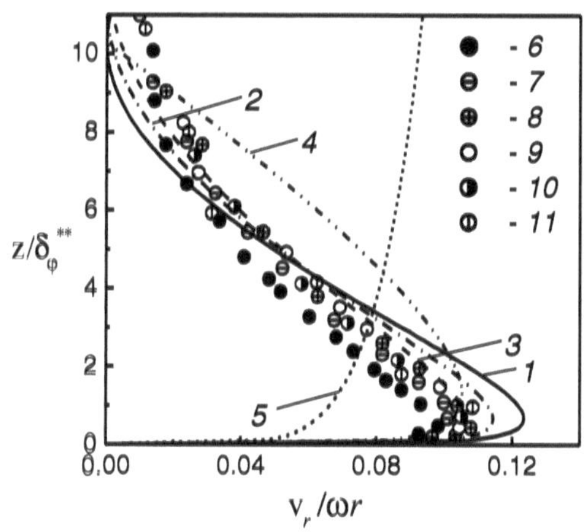

Fig. 3.7 Profiles of the radial velocity component in the turbulent boundary layer over a free rotating disk [1]. *1*—$n = 1/7$, *2*—1/8, *3*—1/9. Equation (2.41), [71]: *4*—$n = 1/7$. Equation (2.44): *5*—$\sigma = 0$, $n = 1/7$. Experiments: *6*—$Re_\omega = 0.4 \times 10^6$, *7*—$0.65 \times 10^6$, *8*—$0.94 \times 10^6$, *9*—$1.6 \times 10^6$ [12], *10*—0.6×10^6, *11*—1.0×10^6 [7]

Table 3.6 Constants in Eqs. (2.77)–(2.80), (3.28) and (3.30)–(3.41) at different σ [1, 89]

Coefficient	n = 1/7	n = 1/8	n = 1/9	n = 1/10
α, $\sigma = 2$	0.2087	0.1842	0.1649	0.1493
α, $\sigma = 1$	0.162	0.143	0.128	0.116
α, $\sigma = 0$	0.0925	0.0818	0.0733	0.0664
γ, $\sigma = 2$	0.530	0.4977	0.4773	0.4597
γ, $\sigma = 1$	0.526	0.497	0.479	0.463
γ, $\sigma = 0$	0.616	0.588	0.571	0.556
ε_M, $\sigma = 2$	0.1466	0.1127	0.0901	0.0734
ε_M, $\sigma = 1$	0.146	0.112	0.09	0.073
ε_M, $\sigma = 0$	0.139	0.107	0.086	0.0704
K_V, $\sigma = 2$	0.203	0.183	0.1661	0.1523
K_V, $\sigma = 1$	0.167	0.15	0.1364	0.125
K_V, $\sigma = 0$	0.111	0.10	0.0909	0.0833
K_3, $\sigma = 2$	0.02683	0.02079	0.01673	0.0137
K_3, $\sigma = 1$	0.0267	0.0207	0.0166	0.0136
K_3, $\sigma = 0$	0.0255	0.0198	0.016	0.0131
n_R	0.8	0.8182	0.8333	0.8462
m	0.6	0.6363	0.6667	0.6923
n_p, $\sigma = 2$	0.5018	0.4894	0.4797	0.4719
n_p, $\sigma = 1$	0.48	0.471	0.463	0.457
n_p, $\sigma = 0$	0.45	0.444	0.44	0.436

Equation (2.70), complemented with Eqs. (3.27)–(3.30), can be analytically solved only for the condition $\Delta \geq 1$ [75, 89]. For $\sigma = 2$, such a solution degenerates to Eqs. (2.89)–(2.91) at $N = 0$ and $\beta = 0$.

Thus, for $\Delta \geq 1$, the coefficients in Eq. (3.4) can be written as

$$n_R = (n + 1)/(3n + 1), \tag{3.31}$$

$$K_1 = K_3 \Delta^{-n} Pr^{1-n_p}, \tag{3.32}$$

$$\Delta^{-n} = \left[\frac{4 + m}{2 + m + n_*} K_V Pr^{-n_p} + (1 - K_V) \right]^{-1}, \tag{3.33}$$

$$K_1 = K_3 Pr \left[\frac{4 + m}{2 + m + n_*} K_V + (1 - K_V) Pr^{n_p} \right]^{-1}, \tag{3.34}$$

$$K_2 = K_1 (n_* + 2)/(2 + n_* + m), \tag{3.35}$$

$$K_3 = A_c (1 + \alpha^2)^{1/2} = C_n^{-2/(n+1)} \gamma^{-2n/(n+1)} (1 + \alpha^2)^{0.5(1-n)/(n+1)}, \tag{3.36}$$

$$K_V = 1 - D_{2*}/A_{1*}. \tag{3.37}$$

In Eq. (3.37), the constants are defined as [1]: $D_{2*} = D_2$ and $A_{1*} = A_1$ for $\sigma = 2$; $D_{2*} = 1/(2n + 1) - 1/(2n + 2)$ and $A_{1*} = 1/(n + 1) - 1/(n + 2)$ for $\sigma = 1$; $D_{2*} = 1/(2n + 1)$ and $A_{1*} = 1/(n + 1)$ for $\sigma = 0$. Parameters K_3 and K_V presented in Table 3.6 do not depend on n_*. In view of the relation $2n_R = 1 + m$ [which follows from Eqs. (2.78) and (3.31)], Eqs. (3.35) and (3.26) become identical.

Let us denote $G = Pr^{n_p}(2 + m + n_*)/(4 + m)$, use the Taylor's series expansion of the entire term in brackets in Eq. (3.34) for $Pr \to 1$, $n_* \to 2$ in the neighborhood of $G = 1$ and neglect summands of an infinitesimal order. As a result, one can obtain [74, 75, 89]

$$Nu = K_3 \left(\frac{2 + m + n_*}{4 + m} \right)^{K_V} Re_{\omega}^{n_R} Pr^{1 - n_p(1 - K_V)}. \tag{3.38}$$

Let us further consider a solution for the most widely used value $n = 1/7$ and round down: $K_V = 0.2$. In doing so, Eq. (3.38) becomes identical to the Dorfman's solution (3.21), if one sets the overall exponent for the Pr number equal to 0.6 and keeps this unchanged and independent of n. Based on this, the unknown parameter n_p can be determined as

$$n_p = 0.4/(1 - K_V). \tag{3.39}$$

Table 3.6 contains numerical values of the parameter n_p calculated by Eq. (3.39). Equation (3.38) coincides with Eq. (3.34) solely for $Pr \to 1$ and $n_* \to 2$. A mathematical interpretation of this fact is that Eq. (3.38) represents a particular case of Eq. (3.34).

For $\Delta \le 1$ and $N = 0$ and $\beta = 0$ (a single rotating disk), Eq. (2.89) remains transcendental

$$\Delta^{2n+1}\left(a_* - 2b_*\Delta + c_*\Delta^2\right) = \frac{4 + m}{2 + m + n_*}(a_* - 2b_* + c_*)Pr^{-n_p}. \tag{3.40}$$

In Eq. (3.40), $n_T = n$ has been set and the subscript "T" at the coefficients a_*, b_* and c_* has been omitted. Equation (3.40) holds for the case $Pr \ge 1$ (or $Sc \ge 1$): naphthalene sublimation in air, flows of liquids etc. Therefore, an analysis of Eq. (3.40) is relegated to Chap. 6.

Table 3.7 and Fig. 3.8 illustrate the influence of the parameter σ on the coefficients K_1 and K_2 in Eqs. (3.34) and (3.35) [as compared to Eqs. (3.21) and (3.22)] in a form of a dependence on the parameter n_* for Pr numbers 0.72 (air) and 1.0. Equation (3.34) at $\sigma = 2$ demonstrates the best agreement with experiments, especially for $n_* \le 0$, which **confirms the choice of the value $\sigma = 2$ as a major one in the present integral method**.

Table 3.7 Constants in Eqs. (3.34) and (3.35) for different values σ [75, 89]

Coefficient	Equation	$n_* = -0.6$	$n_* = 0$	$n_* = 2$	$n_* = 6$
Calculation for $Pr = 0.72$					
K_1	Equation (3.34), $\sigma = 2$	0.0169	0.0187	0.022	0.0246
K_2	Equation (3.35), $\sigma = 2$	0.0118	0.0144	0.0191	0.0229
K_1	Equation (3.34), $\sigma = 1$	0.0176	0.0191	0.0219	0.0240
K_1	Equation (3.34), $\sigma = 0$	0.0180	0.0191	0.0209	0.0222
Calculation for $Pr = 1$					
K_1	Dorfman, Eq. (3.21)	0.0226	0.0238	0.0267	0.0303
K_2	Dorfman, Eq. (3.22)	0.0158	0.0183	0.0232	0.0282
K_1	Equation (3.34), $\sigma = 2$	0.0212	0.0232	0.0268	0.0296
K_2	Equation (3.35), $\sigma = 2$	0.0149	0.0178	0.0233	0.02755
K_1	Equation (3.34), $\sigma = 1$	0.0219	0.0237	0.0267	0.0289
K_1	Equation (3.34), $\sigma = 0$	0.0222	0.0235	0.0255	0.0269

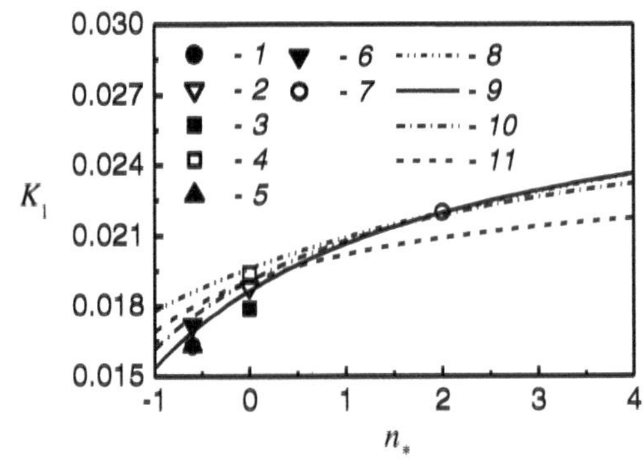

Fig. 3.8 Effect of the exponents n_* and σ on the constant K_1 in turbulent air flow ($Pr = 0.72$) [1]. Experiments: *1*—[6, 27, 28]; *2*—[15, 23, 62, 63]; *3*—[84]; *4*—[83]; *5*—[85], K_1, lower limit; *6*—[85], K_1, upper limit; *7*—[2, 3, 4]. Calculations: *8*—Eq. (3.21) [2]; *9*—Eq. (3.34), $\sigma = 2$; *10*—Eq. (3.34), $\sigma = 1$; *11*—Eq. (3.34), $\sigma = 0$

Data in Tables 3.5 and 3.7 elucidate also inaccuracies of the Dorfman's Eqs. (3.21) and (3.22), which amplify as soon as the Prandtl number diverges from unity.

Figure 3.3 demonstrates that the local Nusselt numbers by Eq. (3.34) at $K_1 = 0.0169$ for q_w = const. and $K_1 = 0.0187$ for T_w = const. agree well with experiments [6].

Using the definition of the Nu number and Eq. (3.4), one can determine the exponent n_* in Eq. (2.30) for the boundary condition q_w = const. [26]

$$\frac{T_w - T_\infty}{T_\infty} = \frac{q_w b}{\lambda T_\infty K_1} Re_\varphi^{-(n+1)/(3n+1)} x^{(n-1)/(3n+1)}. \quad (3.41)$$

For the case $q_w \equiv$ const., Eq. (3.41) yields the value $n_* = (n-1)/(3n+1) = -m$ [26]. Table 3.6 summarizes the values of the parameter m calculated at different values of n.

Experimental disk temperature distributions measured in [6] are depicted in Fig. 3.9. Flow was turbulent at $Re_\omega > 3.6 \times 10^5$ [6]. For $Re_\varphi = 1.6 \times 10^6$ illustrated in Fig. 3.9, the turbulent region was localized over the span $x = 0.474$–1.0. The disk itself comprised three annular regions [6]: an unheated region at $x = 0$–0.15; regions $x = 0.15$–0.4 and 0.4–0.96 heated via two separate heaters having the same power. Data 1 and 2 correspond to the cases with both heaters or only the external one switched on, respectively.

For the case 1, calculations have only been performed for the region $x > 0.5$, where the disk temperature predicted by Eq. (3.41) at $q_w =$ const. (i.e. $n_* = -0.6$) fairly well matches the measurements. Here the experimental data are: $K_1 = 0.0163$, $q_w = 710$ W/m^3, $T_\infty = 298.9$ K, $Re_\varphi = 1.6 \times 10^6$, $b = 0.5$ m; $\lambda = 0.02624$ W/(m K) for air at $T_\infty \approx 300$ K [90].

The wall temperature distribution in the heated region $x > 0.5$ in case 2 (internal heater off) was practically constant (see Fig. 3.9).

Thus, for the case $q_w =$ const. (or $n_* = -0.6$), the constant $K_1 = 0.0169$ calculated by Eq. (3.34) at $n = 1/7$, $n_R = 0.8$ and $\sigma = 2$ differs from the measured value $K_1 = 0.0163$ [6, 27, 28] by only 3.7 %. Dorfman's constant $K_1 = 0.0186$ by Eq. (3.21) diverts from the measurements by 14.1 %. For $1/n = 8.7647$ and $n_R = 0.83$, Eq. (3.34) yields the value $K_1 = 0.0115$, which means only 3.6 % mismatch to the value $K_1 = 0.0111$ obtained in simulations [85].

For the case $T_w =$ const. (or $n_* = 0$), the value $K_1 = 0.0187$ ($T_w =$ const.) calculated by Eq. (3.4) at $n = 1/7$, $n_R = 0.8$ and $\sigma = 2$ much better than $K_1 = 0.0196$ by Dorfman's formula (3.21) agrees with experimental value $K_1 = 0.0188$ [6, 15, 23, 62, 63] (deviation 0.5 %).

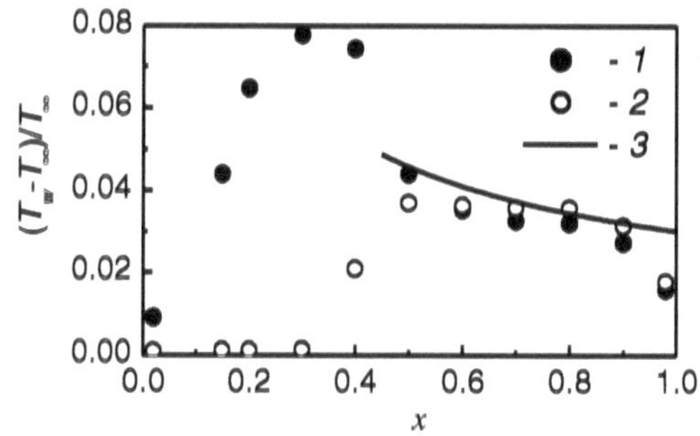

Fig. 3.9 Experimental temperature distribution [6] and its computation over the surface of a rotating disk [1]. *1*—$q_w \equiv$ const.; *2*—$T_w \approx$ const. Computation [26]: *3*—Eq. (3.41)

3.3.4 Arbitrary Distribution of the Wall Temperature

The disk temperature distributions T_w measured in experiments [86, 87], Fig. 3.10, do not agree with the analytical approximations by Eqs. (2.29) and (2.30). Reliability of these experimental data was proved in [50] by a numerical solution of the differential Eqs. (2.9)–(2.15) using the Cebeci–Smith model of turbulence [90]. A numerical version of Dorfman's method ($n = 1/7$) was employed in [86, 87] to model these experimental conditions. Numerically computed Nusselt numbers, similarly to those obtained by the analytical Dorfman's method, agreed well with the experiments for $dT_w/dr > 0$ and noticeably exceeded them for $dT_w/dr \approx 0$ and $dT_w/dr < 0$ [86, 87].

The numerical version of the present integral method, Eqs. (2.72) and (2.73), was used by the author [91] to simulate the experimental conditions [86, 87].

Experimental disk temperature distributions were divided in [86, 87] into four groups complying with Eq. (2.30) at positive ($n_* = 0.4$ and 0.6), approximately constant ($n_* = 0.1$), and negative ($n_* = -0.2$) gradients of the wall temperature T_w. Scatter of the distributions of T_w within each group was less than 10–15 % for different values of Re_φ.

Agreement of Eq. (2.30) with the measurements at the n_* values mentioned above is rather conventional. Obviously, Eq. (2.30) does not assume maxima, minima and inflection points visible in curves depicted in Fig. 3.10 within the region of determination of T_w. However, for convenience, this classification was left unchanged here.

Computed local Nusselt numbers are depicted in Figs. 3.11, 3.12 and 3.13. Experimental distributions of T_w used as the boundary conditions were

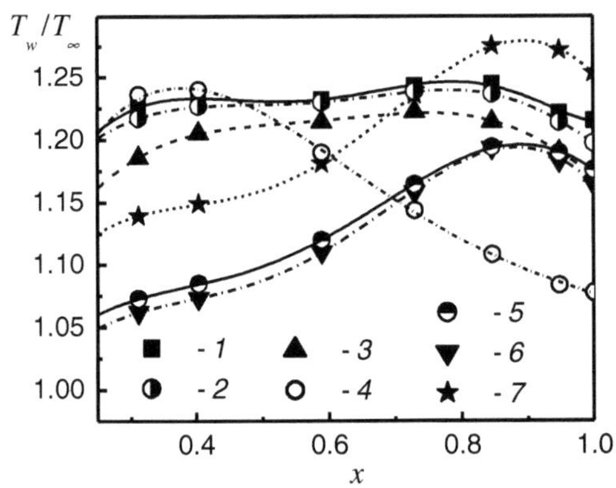

Fig. 3.10 Disk temperature variation: symbols—experiments [87], lines—polynomial approximations of the experiments by [1]. *1*—$n_* = 0.1$, $Re_\varphi = 1.135 \times 10^6$; *2*—$n_* = 0.1$ and $Re_\varphi = 1.19 \times 10^6$; *3*—$n_* = 0.1$ and $Re_\varphi = 3.2 \times 10^6$; *4*—$n_* = -0.2$ and $Re_\varphi = 2.65 \times 10^6$; *5*—$n_* = 0.4$ and $Re_\varphi = 2.67 \times 10^6$; *6*—$n_* = 0.4$ and $Re_\varphi = 3.14 \times 10^6$; *7*—$n_* = 0.6$ and $Re_\varphi = 1.59 \times 10^6$; $x = r/b$

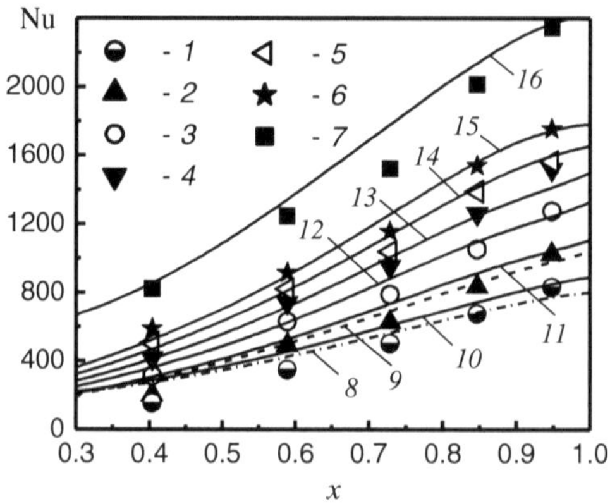

Fig. 3.11 Variation of the Nusselt number for $n_* = 0.1$: symbols *1–7*—experiments [87]. Lines—simulations [1, 91]: *8*—$n = n_T = 1/5$; *9*—$n = 1/6$, $n_T = 1/5$; *10–16*—$n = n_T = 1/6$. *1, 8, 10*—$Re_\varphi = 0.819 \times 10^6$; *2, 9, 11*—$Re_\varphi = 1.08 \times 10^6$; *3, 12*—$Re_\varphi = 1.35 \times 10^6$; *4, 13*—$Re_\varphi = 1.6 \times 10^6$; *5, 14*—$Re_\varphi = 1.88 \times 10^6$; *6, 15*—$Re_\varphi = 2.14 \times 10^6$; *7, 16*—$Re_\varphi = 3.2 \times 10^6$; $x = r/b$

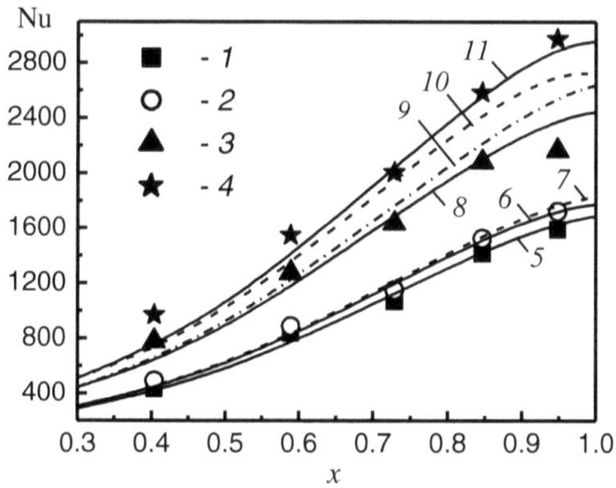

Fig. 3.12 Radial variation of the Nusselt number [1]. Case $n_* = 0.6$: *1, 2*—experiments [87]; *5, 6*—calculations, $n = n_T = 1/6.5$; *7*—calculation, $n = n_T = 1/7$. Case $n_* = 0.4$: *3, 4*—experiments [87]; *8, 10*—calculations, $n = n_T = 1/6$; *9, 11*—calculations, $n = n_T = 1/7$. Reynolds numbers Re_φ: *1, 5*—1.59×10^6; *2, 6, 7*—1.71×10^6. *3, 8, 9*—2.67×10^6; *4, 10, 11*—3.14×10^6

approximated by a 7th-order polynomial, see Fig. 3.10 [1, 91]. Comparisons of the simulations with the experimental data enabled developing hints for the choice of the n and n_T values.

Results for the case $n_* = 0.1$ are plotted in Fig. 3.11 [1, 91]. Computations and experiments for $Re_\varphi = 1.08 \times 10^6$–$3.2 \times 10^6$ match well for $n = n_T = 1/6$, though values $n = n_T = 1/5$ provide a better agreement for the smaller value $Re_\varphi = 0.819 \times 10^6$.

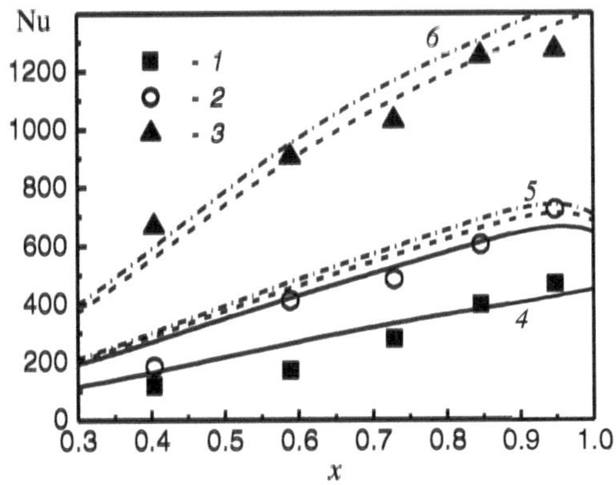

Fig. 3.13 Radial variation of the Nusselt number [1]. Case $n_* = -0.2$: *1–3*—experiments [87]. Calculations [91]: *solid lines 4, 5*—$n = 1/6$, $n_T = 1/4$; *dashed lines 5, 6*—1/6 and 1/5; *dash-dotted lines 5, 6*—both 1/6. Reynolds numbers Re_φ: *1, 4*—0.548×10^6; *2, 5*—1.08×10^6; *3, 6*—2.65×10^6

Figure 3.12 depicts calculations for similar cases $n_* = 0.4$ and $n_* = 0.6$ [1, 91]. At smaller Reynolds numbers $Re_\varphi < 1.71 \times 10^6$, a better agreement with experiments is yield using the exponents $n = n_T = 1/6.5$. For the larger values $Re_\varphi = 2.67 \times 10^6$ and 3.14×10^6, exponents $n = n_T = 1/6$ and 1/7, respectively, are required. Exponents n and n_T for the simulations in Fig. 3.12 are slightly smaller as compared to the case $n_* = 0.1$, where $dT_w/dr \approx 0$ (Fig. 3.11).

In Fig. 3.13, for the negative wall temperature gradient $dT_w/dr < 0$ ($n_* = -0.2$), the values of $n_T = 1/4$ and $n = 1/6$ were needed for $Re_\varphi = (0.548–1.08) \times 10^6$. Exponents $n_T = 1/5$ and $n = 1/6$ were used for a larger value $Re_\varphi = 2.65 \times 10^6$. Thus, for the same value of the Reynolds number Re_φ, the negative gradient $dT_w/dr < 0$ entails the need to use a larger value n_T, whereas the value $n = 1/6$ remains the same (in comparison with the data in Fig. 3.11).

Profiles of the velocity and temperature were not obtained in the measurements [86, 87]. Because of this, an estimation of the exponents n and n_T is made based on the distribution of the Nusselt number. The lower rate of the radial variation in the Nu numbers in Fig. 3.13 (case $n_* = -0.2$) results in a smaller value of the exponent n_R in Eq. (3.4) and, hence, larger exponents n and n_T. Numerical simulations confirm this trend. The experimental investigation [6], among other results, revealed that temperature profiles for q_w = const. ($dT_w/dr < 0$, $n_* \approx -0.6$) and $Re_\omega = 10^6$ were characterized by the exponents $n_T = 1/4–1/5$ (see Fig. 2.5). These exponents correlate with our simulations for the case of $n_* = -0.2$.

To conclude, in case of an arbitrary variation of the disk temperature, predictions of turbulent heat transfer of a rotating disk using a numerical version of the present integral method match well with the experiments [87], whereas the exponent n_T in the temperature profile approximation depends on the thermal boundary conditions.

3.4 Generalized Analytical Solution for Laminar and Turbulent Flow

As said above, Dorfman's Eq. (3.6) for the Nusselt number in laminar regime at $Pr = 1–0.1$ by up to 238 % exceeds the self-similar solution, Table 3.1. The more accurate Eqs. (3.7) and (3.8) [31, 32] are valid only for $T_w = \text{const.}$ ($n_* = 0$). To improve this situation, an approximate solution for the Nusselt number valid over the range $Pr = 1–0.1$ and possessing significantly higher accuracy than Eq. (3.6) was derived in [1, 29, 30].

Equation (2.18) of the velocity boundary layer and Eq. (2.20) of the thermal boundary layer were rewritten as

$$\frac{\mathrm{d}}{\mathrm{d}r}\left[r^4 \delta K_V K_m\right] = \frac{c_f}{2} r^4 (1+\alpha^2)^{1/2}, \tag{3.42}$$

$$\frac{\mathrm{d}}{\mathrm{d}r}\left[r^2 \delta K_H K_m (T_w - T_\infty)\right] = \chi \frac{c_f}{2} r^2 (1+\alpha^2)^{1/2} (T_w - T_\infty). \tag{3.43}$$

Here χ is the Reynolds analogy parameter defined by Eq. (2.52).

The parameters for the turbulent boundary layer are given by Eqs. (2.77)–(2.82); additional conditions are $K_V = \text{const.}$ and $K_m = \text{const.}$ The majority of the constants in Eqs. (2.77)–(2.82) for turbulent flow are listed in Sect. 2.5, while $K_m = \alpha A_1$, $K_V = 1 - D_2/A_1$.

By setting $n = 1$ and $m = 0$, Eqs. (2.77)–(2.82) can be used also for laminar flow. The constants in Eqs. (2.77)–(2.82) for laminar flow were first obtained in [3, 4] by solving and integrating the self-similar Eqs. (2.32)–(2.35)

$$a = 0.8284, \quad K_V = 0.3482, \quad K_m = \frac{I_\infty \alpha}{\delta(\omega/\nu)^{1/2}} = \frac{I_\infty \alpha}{\gamma}, \tag{3.44}$$

$$I_\infty = \alpha^{-1} \int\limits_0^\infty \frac{v_r}{\omega r} d(z\sqrt{\omega/\nu}) = 0.5338, A_c = 0.6159(1+\alpha^2)^{-1/2}. \tag{3.45}$$

If boundary condition (2.30) holds, the dimensionless temperature θ and shape-factor of the temperature profile K_H are also self-similar, i.e. independent of the coordinate r. Substituting Eqs. (2.30), (2.77)–(2.82), (3.44) and (3.45) into Eqs. (3.42) and (3.43) yields

$$(4+m)\gamma K_V K_m = A_c (1+\alpha^2)^{1/2}, \tag{3.46}$$

$$(2+m+n_*)\gamma K_H K_m = \chi A_c (1+\alpha^2)^{1/2}. \tag{3.47}$$

Equation (3.47) contains two unknown quantities: K_H and χ, which can be connected by Eq. (2.51) of Dorfman [2]. The present integral method for turbulent flow validated in Sect. 3.3 enables elaborating a novel model linking K_H and χ, which is much more accurate than Eq. (2.51) and originates automatically from our boundary layer model [29]

$$b_2 K_H = 1 - \chi Pr^{n_p}(1 - K_V)b_1. \tag{3.48}$$

Correction multipliers b_1 and b_2 take account of laminar flow. These multipliers are equal to unity for turbulent flow, and the exponent n_p is defined by Eq. (3.39).

Solving Eq. (3.47) with account for Eq. (2.52), one can assure that the first of Eq. (3.4) again describes the Nusselt number at $n_R = (n+1)/(3n+1)$ and

$$K_1 = (2 + m + n_*)\gamma K_H K_m Pr. \tag{3.49}$$

If one equates the relations for χ resulting from Eqs. (3.47) and (3.48), derives K_H from this equation and substitutes it into Eq. (3.49), this yields

$$K_1 = A_c(1+\alpha^2)^{1/2} Pr\left[\frac{4+m}{2+m+n_*}K_V b_2 + (1-K_V)Pr^{n_p}b_1\right]^{-1}, \tag{3.50}$$

$$K_3 = A_c(1+\alpha^2)^{1/2}. \tag{3.51}$$

Equations (3.50) and (3.34) coincide for turbulent flow, where $b_1 = 1$ and $b_2 = 1$. Setting $n = 1$, $m = 0$, $n_R = 1/2$ for laminar flow, one can obtain from Eq. (3.45) for A_c

$$A_c(1+\alpha^2)^{1/2} = 0.6159. \tag{3.52}$$

In the asymptotic case of $Pr \to 0$, Eq. (3.50) reduces to its asymptotic form, which looks for $b_2 = 1$ identically to the solution derived in [3]

$$K_1 = 0.6159 Pr(2 + n_*)/(4K_V). \tag{3.53}$$

Comparing Eq. (3.50) with the self-similar solution, Table 3.1, one can find the coefficients b_1 and b_2. To satisfy Eq. (3.53), b_1 must be finite and $b_2 = 1$ at $Pr \to 0$. Let us set

$$b_2 = 1 + b_3 Pr^{n_{p1}}, b_1 = \text{const.}, b_3 = \text{const.} \tag{3.54}$$

The coefficients b_1 and b_3 are independent of the Prandtl number and are determined at $Pr = 1$. Further, exponents n_{p1} and n_p were found separately for every Pr number from Table 3.1. In this procedure, the coefficient K_1 from the self-similar solution has been substituted into Eq. (3.50) for $n_* = 0$ and $n_* = 2$. Finally, we obtained [29]

Table 3.8 Values of n_p and n_{p1} depending on the Prandtl number [1, 29, 30]

Pr	0.9	0.8	0.72	0.71	0.6	0.5	0.4	0.3	0.2	0.1	0.01
n_p	0.7290	0.7349	0.7436	0.7435	0.7529	0.7608	0.7721	0.7860	0.8036	0.8338	0.9156
n_{p1}	0.9349	0.9354	0.9316	0.9334	0.9366	0.9439	0.9519	0.9602	0.9765	0.9893	0.999

$$b_1 = 0.6827, \quad b_3 = 0.5939, \tag{3.55}$$

whereas the values of the exponents are listed in Table 3.8.

For computational purposes, the exponents n_{p1} и n_p were described as polynomials

$$n_{p1} = \sum_0^7 a_i Pr^i, n_p = \sum_0^7 c_i Pr^i, \tag{3.56}$$

where $a_0 = 1$, $a_1 = -0.008073$, $a_2 = -0.3558$, $a_3 = 0.5485$, $a_4 = 1.799$, $a_5 = -6.432$, $a_6 = 7.354$, $a_7 = -2.915$, $c_0 = 0.925$, $c_1 = -1.171$, $c_2 = 3.532$, $c_3 = -3.376$, $c_4 = -5.725$, $c_5 = 15.59$, $c_6 = -12.28$, $c_7 = 3.208$.

The coefficient K_1 computed by Eq. (3.50) is presented in Table 3.9. It diverges from the self-similar solution, Table 3.1, by maximum 3.1 % (for $n_* = -1.5$ and $Pr = 1$); the errors become negligible for $Pr \leq 0.1$ [1, 29, 30].

On both sides of Eq. (3.43), the exponents for the r-coordinate must be the same, which yields the expression $1 + m + n_* = 1$ for q_w = const. For laminar flow, $m = 0$, hence, $n_* = 0$. Therefore, the Nusselt number for laminar flow for q_w = const. is exactly the same as that for T_w = const. In turbulent flow, as shown above, $n_* = -m$ for q_w = const.

To conclude, as can be seen from Fig. 3.14, the exact and approximate solutions for air ($Pr = 0.72$) are in good agreement with each other as well as with experimental results.

Table 3.9 Values of the constant K_1 by Eqs. (3.50) and (3.54)–(3.56) [1]

Pr	$n_* = -2$	$n_* = -1.5$	$n_* = -1$	$n_* = -0.5$	$n_* = 0$	$n_* = 1$	$n_* = 2$	$n_* = 3$	$n_* = 4$
1.0	0.0	0.1261	0.2311	0.3199	0.3961	0.5197	0.6159	0.6928	0.7557
0.72	0.0	0.1019	0.1887	0.2635	0.3286	0.4365	0.5223	0.5921	0.6500
0.71	0.0	0.1009	0.187	0.2612	0.3259	0.4332	0.5185	0.5880	0.6457
0.5	0.0	0.0788	0.1477	0.2084	0.2623	0.3539	0.4287	0.4910	0.5437
0.1	0.0	0.0204	0.0399	0.0586	0.0766	0.1104	0.1417	0.1707	0.1977
0.01	0.0	0.00219	0.00438	0.00655	0.00871	0.01301	0.01726	0.02148	0.02565

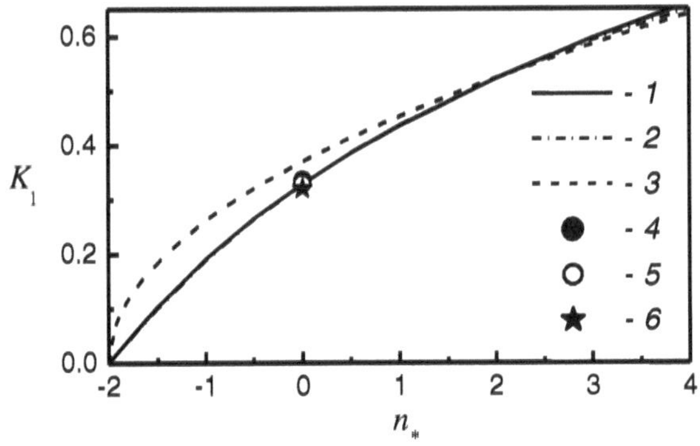

Fig. 3.14 Effect of the exponent n_* on the constant K_1 in Eq. (3.4) for laminar flow at $Pr = 0.72$ [1]. *1*—Exact solution [29, 30], *2*—Eq. (3.50), *3*—Eq. (3.6). Experiments for $n_* = 0$: *4*—0.335 [6, 18, 22], *5*—0.33 [15, 17, 23], *6*—0.32 [16]

3.5 Finding a Wall Temperature Distribution for Arbitrary Nusselt Numbers

3.5.1 Solution of the Problem

Usually the thermal boundary layer equation is solved as a direct problem of searching the Nusselt number at a given wall temperature, Eq. (2.30), or in a modified form

$$\overline{\Delta T} = x^{n_*}, \tag{3.57}$$

where $\overline{\Delta T} = \Delta T / \Delta T_{x=1}$ is the relative non-dimensional temperature difference on a surface.

The solution for the Nusselt number (3.4) found in such a way can be rewritten as

$$\begin{aligned} Nu &= K_1 Re_\varphi^{(n+1)/(3n+1)} x^{2(n+1)/(3n+1)} \\ &= K_1 Re_\varphi^{(n+1)/(3n+1)} x^{1+m}, 2(n+1)/(3n+1) = 1+m. \end{aligned} \tag{3.58}$$

An inverse problem searches for a distribution of ΔT, if the Nusselt number is given

$$Nu = K_1 Re_\varphi^{(n+1)/(3n+1)} x^{m_x}, \quad \mathrm{Nu}_b = K_1 Re_\varphi^{(n+1)/(3n+1)} x^{m_x - 1}, \tag{3.59}$$

where m_x = const. is an arbitrarily selected exponent not equal to $1 + m$ as in Eq. (3.58). An application in practice is e.g. a transient technique of experimental finding of time-independent Nusselt numbers, whereas the measured distributions of ΔT are unsteady. Here a steady-state distribution of ΔT may be found only via solving an inverse problem.

To solve this problem [1, 92], the thermal boundary layer Eq. (3.43) in the integral form can be transformed to

$$\frac{\mathrm{d}}{\mathrm{d}x}\left[Re_{\omega}\bar{\delta}K_H K_m \overline{\Delta T}\right] = \frac{Nu}{Pr}\overline{\Delta T}. \tag{3.60}$$

Equation (2.77) can be presented in the form valid for laminar and turbulent flow

$$\bar{\delta} = C_{\delta}^{*}x^{m}, \quad C_{\delta}^{*} = \gamma Re_{\varphi}^{-2n/(3n+1)}. \tag{3.61}$$

To remind, in laminar flow $m = 1$ and $K_m\bar{\delta} = I_{\infty}\alpha Re_{\varphi}^{-1/2}$ [see Eq. (3.44)]. Equation (2.52) can be used to evaluate the Reynolds analogy parameter χ

$$\chi = \frac{Nu}{\frac{c_f}{2}Re_{\omega}(1+\alpha^2)^{1/2}Pr} = C_{\chi}x^{m_x-m-1}, \tag{3.62}$$

$$C_{\chi} = \frac{K_1}{A_c(1+\alpha^2)^{1/2}Pr}. \tag{3.63}$$

Let us use the notation

$$m_x^{*} = m_x - m. \tag{3.64}$$

In terms of Eq. (3.64), one can make sure that Eq. (3.58) holds at $m_x^{*} = 1, m_x = 1 + m$.

In view of Eqs. (3.62)–(3.64), one can rewrite Eq. (3.48) such as

$$K_H = a_{*} + b_{*}x^{m_x^{*}-1}, \tag{3.65}$$

$$a_{*} = \frac{1}{b_2}, b_{*} = -Pr^{n_p}(1-K_V)\frac{b_1}{b_2}C_{\chi}. \tag{3.66}$$

An integration of Eq. (3.60) in view of Eqs. (3.61)–(3.66) yields [1]

$$\overline{\Delta T} = \frac{a_{*}+b_{*}}{a_{*}+b_{*}x^{m_x^{*}-1}}x^{-2-m}\left[\left(\frac{a_{*}x^{1-m_x^{*}}+b_{*}}{a_{*}+b_{*}}\right)^{\frac{1}{1-m_x^{*}}}x^{-1}\right]^{-\frac{K_1}{PrK_m\gamma b_{*}}} \tag{3.67}$$

$$\overline{\Delta T} = \left[\frac{K_H}{K_{H_{x=1}}}\right]^{-\frac{K_1}{PrK_m\gamma b_{*}(1-m_x^{*})}-1}x^{-2-m}. \tag{3.68}$$

3.5.2 The Limiting Case of the Solution

For $m_x^* \to 1$, Eq. (3.68) degenerates to [1]

$$\overline{\Delta T} = x^{\frac{K_1}{PrK_m\gamma(a_*+b_*)}-2-m}. \tag{3.69}$$

Equations (3.57) and (3.69), being combined and transformed, yield

$$n_* = \frac{K_1}{PrK_m\gamma(a_*+b_*)} - 2 - m. \tag{3.70}$$

Specifying the value of n_* and keeping in mind that $a_* + b_* = K_H$ at $m_x^* = 1$, one can obtain Eq. (3.50) for K_1 valid under the conditions (2.30) or (3.57). In doing so, Eq. (3.57) can be treated as a specific case of Eq. (3.68) at $m_x^* = 1$.

3.5.3 Properties of the Solution for the Temperature Difference on the Wall

The point of extremum x_{ext} of the Eq. (3.68) for $\overline{\Delta T}$ is located at [1, 93]

$$x_{\text{ext}} = \left[\frac{\frac{K_1}{PrK_m\gamma} - b_*(m_x+1)}{a_*(2+m)}\right]^{\frac{1}{1-m_x^*}}. \tag{3.71}$$

Expression $\chi = \Delta^{-n}Pr^{-n_p}$ and Eq. (3.62) for the Reynolds analogy parameter χ yield a relation for the normalized thermal boundary layer thickness Δ [1]

$$\Delta = (\chi Pr^{n_p})^{-1/n} = (C_\chi x^{m_x^*-1}Pr^{n_p})^{-1/n} = (C_\chi Pr^{n_p})^{-1/n}x^{(1-m_x^*)/n}. \tag{3.72}$$

For the boundary condition (3.57) with $m_x^* = 1$, Eq. (3.72) results in the relation Δ = const. The function $\Delta(x)$ is increasing or decreasing at $m_x^* < 1$ or $m_x^* > 1$, respectively.

In view of its physical nature, parameter K_H may be only positive. Equation (3.65) for K_H exhibits a limiting point where $K_H = 0$, whose parameters are

$$\chi_{\text{crit}} = \frac{1}{b_1 Pr^{n_p}(1-K_V)}, \tag{3.73}$$

$$\Delta_{\text{crit}} = (\chi_{\text{crit}}Pr^{n_p})^{-1/n} = \left[\frac{1}{b_1(1-K_V)}\right]^{-1/n}, \tag{3.74}$$

$$x_{\text{crit}} = \left[\frac{\chi_{\text{crit}}}{C_\chi}\right]^{1/(m_x^*-1)}. \quad (3.75)$$

Equations (3.71)–(3.75) can be used in the analysis of the behavior of $\overline{\Delta T}$.

3.5.4 Analysis of the Solution

The novel solution described above in Sect. 3.5 holds for a much wider range of thermal boundary conditions at the wall and incorporates previously known solutions as a specific case, provided that one of the parameters degenerates to unity [1, 92].

Indeed, Eq. (3.67) [or (3.68)] for the temperature difference and Eq. (3.59) for the Nusselt number contain two independent parameters: K_1 and m_x. At $m_x = 1 + m$ (or $m_x^* = 1$), Eqs. (3.67), (3.68) and (3.59) reduce to the known Eq. (3.57) for $\overline{\Delta T}$ and Eq. (3.58) [or (3.4)] for Nu with only one independent parameter: K_1 or n_*. Equations (3.57), (3.67) and (3.68) for $\overline{\Delta T}$ do not depend on the Reynolds number Re_φ.

Both one-parameter Eq. (3.58) [or (3.4)] and two-parameter Eq. (3.59) represent monotonic distributions of the Nusselt number, i.e. the sign of the derivative dNu/dx is constant over the entire range of variation of the radial coordinate x.

At $m_x^* = 1$, the sign of the derivative $d\overline{\Delta T}/dx$ in Eq. (3.57) is also constant. On the contrary, at $m_x^* \neq 1$, functions of $\overline{\Delta T}$ given by Eq. (3.67) or (3.68) are non-monotonic and enable predicting curves of $\overline{\Delta T}$ exhibiting points of maxima and minima.

Application to laminar flow. For validation of the model, experiments [94–96] (laminar air flow for $Re_\varphi = 53500$, $Pr = 0.71$) were chosen. Let us assume that the temperature difference $\overline{\Delta T}$ is determined by one-parameter Eq. (3.57) at $n_* = -1$, while the Nusselt number Nu_b is constant and determined by Eq. (3.58) with $K_1 = 0.187$ (here $m = 0$, $m_x^* = m_x$).

Setting the value $K_1 = 0.187$ and replacing $m_x = 1$ by $m_x = 0.6$ in Eq. (3.59) yields a radial dependence for Nu_b starting at infinity at $x \to 0$ and further monotonically subsiding (curve 1 in Fig. 3.15). Based on Eq. (3.68), the temperature difference $\overline{\Delta T}$ is a non-linear function of the coordinate x being zero at $x = 0$, having a maximum $\overline{\Delta T} = 2.55$ at $x \approx 0.2$ and further diminishing as a function similar to x^{-1} (curve 1 in Fig. 3.16). Obviously, $\overline{\Delta T} = 1$ at $x = 1$. At the point of maximum, Eqs. (3.73)–(3.75) hold.

If $n_* = 2$ in Eq. (3.57), the function $\overline{\Delta T}(x)$ is increasing; at the same time, the Nusselt number Nu_b is constant with $K_1 = 0.5185$ (and $m_x = 1$). Provided that $K_1 = 0.5185$ and $m_x = 1.6$ in Eq. (3.59), the function $Nu_b(x)$ is monotonically increasing.

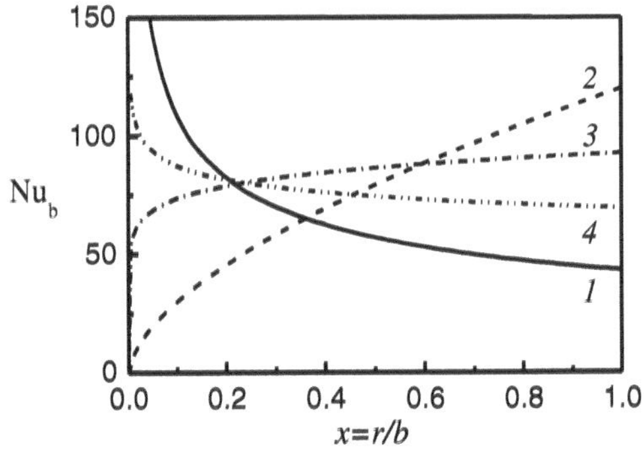

Fig. 3.15 Radial distribution of the Nusselt number Nu_b by Eq. (3.59) for laminar flow ($m = 0$) at $Re_\varphi = 53{,}500$, $Pr = 0.71$ [1]. *1*—$m_x = 0.6$, $K_1 = 0.187$; *2*—$m_x = 1.6$, $K_1 = 0.5185$; *3*—$m_x = 1.1$, $K_1 = 0.4$; *4*—$m_x = 0.9$, $K_1 = 0.3$

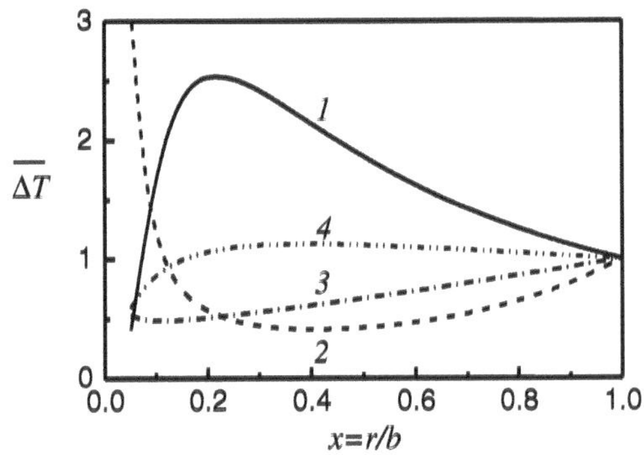

Fig. 3.16 Radial distribution of the temperature difference $\overline{\Delta T} = (T_w - T_\infty)/(T_w - T_\infty)_{x=1}$ by Eq. (3.68) for laminar flow ($m = 0$) at $Re_\varphi = 53500$, $Pr = 0.71$ [1]. *1*—$m_x = 0.6$, $K_1 = 0.187$; *2*—$m_x = 1.6$, $K_1 = 0.5185$; *3*—$m_x = 1.1$, $K_1 = 0.4$; *4*—$m_x = 0.9$, $K_1 = 0.3$

Given this combination of K_1 and m_x, the temperature difference decreases from $\overline{\Delta T} \to \infty$ and reaches a point of minimum at $x \approx 0.4$ with $\overline{\Delta T} = 0.4075$. Further at $x > 0.4$, the function $\overline{\Delta T}$ demonstrates a trend of increase as a function similar to x^2.

Curves 3 and 4 in Figs. 3.15 and 3.16 were plotted to outline the heat transfer regimes situated between the cases 1 and 2 discussed above.

Application to turbulent flow. For our analysis, we selected the experiments [86, 87], while the equality $n = n_T$ was always held while using Eqs. (3.59) and (3.68) [93, 97]. The results of simulations for the case $n_* = 0.1$ (performed at $n = n_T = 1/6$) are shown in Figs. 3.17 and 3.18. In the region $x \approx 0.3$–0.85, an approximation $\overline{\Delta T} = c_{0*}x^{n_*}$ at $n_* = 0.06$, $c_{0*} = 1.16$ for $Re_\varphi = (1.08$–$1.35) \times 10^6$ and at $n_* = 0.06$, $c_{0*} = 1.26$ for $Re_\varphi = (2.14$–$3.2) \times 10^6$ locally correlates with experimentally measured function $\overline{\Delta T}$.

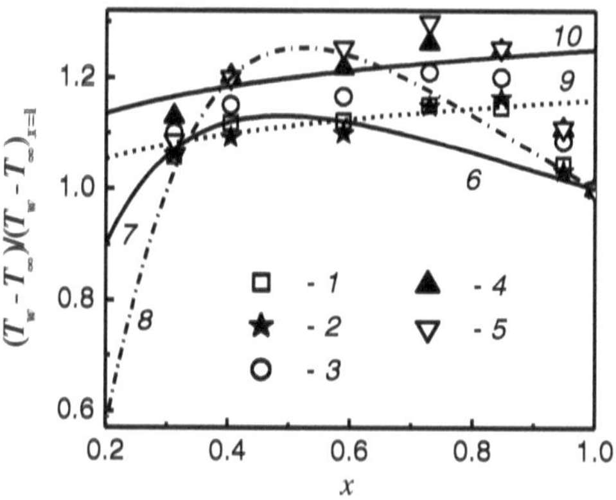

Fig. 3.17 Radial variation of the temperature difference $\overline{\Delta T}$ for $Pr = 0.72$, case conventionally $n_* = 0.1$ [1]. Experiments [87]: *1*—$Re_\varphi = 1.08 \times 10^6$; *2*—$1.6 \times 10^6$; *3*—$1.88 \times 10^6$; *4*—$2.14 \times 10^6$; *5*—$3.2 \times 10^6$. Calculations by Eq. (3.68) for $n = 1/6$: *6*—$K_1 = 0.0232$ and $m_x = 1.48$; *7*—0.0229 and 1.43; *8*—0.0224 and 1.38. Approximation $\overline{\Delta T} = c_{0*}x^{n_*}$: *9*—$c_{0*} = 1.16$ and $n_* = 0.06$; *10*—$c_{0*} = 1.25$ and $n_* = 0.06$

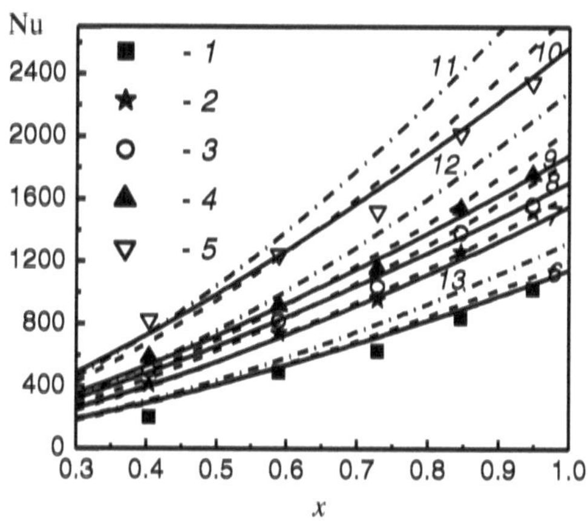

Fig. 3.18 Radial variation of the Nusselt number, case conventionally $n_* = 0.1$ [1]: *1–5*—experiments [87]; *6–11*—calculations by Eqs. (3.58) and (3.59) for $n = n_T = 1/6$. *Solid lines*, Eq. (3.59): *6, 7*—$K_1 = 0.0232$, $m_x = 1.48$; *8–10*—$K_1 = 0.0224$, $m_x = 1.38$. *Dashed lines 6–10*—Eq. (3.58), $K_1 = 0.0232$, $n_R = 0.778$, $m_x = 1.556$. *Dash-dotted lines 11–13*—Dorfman's Eq. (3.21) for $n_* = 0.1$. *1, 6, 13*—$Re_\varphi = 1.08 \times 10^6$; *2, 7*—$1.6 \times 10^6$; *3, 8*—$1.88 \times 10^6$; *4, 9, 12*—$2.14 \times 10^6$; *5, 10, 11*—$3 \times 10^6$

As a result, calculations by Eq. (3.58) [with K_1 given by Eq. (3.34)] agree well with measured Nu numbers over the same range of x (here the exact value c_{0*} is unimportant).

At the same time, for $x > (0.7–0.85)$ experimental temperature difference $\overline{\Delta T}$ diminishes, while the predicted $\overline{\Delta T}$ continues increasing. Owing to this, predicted

Nusselt numbers surpass the experimental data. This tendency becomes more noticeable at larger Reynolds numbers Re_φ. Dorfman's Eq. (3.21) at $n_* = 0.1$, $n = n_T = 1/7$ (see curves 11–13 in Fig. 3.18 and explanations to them) diverges from experiments even more noticeably.

Using Eq. (3.59) for the Nusselt number, Fig. 3.18, and Eq. (3.68) for $\overline{\Delta T}$ can provide a better agreement between simulations and experiments (see Fig. 3.17). While Eq. (3.57) ensures a positive sign of $\mathrm{d}\overline{\Delta T}/\mathrm{d}x$ for any x, Eq. (3.68), on the contrary, enables rather flexible modeling of the sign of the derivative $\mathrm{d}\overline{\Delta T}/\mathrm{d}x$, which changes here from "plus" to "minus" with increasing x. This ensures a more close agreement between the computed and measured values of the Nusselt number (Fig. 3.18).

Predictions for the case $n_* = -0.2$ (performed for $n = n_T = 1/6$) are depicted in Figs. 3.19 and 3.20. A fair consistency of the computed and measured functions of $\overline{\Delta T}$ is evident for $x \geq 0.6$ (Fig. 3.19). Qualitative agreement of the sign of the predicted and measured function $\mathrm{d}\overline{\Delta T}/\mathrm{d}x$ demonstrates only curve 4 over the range $x = 0.35$–0.45. Important is that the approximation $\overline{\Delta T} = c_{0*}x^{n_*}$ correlates with the experiments for $x > 0.6$ only for $n_* = -1.5$ (and $c_{0*} = 1.14$), though the value suggested in [86, 87] is $n_* = -0.2$ (curve 6).

The Nusselt numbers in Fig. 3.20 were computed for two different values of the Reynolds numbers. Experimental data 1 and *upper lines* 3–6 relate to the value $Re_\varphi = 2.65 \times 10^6$, whereas experiments 2 and *lower lines* 3–6 correspond to $Re_\varphi = 1.08 \times 10^6$. Curves for the Nu number depend noticeably on the values for K_1 and m_x.

The lower curve 4 of the Nu number in Fig. 3.20 computed by Eq. (3.58) for $Re_\varphi = 1.08 \times 10^6$ conforms to the experiments 2. Nevertheless, at $x \approx 0.4$ the flow is laminar, thus an agreement with it testifies that the computed curve 4 lies lower than

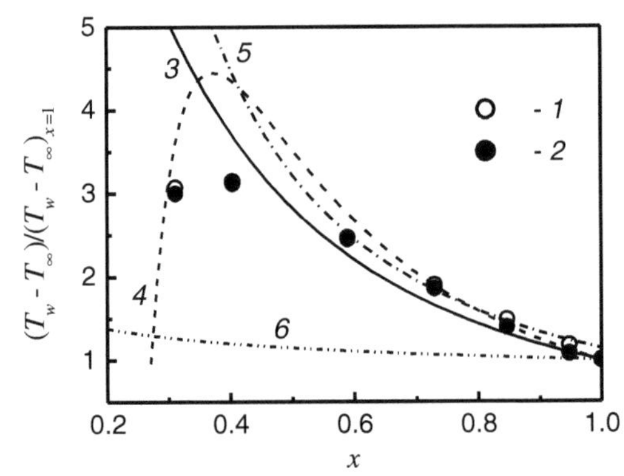

Fig. 3.19 Radial variation of the temperature difference $\overline{\Delta T} = (T_w - T_\infty)/(T_w - T_\infty)_{x=1}$ for $Pr = 0.72$, $n_* = -0.2$ [1]. Experiments [87]: *1*—$Re_\varphi = 1.08 \times 10^6$; 2–2.65 × 10^6. Calculations by Eq. (3.68) for $n = 1/6$: *3*—$K_1 = 0.0157$, $m_x = 1.3$; *4*—$K_1 = 0.0137$, $m_x = 0.775$. Approximation $\overline{\Delta T} = c_{0*}x^{n_*}$: *5*—$c_{0*} = 1.14$, $n_* = -1.5$; *6*—$c_{0*} = 1.0$, $n_* = -0.2$

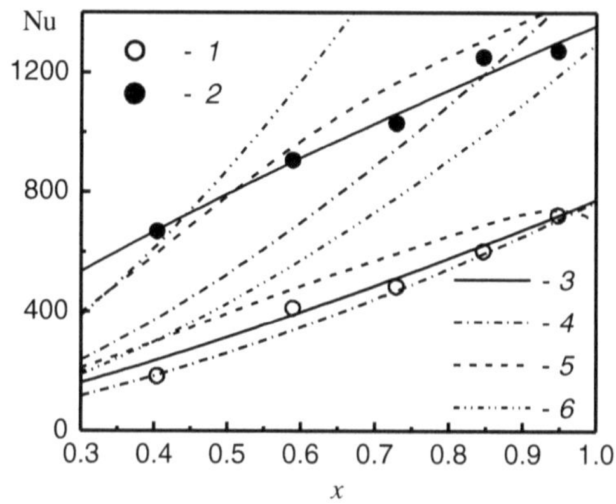

Fig. 3.20 Variation of the Nusselt number, $n_* = -0.2$ [1]: *1*, *2*—experiments [87]; *3–5*—Eq. (3.58), $n = n_T = 1/6$; *6*—Dorfman's Eq. (3.21), $n_* = -0.2$. *Lower lines 3–6* and symbol *1*—$Re_\varphi = 1.08 \times 10^6$; *upper lines 3–6* and symbol *2*—$Re_\varphi = 2.65 \times 10^6$. *Lower line 3*—Eq. (3.59), $K_1 = 0.0157$, $m_x = 1.3$; *upper line 3*—Eq. (3.59), $K_1 = 0.0137$, $m_x = 0.775$. Lines *4*—Eq. (3.58), $K_1 = 0.0156$, $n_R = 0.778$, $m_x = 1.556$, $n_* = -1.5$. Lines *5*—numerical modelling (see Fig. 3.13)

it should be expected. Therefore, for $Re_\varphi = 2.65 \times 10^6$ experimental points 1 significantly exceed the upper curve 4 at $x \leq 0.7$, where turbulent flow is developed.

Using Eq. (3.59) with $\overline{\Delta T}$ predicted by Eq. (3.68), one can attain a good agreement with the experiments for the Nu number, if the coefficients K_1 and m_x are properly selected.

Dorfman's Eq. (3.21) at $n_* = -0.2$ predicts much too high Nusselt numbers as compared to the experimental data (curves 6 in Fig. 3.20).

Predictions and measurements for $n_* = 0.4$ and $n_* = 0.6$ are depicted in Figs. 3.21 and 3.22. Values $n = n_T = 1/6.5$ and $n = n_T = 1/6$ were used for $Re_\varphi = 1.59 \times 10^6$ and 2.67×10^6, accordingly. The value $n_* = 0.6$ in Eq. (3.57) indeed conforms to the experiments at $x = 0.3$–0.6 and $Re_\varphi = 1.59 \times 10^6$ (Fig. 3.21). However, experiments for $Re_\varphi = 2.67 \times 10^6$ can be better simulated with $n_* = 0.7$ rather than with $n_* = 0.4$ [87].

In addition, Eq. (3.68) ensures a variation of the absolute value of $\mathrm{d}\overline{\Delta T}/\mathrm{d}x$ in the radial direction and a wide range of selection of the parameters K_1 and m_x (Fig. 3.21).

The local Nusselt numbers in Fig. 3.22 show that Eqs. (3.59) and (3.68) with carefully selected values K_1 and m_x enable a better agreement with experiments than Eqs. (3.57) and (3.58), which can be seen for the larger value $Re_\varphi = 2.67 \times 10^6$. As the curves 6 and 7 in Fig. 3.22 prove, the deviation of the Dorfman's Eq. (3.21) is not significant.

Thus, Eqs. (3.59) and (3.68) essentially expand the possibilities for analytical predictions of heat transfer over a rotating disk subject to arbitrary thermal boundary conditions.

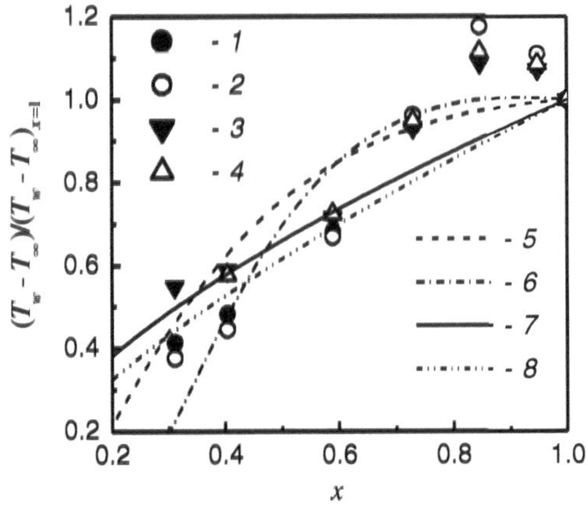

Fig. 3.21 Radial variation of the temperature difference $\overline{\Delta T} = (T_w - T_\infty)/(T_w - T_\infty)_{x=1}$ for $Pr = 0.72$, cases $n_* = 0.4$ (data *1*, *2*, *6*, *8*) and $n_* = 0.6$ (data *3*, *4*, *5*, *7*) [1]. Experiments [87]: *1*—$Re_\varphi = 2.67 \times 10^6$; 2–$3.14 \times 10^6$; *3*—$0.615 \times 10^6$; *4*—$1.71 \times 10^6$. Calculations by Eq. (3.68): *5*—$K_1 = 0.0219$, $m_x = 1.48$, $n = 1/6.5$; *6*—$K_1 = 0.0249$, $m_x = 1.34$, $n = 1/6$. Calculations by Eq. (3.57): *7*—$n_* = 0.6$, *8*—$n_* = 0.7$

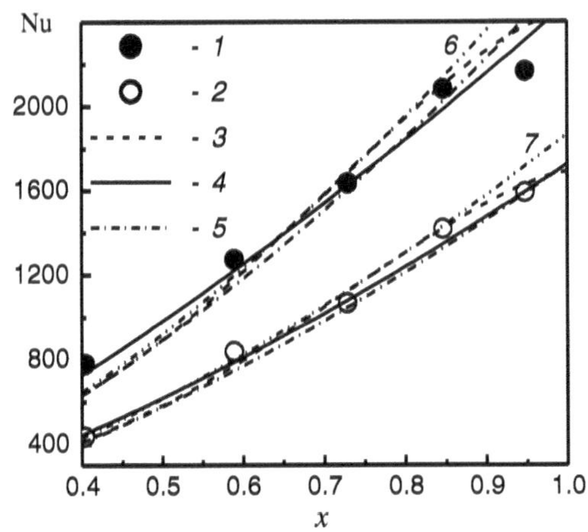

Fig. 3.22 Radial variation of the Nusselt number [1]. *1*, *2*—experiments [87]. *3*—numerical modelling (Fig. 3.12). Case $\underline{n_* = 0.4}$ (upper group of lines, $Re_\varphi = 2.67 \times 10^6$), calculations at $n = n_T = 1/6$: *4*—Eq. (3.59), $m_x = 1.34$, $K_1 = 0.0249$; *5*—Eq. (3.58), $n_* = 0.7$, $K_1 = 0.0262$, $n_R = 0.778$, $m_x = 1.556$. Case $\underline{n_* = 0.6}$ (lower group of lines, $Re_\varphi = 1.59 \times 10^6$), calculations at $n = n_T = 1/6.5$: *4*—Eq. (3.59), $m_x = 1.48$, $K_1 = 0.0219$; *5*—Eq. (3.58), $m_x = 1.58$, $K_1 = 0.02265$, $n_R = 0.7896$, $n_* = 0.6$. Line *6*—Dorfman's Eq. (3.21), $n_* = 0.4$; line *7*—Eq. (3.21), $n_* = 0.6$

3.6 Theory of Local Modelling

The bedrock of the theory of local modelling is Eq. (2.51) at M_s = const. and $n_* = 2$ (see Sect. 2.3.2), which results in the Dorfman's solutions (3.6) and (3.21). An improvement of this model can be attained via setting a variable value of M_s dependent on the parameter n_* [1]. Let us rewrite the thermal boundary layer Eq. (3.60) such as

$$\frac{1}{\Delta T}\frac{\mathrm{d}}{\mathrm{d}r}\left[Re_T^{**}r\Delta T\right] = Nu/Pr. \tag{3.76}$$

Let us further substitute Eq. (2.51) and boundary condition (2.30) into Eq. (3.76). As a result, the Nusselt number can be expressed as

$$Nu = M_s^{\frac{1}{1+\sigma}}(2n_R + n_* + 1)^{\frac{\sigma}{1+\sigma}}(1+\alpha^2)^{\frac{1}{2(1+\sigma)}}Re_\omega^{\frac{1}{1+\sigma}}Pr^{\frac{1+\sigma-n_s}{1+\sigma}}. \tag{3.77}$$

We wish to elucidate just the basic features of the theory; hence, listed below are the solutions for only a single value $Pr = 0.72$. As a result, Eq. (3.34), together with the equations for M_s and St for *turbulent flow* (n = 1/7) can be written as

$$\frac{1}{K_1} = 34.99 + \frac{48.33}{2.6+n_*}, \tag{3.78}$$

$$M_s = \frac{1.252}{(34.99 + 48.33/(2.6+n_*))^{1.25}(2.6+n_*)^{0.25}}, \tag{3.79}$$

$$St = \frac{1.475 \cdot Re_T^{**-0.25}}{(34.99 + 48.33/(2.6+n_*))^{1.25}(2.6+n_*)^{0.25}}. \tag{3.80}$$

For *laminar flow*, the constant K_1 given by Eq. (3.50) and the Stanton number for Pr = 0.72 can be expressed as

$$K_1 = \frac{0.4435}{0.3486 + 2.002/(2+n_*)}, \tag{3.81}$$

$$St = \frac{0.2922 \cdot Re_T^{**-1.0}}{(2+n_*)(0.3486 + 2.002/(2+n_*))^2}. \tag{3.82}$$

To conclude, Eqs. (3.80) and (3.82) essentially improve agreement with experiments, however, at the expense of ignoring the basic postulate of the theory: M_s = const. The mathematical formulation of Eqs. (3.80) and (3.82) was suggested in analogy to Eq. (3.34) obtained by means of a fundamentally different theoretical model.

3.7 Unsteady Heat Transfer

3.7.1 *Transient Experimental Technique*

Thermochromic liquid crystals are used for experimental measurements of steady-state heat transfer rate over a surface using a transient technique. The following physical phenomenon lies behind this technique: after a short time from the onset of unsteady heating/cooling, the heat transfer coefficient accepts a time-independent value equivalent to that for steady-state heat transfer subject to identical thermal boundary conditions.

Experimental data reduction operates by employing a one-dimensional heat conduction solution for a semi-infinite wall subject to a convective boundary condition for a step change in the fluid temperature T_∞ [98–105]

$$F_t(t) = \frac{T_w(t) - T_\infty}{T_{w,i} - T_\infty} = \exp(\gamma^2) \cdot \mathrm{erfc}(\gamma), \quad \gamma = \alpha\sqrt{a_w t}/\lambda_w \tag{3.83}$$

where $T_{w,i}$ and T_∞ are constants; a_w and λ_w denote the thermal diffusivity and conductivity of the body, accordingly. Given a measured curve of the surface temperature $T_w(t)$, Eq. (3.83) can be solved to find the heat transfer coefficient α. The semi-infinite-plate model holds, if conduction heat transfer does not penetrate deeply into the body.

A solution for a plate with a thickness s having identical heat transfer coefficients at both sides [13, 106] replaces Eq. (3.83), if a plate is relatively thin

$$F_t(t) = \vartheta(t, y = 1), \vartheta(t, y) = \sum_{m=1}^{\infty} E_m \cos(\mu_m y) \exp(-\mu_m^2 \mathrm{Fo}), \tag{3.84}$$

$$E_m = \frac{2\sin(\mu_m)}{\mu_m + \sin(\mu_m)\cos(\mu_m)}, \quad \cot(\mu_m) = \mu_m/Bi, \tag{3.85}$$

where Eq. (3.85) defines eigenvalues μ_n; $\vartheta(t, y) = (T(t, y) - T_\infty)/(T_{w,i} - T_\infty)$; $y = z/(0.5s)$.

It will be shown below that the transient technique for measurements of surface heat transfer coefficients described above can be applied for a Plexiglas® disk subject to even very strongly non-uniform initial temperature distribution. A disk made of aluminium is not usable for this purpose [1], because of strong radial heat conduction effects.

3.7.2 Self-similar Equations for Unsteady Convective Heat Transfer

Self-similar functions and independent variables for unsteady heat transfer in stationary flow over a rotating disk were obtained in [1, 107, 108]. As a result, Eq. (2.12) of the thermal boundary layer and the boundary condition (2.29) take the form

$$\theta'' = Pr[g_*\theta + \theta'(H - \eta/2) + n_* F\theta], \tag{3.86}$$

$$\eta = z\Big/(\nu t)^{1/2}, H(\eta) = v_z(t/\nu)^{1/2} \tag{3.87}$$

$$g_* = \frac{t}{F_t}\frac{\mathrm{d}F_t}{\mathrm{d}t}, \tag{3.88}$$

$$\theta = 1 \text{ for } \eta = 0 \text{ and } \theta = 0 \text{ for } \eta \to \infty. \tag{3.89}$$

Here primes denote derivatives with respect to the similarity variable η.

The solution of Eq. (3.86) requires specifying the function $H(\eta)$. For this purpose, the time-independent Eqs. (2.32)–(2.35) (for $N = 0$, $\beta = 0$) must be non-dimensionalized and solved with respect to the similarity variable η instead of $\zeta = z\sqrt{\omega/\nu}$. As a result, at the non-dimensionalization of the velocity components and the static pressure in Eq. (2.26), parameter $1/t$ replaces the angular velocity ω

$$F(\eta) = v_r r/t, G(\eta) = v_\varphi t/r, \text{ and } P(\eta) = -pt/(\rho\nu), \tag{3.90}$$

whereas Eqs. (2.32)–(2.35) and functions F, G, H and P still do not depend on time.

As soon as $v_\varphi = \omega r$ at $z = 0$, the new function $G(\eta) = \omega t$ at $\eta = 0$. Therefore, the boundary conditions (2.27) and (2.28) can be finally rewritten to a new self-similar form

$$\eta = 0: \quad F = H = 0, \quad G = \omega t, \tag{3.91}$$

$$\eta \to \infty: \quad G = F = 0. \tag{3.92}$$

To calculate the Nusselt number Nu_b, the following relations are used

$$Nu_b = K_1 Re_\varphi^{1/2}, \quad K_1 = -\left(\frac{\mathrm{d}\theta}{\mathrm{d}\zeta}\right)_{\zeta=0} = \frac{1}{\sqrt{\omega t}}\left(\frac{\mathrm{d}\theta}{\mathrm{d}\eta}\right)_{\eta=0}. \tag{3.93}$$

The new non-dimensional parametric variable ωt arises in Eqs. (3.91) and (3.93).

Equations (3.87) and (3.90) were derived with the help of the group theory [90]. Authors of the work [31] employed self-similar functions almost identical to

Eqs. (3.87) and (3.90) with a difference to within a constant; however, this model was not formally grounded from the mathematical point of view.

3.7.3 Cooling of an Isothermal Rotating Disk

Mathcad software was used to numerically solve Eqs. (3.86) and (2.32)–(2.35) [in view of Eqs. (3.90)–(3.92)] for the condition T_w = const. (or $n_* = 0$) [1, 107, 108].

Equation (3.88) with account for Eq. (3.83) transforms to

$$g_* = \gamma^2 - \gamma/(\pi^{1/2}F_t), \quad \gamma = K_1 Pr^{-1/2}(a_w/a)^{1/2}(\lambda/\lambda_w)\sqrt{\omega t}. \tag{3.94}$$

In simulations, physical properties and geometric parameters were: for Plexiglas® [96] $\lambda_w = 0.19$ W/(m^2 K), $a_w = 1.086 \times 10^{-7}$ m^2/s; for air [90] $\lambda = 0.02624$ W/(m^2 K), $a = 2.216 \times 10^{-5}$ m^2/s; $Pr = 0.71$; thickness of the disk $s = 0.01$ m; $Re_\varphi = 5.35 \times 10^4$ that means $\omega = 52.36$ 1/s (500 r.p.m.) [96]. The value of $K_1 = 0.326$ at T_w = const. was used in calculation of the Biot number in Eq. (3.84) and parameter γ that gives $\gamma = 0.0768\sqrt{\omega t}$, $Bi = 0.395$.

The constant K_1, together with the Nusselt number Nu_b, become very fast time-independent (see Fig. 3.23): at $\omega t \approx 130$ or $t \approx 2.5$ s (setting 1 % deviation from the steady-state as a threshold), whereas $F_t(t) \approx 0.96$ [1]. The curves for $K_1(t)$ predicted using Eqs. (3.83) and (3.84) practically coincide (see Fig. 3.23).

If heat transfer coefficients of an isothermal disk of a thickness s (Fig. 2.1) are identical at $z = 0$ and $z = -s$, solution (3.84) testifies that already for Fo = 0.3 (or t = 69 s) the function $F_t(t)$ with an inaccuracy of 0.37 % is represented by just the first term of Eq. (3.84). In other words, a so-called regular regime of heat transfer [106] is established.

As seen in Fig. 3.23, during the initial time period, dimensionless surface temperatures $F_t(t)$ calculated by Eqs. (3.83) and (3.84) practically coincide and start diverging at Fo = 0.456, when the process of cooling involves the entire thickness of the thin disk, which afterwards cools down much faster than the disk with a

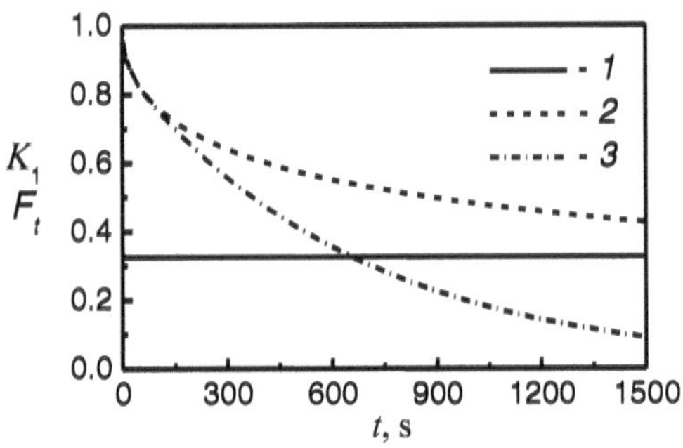

Fig. 3.23 Variation of K_1 and F_t with time [1]. *1*—K_1; *2*—$F_t(t)$, Eq. (3.83); *3*—$F_t(t)$, Eq. (3.84)

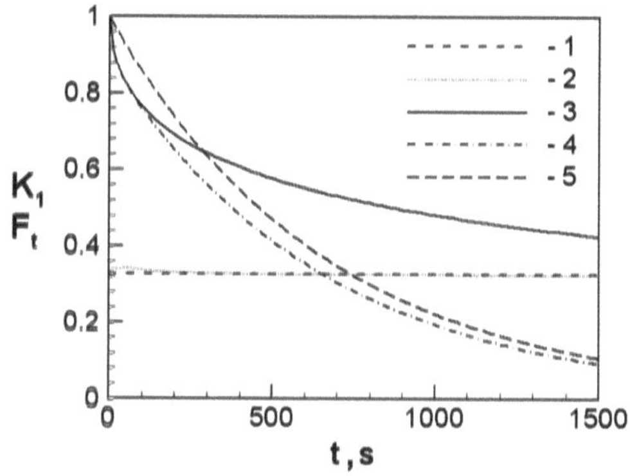

Fig. 3.24 Variation of the parameters K_1 and F_t in time according to the self-similar solution and simulations using the CFX-5 [111]. *1*—K_1, self-similar solution; *2*—K_1, CFX-5; *3*—$F_t(t)$, Eq. (3.83); *4*—$F_t(t)$, Eq. (3.84); *5*—$F_t(t)$, CFX-5

semi-infinite thickness. In [109, 110], this limiting Fourier number equals to 0.25 and 1.0, respectively.

The problem of unsteady cooling of a finite-thickness disk considered here was simulated also as a conjugate problem using the commercial CFD software CFX-5 [111]. The temporal curve of the cooling rate of the disk surface obtained from the CFD simulations and presented in Fig. 3.24 agrees rather well with that predicted by Eq. (3.84).

3.7.4 Unsteady Two-Dimensional Heat Conduction in a Non-uniformly Heated Disk

Differential equation of unsteady 2D heat conduction in the plate together with the boundary conditions can be written as [108]

$$\frac{\partial \vartheta}{\partial \mathrm{Fo}} = \frac{1}{H^2}\left(\frac{\partial^2 \vartheta}{\partial x^2} + \frac{1}{x}\frac{\partial \vartheta}{\partial x}\right) + \frac{\partial^2 \vartheta}{\partial y^2}, \tag{3.95}$$

$$\mathrm{Fo} = 0: \quad \vartheta = x^{n_*}, \tag{3.96}$$

$$x = 0: \quad \frac{\partial \vartheta}{\partial x} = 0, \quad x = 1: \quad \left(\frac{\partial \vartheta}{\partial x}\right)_{x=1} = -Bi_1 \vartheta_{x=1}, \tag{3.97}$$

$$y = 0: \quad \frac{\partial \vartheta}{\partial y} = 0, \quad y = 1: \quad \left(\frac{\partial \vartheta}{\partial y}\right)_{y=1} = -Bi_2 \vartheta_{y=1}, \tag{3.98}$$

where $x = r/b$ and $y = z/(0.5s)$. In fact, Eq (3.96) is Eq. (2.30) reformatted with the help of the new variables. Here the Biot numbers Bi_1 and Bi_2 characterize convective heat transfer from the cylindrical and flat surfaces of the disk, respectively.

The method of separation of variables was used to solve Eqs. (3.95)–(3.98) analytically [108], where the final solution can be presented as

$$\vartheta(\text{Fo}, x, y) = \sum_{n=1}^{\infty}\sum_{m=1}^{\infty} D_n E_m J_0(\mu_{xn}x)\cos(\mu_{ym}y)\exp\left[-(\mu_{xn}^2/H^2 + \mu_{ym}^2)\text{Fo}\right], \quad (3.99)$$

$$D_n = \frac{{}_1F_2(1 + n_*/2; 1, 2 + n_*/2; -\mu_{xn}^2/4)/(2 + n_*)}{0.5[J_0^2(\mu_{xn}) + J_1^2(\mu_{xn})]}, \quad (3.100)$$

$$\frac{J_1(\mu_{xn})}{J_0(\mu_{xn})} = \frac{Bi_1}{\mu_{xn}}. \quad (3.101)$$

Here, E_m and μ_{ym} are defined by Eq. (3.85); J_0 and 102 are Bessel functions of the first kind and zero/first order, accordingly. Ignoring radial heat conduction, Eq. (3.99) simplifies to Eq. (3.84). In Eq. (3.100), ${}_1F_2$ is a hypergeometric function of $-\mu_{xn}^2/4$ [112].

Figures 3.25 and 3.26 depict radial variations of the disk temperature in cases with $n_* = -1$ and 2, where n_* is strongly different from zero. Figures 3.25 and 3.26 demonstrate that during cooling of a Plexiglas® disk the surface temperature $\vartheta_w/\vartheta_w(x=1)$ in fact repeats the initial functions (3.96) over practically entire disk surface.

Strictly saying, Eq. (3.96) contradicts with the steady-state boundary conditions (3.97) for $x = 0$ and $x = 1$. Hence, time-dependent distributions $\vartheta_w/\vartheta_w(x=1)$ are distorted in the neighborhood of the locations $x = 0$ and $x = 1$; therefore the behavior of the heat transfer coefficient $\alpha_{2,t}$ is studied within a region of $x = 0.2$–0.9.

To conclude, the transient technique for measurements of surface heat transfer can be used together with a Plexiglas® disk subject to any initial temperature distribution.

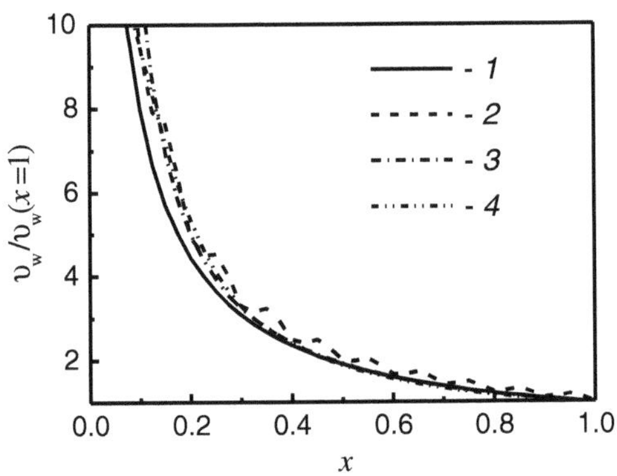

Fig. 3.25 Variation of the function $\vartheta_w/\vartheta_w(x=1)$ with x according to Eq. (3.99) at $n_* = -1$ [1]. *1*—Eq. (3.96); *2*—Fo = 0.00652; *3*—Fo = 0.869; *5*—Fo = 2.607

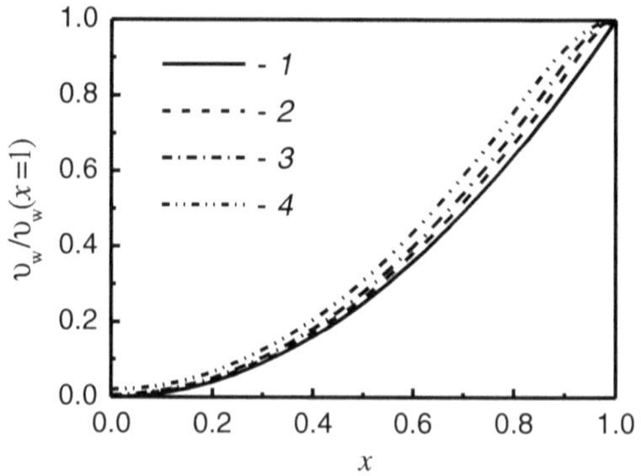

Fig. 3.26 Variation of the function ϑ_w/x^{n_*} with x according to Eq. (3.99) at $n_* = 2$ (*dash-dotted lines*) and $n_* = 0.5$ (*solid lines*) for different values of Fo [1]. *1*—Eq. (3.96); *2*—Fo = 0.0261; *3*—Fo = 0.261; *4*—Fo = 0.869; *5*—Fo = 2.607

References

1. Shevchuk IV (2009) Convective heat and mass transfer in rotating disk systems. Springer, Berlin, Heidelberg
2. Dorfman LA (1963) Hydrodynamic resistance and the heat loss of rotating solids. Oliver and Boyd, Edinburgh, UK
3. Owen JM, Rogers RH (1989) Flow and heat transfer in rotating-disc systems. Volume 1: rotor-stator systems. Research Studies Press Ltd., Taunton, Somerset, England
4. Owen JM, Rogers RH (1995) Flow and heat transfer in rotating-disc systems Volume 2: rotating cavities. Research Studies Press Ltd., Taunton, Somerset, England
5. Schlichting G (1968) Boundary-layer theory. McGraw-Hill Book Company, New York
6. Elkins CJ, Eaton JK (1997) Heat transfer in the rotating disk boundary layer. Stanford University, Department of Mechanical Engineering, Thermosciences Division Report TSD–103. Stanford University, USA
7. Itoh M, Hasegawa I (1994) Turbulent boundary layer on a rotating disk in infinite quiescent fluid. JSME Int J Ser B 37(3):449–456
8. Lingwood RJ (1996) An experimental study of absolute instability of the rotating-disk boundary-layer flow. J Fluid Mech 314:373–405
9. Theodorsen Th, Regier A (1944) Experiments on drag of revolving discs, cylinders and streamline rods at high speeds. NASA Report No 793
10. Kempf G. (1924) Über Reibungswiderstand rotierender Scheiben. Vorträge auf dem Gebiet der Hydro- und Aerodynamik, Innsbrucker Kongress, 1922. Berlin, p 22
11. Schmidt W (1921) Ein einfaches Messverfahren für Drehmomente. Z VDI 65:411–444
12. Littel HS, Eaton JK (1994) Turbulence characteristics of the boundary layer on a rotating disk. J Fluid Mech 266:175–207
13. Elkins CJ, Eaton JK (2000) Turbulent heat and momentum transfer on a rotating disk. J Fluid Mech 402:225–253
14. Parthasarathy RN (2002) Analysis of the turbulent boundary layer on a rotating disk. Microsyst Technol 8(4–5):278–281
15. Bogdan Z (1982) Cooling of a rotating disk by means of an impinging jet. In: Proceedings of 7th IHTC heat transfer 1982, vol 3. Munich, Germany, pp 333–336

16. Chen Y-M, Lee W-T, Wu S-J (1998) Heat (mass) transfer between an impinging jet and a rotating disk. Heat Mass Transfer 34(2–3):101–108
17. Cobb EC, Saunders OA (1956) Heat transfer from a rotating disk. In: Proceedings of the Royal Society A, vol 236, pp 343–349
18. Janotková E, Pavelek M (1986) A naphthalene sublimation method for predicting heat transfer from a rotating surface. Strojnícky časopis 37(3):381–393 (in Czech)
19. Janotková E, Pavelek M (1993) Application of interferometry to temperature field measurements in neighbourhood of rotating discs. In: Proceedings of 11th CHISA 93. Paper No 0015, Prague, Czech Republic
20. Kreith F, Taylor JH, Chong JP (1969) Heat and mass transfer from a rotating disk. Trans ASME J Heat Transfer 81:95–105
21. Pavelek M, Janotková E (1989) Surface temperature calculation for applying naphthalene sublimation method on rotating disks. In: Proceedings of 6th conference thermogrammetry thermal engineering. Budapest, Hungary, pp 111–115
22. Pavelek M, Janotková E (1996) An experimental study of heat transfer from co-rotating disks. Proceedings of 12th CHISA 96, Paper G 8.48, Prague, Czech Republic
23. Popiel CO, Boguslawski L (1975) Local heat-transfer coefficients on the rotating disk in still air. Int J Heat Mass Transfer 18(1):167–170
24. Shimada R, Naito S, Kumagai S, Takeyama T (1987) Enhancement of heat transfer from a rotating disk using a turbulence promoter. JSME Int J Ser B 30(267):1423–1429
25. Sparrow EM, Chaboki A (1982) Heat transfer coefficients for a cup-like cavity rotating about its own axis. Int J Heat Mass Transfer 9(25):1334–1341
26. Shevchuk IV (2000) Turbulent heat transfer of rotating disk at constant temperature or density of heat flux to the wall. High Temp 38(3):499–501
27. Cardone G, Astarita T, Carlomagno GM (1996) Infrared heat transfer measurements on a rotating disk. Opt Diagn Eng 1(2):1–7
28. Cardone G, Astarita T, Carlomagno GM (1997) Heat transfer measurements on a rotating disk. Int J Rotating Mach 3(1):1–9
29. Shevchuk IV (2001) Effect of the wall temperature on laminar heat transfer in a rotating disk: an approximate analytical solution. High Temp 39(4):637–640
30. Shevchuk IV (2001) An improved enthalpy-thickness model for predicting heat transfer of a free rotating disk using an integral method. In: Proceedings of 35th ASME NHTC'01. Paper NHTC2001-20195, Anaheim, California, USA
31. Cheng W-T, Lin H-T (1994) Unsteady and steady mass transfer by laminar forced flow against a rotating disk. Wärme und Stoffübertragung 30(2):101–108
32. Lin H-T, Lin L-K (1987) Heat transfer from a rotating cone or disk to fluids at any Prandtl number. Int Commun Heat Mass Transfer 14(3):323–332
33. Sparrow EM, Gregg JL (1959) Heat transfer from a rotating disc to fluids at any Prandtl number. Trans ASME J Heat Transfer 81:249–251
34. Smith NH (1946) Exploratory investigation of laminar boundary layer oscillations on a rotating disc. NACA Tech Note 1227:1–21
35. Gregory N, Stuart JT, Walker WS (1955) On the stability of three-dimensional boundary layers with application to the flow due to a rotating disk. Phil Trans Roy Soc Ser A 248:155–199
36. Corke TC, Knasiak KF (1998) Stationary travelling cross-flow mode interactions on a rotating disk. J Fluid Mech 355:285–315
37. Fedorov BI, Plavnik GZ, Prokhorov IV, Zhukhovitskii LG (1976) Transitional flow conditions on a rotating disk. J Eng Phys 31:1448–1453
38. Jarre S, Le Gal P, Chauve MP (1996) Experimental study of rotating disk instability. I. Natural flow. Phys Fluids 8(2):496–508
39. Jarre S, Le Gal P, Chauve MP (1996) Experimental study of rotating disk instability II. Forced flow. Phys Fluids 8(11):2985–2994
40. Malik MR, Wilkinson SP, Orzag SA (1981) Instability and transition in a rotating disc. AIAA J 19(9):1131–1138

41. Wilkinson SP, Malik MR (1985) Stability experiments in the flow over a rotating disk. AIAA J 23(4):588–595
42. Chin D, Litt M (1972) An electrochemical study of flow instability on rotating disk. J Fluid Mech 54:613–625
43. Cochran WG (1934) The flow due to a rotating disk. In: Proceedings of Cambridge philosophical society, vol 30, pp 365–375
44. Cooper AJ, Carpenter PW (1997) The stability of rotating-disc boundary-layer flow over a compliant wall. Part 1. Type I and II instabilities. J Fluid Mech 350:231–259
45. Hall P (1986) An asymptotic investigation of the stationary modes of instability of the boundary layer on a rotating disk. In: Proceedings of royal society of London A, vol 406, no 1830, pp 93–106
46. Kohama Y (1983) Study on boundary layer transition of a rotating disk. Acta Mech 50(3–4):193–199
47. Lingwood RJ (1995) Absolute instability of the boundary layer on a rotating disk. J Fluid Mech 299:17–33
48. Lingwood RJ (1997) Absolute instability of the Ekman layer and related rotating flows. J Fluid Mech 331:405–428
49. Malik MR (1986) The neutral curve for stationary disturbances in rotating-disk flow. J Fluid Mech 164:275–287
50. Ong CL, Owen JM (1991) Computation of the flow and heat transfer due to a rotating disc. Int J Heat Fluid Flow 12(2):106–115
51. Pier B (2003) Finite-amplitude crossflow vortices, secondary instability and transition in the rotating-disk boundary layer. J Fluid Mech 487:315–343
52. Sahin I (1988) Applications of various coordinate transformations for rotating disk flow stability. AIAA J 26(3):368–370
53. Szeri AZ, Giron A (1984) Stability of flow over a rotating disk. Int J Numer Methods Fluids 4 (10):989–996
54. Watanabe T (1985) Stability of boundary layers along a rotating disk. Trans JSME Ser B 51 (470):3344–3347
55. Isachenko VP, Osipova VA, Sukomel AS (1977) Heat transfer. Mir Publishers, Moscow, USSR
56. Kobayashi R, Kohama Y, Takamadate Ch (1980) Spiral vortices in boundary layer transition regime on a rotating disk. Acta Mech 35(1–2):71–82
57. Zoueshtiagh F, Ali R, Cooley AJ, Thomas PJ, Carpenter PW (2003) Laminar-turbulent boundary-layer transition over a rough rotating disk. Phys Fluids 15(8):2441–2444
58. Watanabe T (1989) Effect of surface roughness on boundary layer transition on a rotating disk. Trans JSME Ser B 55(515):1842–1846
59. Gregory N, Walker WS (1960) Experiments on the effect of suction on the flow due to a rotating disk. J Fluid Mech 9:225–234
60. Clarkson MH, Chin SC, Shacter P (1980) Flow visualization of inflexional instabilities on a rotating disk. AIAA 80–0279:1–14
61. Dennis RW, Newstead C, Ede AJ (1970) The heat transfer from a rotating disc in an air crossflow. In: Proceedings of IV IHTC, vol 3. Paper no FC 7.1, Paris-Versailles, France
62. Trinkl CM, Bardas U, Weyck A, aus der Wiesche S (2011) Experimental study of the convective heat transfer from a rotating disc subjected to forced air streams. Int J Thermal Sci 50(1):73–80
63. Pellé J, Harmand S (2007) Heat transfer study in a discoidal system: the influence of an impinging jet and rotation. Experim Heat Transfer 20(4):337–358
64. Tien CL, Campbell DT (1963) Heat and mass transfer from rotating cones. J Fluid Mech 17:105–112
65. Daguenet M (1968) Etude du transport de matière en solution, a l'aide des électrodes a disque et a anneau tournants. Int J Heat Mass Transfer 11(11):1581–1596
66. Dossenbach O (1976) Simultaneous laminar and turbulent mass transfer at a rotating disk electrode. Berichte der Bunsen-Gesellschaft Phys Chem Chem Phys 80(4):341–343

67. Mohr CM, Newman J (1976) Mass transfer to a rotating disk in transitional flow. J Electrochem Soc 123(11):1687–1691
68. Deslouis C, Tribollet B, Viet L (1980) Local and overall mass transfer rates to a rotating disk in turbulent and transition flows. Electrochim Acta 25(8):1027–1032
69. Le Palec G (1989) Numerical study of convective heat transfer over a rotating rough disk with uniform wall temperature. Int Commun Heat Mass Transfer 16(1):107–113
70. Le Palec G, Nardin P, Rondot P (1990) Study of laminar heat transfer over a sinusoidal—shaped rotating disk. Int J Heat Mass Transfer 33(6):1183–1192
71. von Karman Th (1921) Über laminare und turbulente Reibung. Z Angew Math Mech 1 (4):233–252
72. Cham T-S, Head TR (1969) Turbulent boundary-layer flow on a rotating disk. J Fluid Mech 37(1):129–147
73. Shvets IT, Dyban EP (1974) Air cooling of gas turbine parts. Naukova Dumka, Kiev, Ukraine (in Russian)
74. Shevchuk IV, Khalatov AA (1997) Integral method for calculating the characteristics of a turbulent boundary layer on a rotating disk: quadratic approximation of the tangent of the flow swirl angle. Heat Transfer Res 28(4–6):402–413
75. Shevchuk IV (1999) Simulation of heat transfer and hydrodynamics over a free rotating disk using an improved radial velocity profile. J Thermal Sci 8(4):243–249
76. Abrahamson S, Lonnes S (1993) An integral method for turbulent boundary layers on rotating disks. Trans ASME J Fluids Eng 115(4):614–619
77. Kapinos VM (1964) Hydraulic resistance and heat transfer of a free disc with a cob. Izvestiya vuzov Energetika 11:85–92 (in Russian)
78. Case P (1966) Measurements of entrainment by a free rotating disk. J Roy Aero Soc 71:124–129
79. Kabkov VIa (1974) Characteristics of turbulent boundary-layer on a smooth disk rotating in a large volume. Teplofizika i Teplotekhnika Naukova Dumka Kiev 28:119–124 (in Russian)
80. Goldstein S (1935) On the resistance to the rotation of a disc immersed in a fluid. Proc Cambridge Phil Soc 31:232–241
81. Buznik VM, Artemov GA, Bandura VN, Kardashev YD, Fedorovskiy AM (1966) Heat transfer of a flat disk rotating in quiescent environment, Izvestiya vuzov. Energetika 1:84–86 (in Russian)
82. Kuznetsov AL (1962) Experimental investigation of heat transfer from a disk rotating in quiescent environment. Trudy Leningrad Korablestroit Inst 38:183–186. (in Russian)
83. Nikitenko NI (1963) Experimental investigation of heat exchange of a disk and a screen. J Eng Phys 6(6):1–11
84. McComas ST, Hartnett JP (1970) Temperature profiles and heat transfer associated with a single disc rotating in still air. In: Proceedings of IV IHTC, vol 3. Paper FC 7.7, Paris-Versailles, France
85. Tadros SE, Erian FF (1983) Heat and momentum transfer in the turbulent boundary layer near a rotating disk. Paper ASME 83–WA/HT–6
86. Northrop A, Owen JM (1988) Heat transfer measurements in rotating-disc systems. Part 1: The free disc. Int J Heat Fluid Flow 9(1):19–26
87. Northrop A (1984) Heat transfer in a cylindrical rotating cavity. D. Phil. thesis. University of Sussex, Brighton, UK
88. Johnson MC (1980) Turbulent heat transfer to a rotating disk: a review and extension of Dorfman. Trans ASME J Heat Transfer 102(4):780–781
89. Shevchuk IV (1998) Simulation of heat transfer in a rotating disk: the effect of approximation of the tangent of the angle of flow swirling. High Temp 36(3):522–524
90. Cebeci T, Bradshaw P (1984) Physical and computational aspects of convective heat transfer. Springer, Berlin, Heidelberg
91. Shevchuk IV (2002) Numerical modeling of turbulent heat transfer in a rotating disk at arbitrary distribution of the wall temperature. J Eng Phys Thermophys 75(4):885–888

92. Shevchuk IV, Saniei N, Yan XT (2001) Effect of the wall temperature distribution in direct and inverse problems of laminar heat transfer over a free rotating disk. In: Proceedings of 35th ASME NHTC'01. Paper NHTC2001–20196, Anaheim, California, USA
93. Shevchuk IV (2001) Turbulent heat transfer over a free rotating disk: analytical and numerical predictions using solutions of direct and inverse problems. In: Proceedings of IMECE'01. New York, USA, pp 1–8
94. Saniei N, Yan X (1998) Effect of jet location on impingement cooling of a rotating disk. Proc ASME HTD 361:203–210
95. Saniei N, Yan XT, Schooley WW (1998) Local heat transfer characteristics of a rotating disk under jet impingement cooling. Proc 11th IHTC Heat Transfer 5:445–450
96. Saniei N, Yan XT (2000) An experimental study of heat transfer from a disk rotating in an infinite environment including heat transfer enhancement by jet impingement cooling. J Enhanced Heat Transfer 7:231–245
97. Shevchuk IV (2005) A new type of the boundary condition allowing analytical solution of the thermal boundary layer equation. Int J Thermal Sci 44(4):374–381
98. Baughn JW, Ireland PT, Jones TV, Saniei N (1989) A comparison of the transient and heated-coating methods for measurement of local heat transfer coefficients on pin fin. Trans ASME J Heat Transfer 111(4):877–881
99. Harmand S, Pellé J, Poncet S, Shevchuk IV (2013) Review of fluid flow and convective heat transfer within rotating disk cavities with impinging jet. Int J Thermal Sci 67:1–30
100. Holman JP (1981) Heat transfer. McGraw-Hill, Inc., New York
101. Kingsley-Rowe JR, Lock GD, Owen JM (2005) Transient heat transfer measurements using thermochromic liquid crystal: lateral-conduction error. Int J Heat Fluid Flow 26(2):256–263
102. Lock GD, Daguze F, Syson BJ, Owen JM (1996) The application of thermal imaging techniques to determine heat transfer coefficients for slow thermal transients. In: Proceedings of 13th Symposium measuring techniques transonic supersonic flows in Cascades and Turbomach. Zurich, Switzerland, pp 1–7
103. Martinez-Botaz RF, Lock GD, Jones TV (1994) Heat transfer measurements in an annular cascade of transonic gas turbine blades using the transient liquid crystal technique. ASME Paper 94-GT-172, pp 1–8
104. Roy RP, Xu G, Feng J (2001) A study of convective heat transfer in a model rotor-stator disk cavity. Trans ASME J Turbomach 123(3):621–632
105. Wagner G, Kotulla M, Ott P, Weigand B, von Wolfersdorf J (2005) The transient liquid crystal technique: influence of surface curvature and finite wall thickness. Trans ASME J Turbomach 127(1):175–182
106. Luikov AV (1968) Analytical heat diffusion theory. Academic Press, New York
107. Shevchuk IV (2004) Unsteady-state laminar heat transfer in a rotating disk: self-similar solution. High Temp 42(4):592–595
108. Shevchuk IV (2006) Unsteady conjugate laminar heat transfer of a rotating non-uniformly heated disk: application to the transient experimental technique. Int J Heat Mass Transfer 49 (19–20):3530–3537
109. Schultz DL, Jones TV (1973) Heat transfer measurements in short duration hypersonic facilities. Aeronautical RD AGARDOGRAPH, NATO Advisory Group, p 165
110. Vogel G, Weigand B (2001) A new evaluation method for transient liquid crystal experiments. In: Proceedings 35th ASME NHTC'01. Paper NHTC2001–20250, Anaheim, California, USA
111. Shevchuk IV, Indinger T (2005) Transient conjugate heat transfer of a disk rotating in still air. In: Proceedings 4th international conference computational heat mass transfer. Paper ICCHMT'05–150, Paris, France
112. Weisstein EW (2002) Hypergeometric function. MathWorld. A Wolfram Web Resource

Chapter 4
Forced External Flow Over a Rotating Disk

4.1 Rotating Disk in a Fluid Rotating as a Solid Body

4.1.1 Turbulent Flow

Disks rotating with an angular velocity ω in a fluid corotating with a different velocity Ω may be found in many technical applications: swirling flows over rotating disks, disks corotating at different angular speeds, etc. [1–4]. If $\omega > \Omega$ (in other words, $\beta = v_{\varphi,\infty}/(\omega r) < 1$), disk rotation engenders centrifugal fluid flow (see Fig. 4.1) qualitatively similar to the flow caused by a single rotating disk at $\beta = 0$.

If $\omega < \Omega$ (or $\beta > 1$), disk rotation causes centripetal flow. For $\beta \rightarrow \infty$, fluid rotates over a stationary surface [2, 4, 5]; this case is outside of the scope of the present work.

Radial variation of the parameter β is different in different problems. Two frequently studied cases are solid body rotation, where $\beta = \Omega/\omega$ = const. (2.27), and free vortex law, where $\beta = \text{const.}/r^2$ or $v_{\varphi,\infty} = (v_{\varphi,\infty} r)_i/r$.

Other forms of the radial distribution of the parameter β were studied, e.g., in the works [2, 3, 7].

For laminar flow, β = const. and $N = 0$, one can find an exact solution of the self-similar Navier–Stokes and energy Eqs. (2.32)–(2.36) (see Sect. 4.1.2). For turbulent flow, a solution of the boundary layer equations can be derived using an integral method at β = const. and $N = 0$. Such a solution has the form of Eqs. (2.77)–(2.87).

The von Karman's integral method [8], based on Eqs. (2.40) and (2.41) for $n = 1/7$, yields a solution for the parameters α, γ_*, ε_m, and ε_M to be substituted into Eqs. (2.77)–(2.81) [2, 9]

$$\alpha^2 = \frac{2300(1+8\beta)}{49(1789-409\beta)}, \quad \gamma_* = \left[\frac{81(1+\alpha^2)^{3/8}}{49(23+37\beta)\alpha}\right]^{4/5}, \tag{4.1}$$

I.V. Shevchuk, *Modelling of Convective Heat and Mass Transfer in Rotating Flows*, Mathematical Engineering, DOI 10.1007/978-3-319-20961-6_4

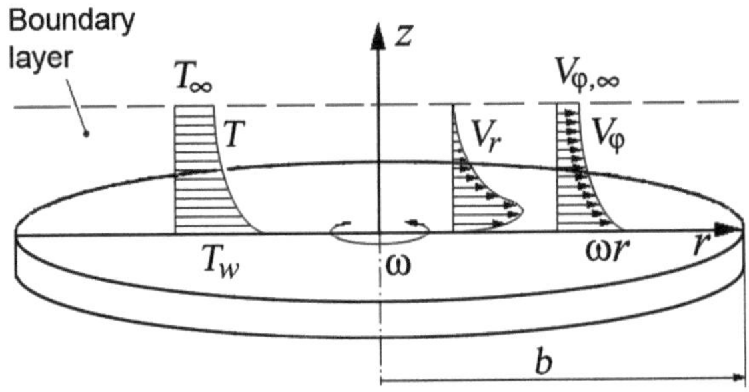

Fig. 4.1 Schematic outline of velocity and temperature profiles on a disk rotating in a rotating fluid for $\beta < 1$ [6]. Tangential velocity component v_φ is turned 90°

$$\varepsilon_m = \frac{49\pi}{60}\mathrm{sgn}(1-\beta)|1-\beta|^{8/5}\alpha\gamma_*,$$
$$\varepsilon_M = \frac{49\pi}{4140}\mathrm{sgn}(1-\beta)|1-\beta|^{8/5}(23+73\beta)\alpha\gamma_*. \tag{4.2}$$

An integral method [4] employing Eqs. (2.40) and (2.42) for n = 1/7 yields

$$\alpha^2 = \frac{0.2222(\beta+0.125)}{0.6374-0.1176}, \quad \gamma_* = 0.04809(1+\alpha^2)^{3/10}[\alpha(0.2425+0.2772\beta)]^{-4/5}, \tag{4.3}$$

$$\varepsilon_m = 1.6332\alpha(1-\beta)\gamma_*, \quad \varepsilon_M = 0.123\left(1+\alpha^2\right)^{0.375}\gamma_*^{-0.25}|\beta-1|^{1.75}. \tag{4.4}$$

The *present integral method* results in a solution, which for β = const. and $N = 0$ (or $\kappa = 0$) has a form of Eqs. (2.77)–(2.87) for the velocity boundary layer, and Eqs. (2.89)–(2.91) for the thermal boundary layer (in fact, Eqs. (4.3) and (4.4) [4] are the same as Eqs. (2.77)–(2.87) simplified for the conditions where $N = 0$ and n = 1/7).

According to Eq. (2.83), a solution for $N = 0$ (or $\kappa = 0$) is valid just at $C_5 \le \beta \le H_8/H_7$ [10], where $C_5 = -0.125$, $H_8/H_7 = 4.412$ for n = 1/7, and $C_5 = -0.1$, $H_8/H_7 = 6.026$ for n = 1/9 (in [2], $C_5 = -0.125$ and $H_8/H_7 = 4.374$ for n = 1/7). The constants α, γ, ε_m^*, and ε_M in Eqs. (2.83)–(2.86) [4, 10] along with their values computed by Eqs. (4.1) and (4.2) [2] are specified in Table 4.1. Values α by the present integral method exceed those obtained in the work [2]; however, this practically does not affect the moment coefficient. Coefficients ε_m^* (mass flow rate) predicted by both methods for $0.6 < \beta < 3$ practically coincide.

The present integral method yields a relation for the enthalpy thickness [11]

$$\overline{\delta}_T^{**} = -(1-\beta)\alpha(-A_1 + \chi Pr^{n_p} D_2). \tag{4.5}$$

Equation (2.89) for the thermal boundary layer for $\Delta \le 1$ and $N = 0$ remains transcendental, whereas the constants E_1, E_2, C_4, and C_5 are accordingly simplified.

Table 4.1 Constants in Eqs. (2.77)–(2.81) for β = const., $N = 0$ [6]

β	Solution (2.83)–(2.86) [4, 10], $n = 1/7$				Solution (4.1), (4.2) [2], $n = 1/7$			
	α	γ	ε_m^*	ε_M	α	γ	ε_m^*	ε_M
0	0.2087	0.5299	0.1806	0.1466	0.162	0.526	0.219	0.146
0.2	0.343	0.3084	0.1727	0.1205	0.267	0.286	0.196	0.1208
0.4	0.4445	0.2232	0.162	0.0845	0.348	0.197	0.1766	0.085
0.6	0.5331	0.1754	0.1527	0.0482	0.42	0.149	0.1603	0.0486
0.8	0.6151	0.1443	0.1449	0.0165	0.487	0.119	0.1446	0.0172
1	0.6936	0.1224	0.1386	0	0.553	0.0983	0.1395	0
2	1.084	0.0683	−0.1208	0.322	0.907	0.0498	0.116	0.326
3	1.564	0.046	−0.1175	1.28	1.445	0.031	−0.1151	1.356
4	2.348	0.0335	−0.1285	3.368	3.182	0.02	−0.1629	4.682

For the case of air flow, where $\Delta \geq 1$, Eq. (2.89) yields a relation for Δ [valid under the boundary condition (2.31)] [10]

$$\Delta^{-n} = \left[\frac{(4+m)K_V Pr^{-n_p} + (2+m+n_*)(1-K_V) - Pr^{-n_p}2\beta/(\beta-1)}{(2+m+n_*) - n_*\beta/(\beta-1)}\right]^{-1}. \quad (4.6)$$

In Eq. (4.6), the parameter K_V is defined by Eq. (3.37). Equations (2.90) and (2.91) for the Stanton and Nusselt numbers remain the same.

Comparing Eqs. (2.52) and (2.66), one can obtain expressions for the Reynolds analogy parameter χ and the coefficient K_1 [11] given by

$$\chi = \Delta^{-n} Pr^{-n_p}, \quad K_1 = \chi A_c \left(1+\alpha^2\right)^{1/2}(1-\beta)Pr, \quad (4.7)$$

$$\chi = \frac{(2+m+n_*) + n_*\beta/(1-\beta)}{(2+m+n_*)Pr^{n_p}(1-K_V) + (4+m)K_V + 2\beta/(1-\beta)}. \quad (4.8)$$

For the boundary condition (2.30) mainly used here, Eq. (4.8) yields [11]

$$\chi = \frac{2+m+n_*}{(2+m+n_*)(1-K_V)Pr^{n_p} + (4+m)K_V + 2\beta/(1-\beta)}. \quad (4.9)$$

The dependence of the Reynolds analogy parameter χ on the swirl parameter β for turbulent flow is elucidated in Fig. 4.2.

Data for the constant K_1 for $Pr = 0.71$ and 1.0, at n_* varying over the range of $-2.6 \leq n_* \leq 4$ are documented in Tables 4.2 and 4.3. As one can see from these tables, heat transfer rate diminishes for higher values of β because of decreasing shear of the tangential velocity. Heat transfer rate enhances with increased values of the exponent n_* [11].

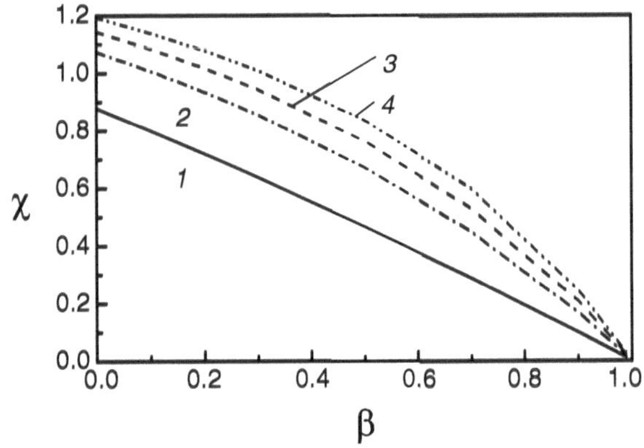

Fig. 4.2 Effect of the swirl parameter β on the Reynolds analogy parameter χ in turbulent flow for $Pr = 0.71$ [6]. *1*—$n_* = -0.6$; *2*—$n_* = 1$; *3*—$n_* = 2$; *4*—$n_* = 3$

4.1.2 Laminar Flow

Exact solution. The self-similar Eqs. (2.32)–(2.36) for $N = 0$ and $\beta < 1$ were solved in [11] by using the Mathcad software under the following boundary conditions:

$$\begin{aligned} &\zeta = 0: && F = H = 0, G = 1, && \theta = 1, \\ &\zeta \to \infty: && G = \beta, F = 0, H \to \text{const.}, && \theta = 0. \end{aligned} \tag{4.10}$$

The velocity components F, G, H and dimensionless temperature θ parametrically depending on β are depicted in Figs. 4.3, 4.4, 4.5, and 4.6. The radial velocity F has a form typical for wall jets (Fig. 4.3). The axial velocity H is directed toward the disk surface (Fig. 4.5) and, hence, is negative. The shear of the tangential velocity G (Fig. 4.4) is the driving force of the flow. Increasing the swirl parameter β (or fluid's angular speed of rotation Ω) results in the diminishing shear of the tangential velocity G, suppresses the radial and axial flow, and reduces the temperature gradient in the vicinity of the disk (Figs. 4.3, 4.4, 4.5, and 4.6) [11].

Tables 4.4 and 4.5 list the constants used to calculate the following quantities:

$$\begin{aligned} &\tau_{w\varphi}/\left(\rho\omega^2 r^2\right) = -G'_w Re_\omega^{-1/2}, \quad \tau_{wr}/\left(\rho\omega^2 r^2\right) = F'_w Re_\omega^{-1/2}, \\ &\dot{m}_d/(\mu r) = -\pi H_\infty Re_\omega^{1/2}, \quad C_M = -2\pi G'_w Re_\varphi^{-1/2}. \end{aligned} \tag{4.11}$$

As follows from the discussion above, higher values of the swirl parameter β are accompanied with lower values of the wall shear stress and a reduced mass flow rate through the boundary layer. Tables 4.4 and 4.5 complement similar archival data presented in [2, 3] (factor $\beta^{-3/2}$ is discarded in the first line of Table 4.5).

Equation (3.4) for $n_R = 1/2$ and Eq. (3.5) define the Nusselt numbers and the constant K_1, accordingly. Thus increasing the swirl parameter β deteriorates the heat transfer (details see below).

Integral method: fluid flow. In frames of the integral method, the modified form of the solution for turbulent flow is used for laminar flow. In doing so, selected

Table 4.2 Constant K_1 in turbulent flow for $Pr = 1$ and $\beta = \text{const.}$ [6]

β	$n_* = -2.6$	$n_* = -0.6$	$n_* = 0$	$n_* = 0.5$	$n_* = 1$	$n_* = 2$	$n_* = 3$	$n_* = 4$
0.0	0.0	0.0212	0.0232	0.0244	0.0254	0.0268	0.0278	0.0286
0.1	0.0	0.0201	0.0222	0.0236	0.0247	0.0264	1.024	0.0284
0.2	0.0	0.0182	0.0204	0.0219	0.0231	0.0249	0.0262	0.0272
0.3	0.0	0.0159	0.0181	0.0196	0.0208	0.0227	0.0241	0.0252
0.5	0.0	0.0107	0.0126	0.0139	0.015	0.0169	0.0183	0.0195
0.7	0.0	0.0054	0.0065	0.00741	0.0082	0.0096	0.0107	0.0117
0.9	0.0	0.001	0.0013	0.00155	0.0018	0.0022	0.0026	0.0029
0.99	0.0	2.8×10^{-5}	3.6×10^{-5}	4.3×10^{-5}	4.0×10^{-5}	6.3×10^{-5}	8.3×10^{-5}	9.0×10^{-5}

Table 4.3 Constant K_1 in turbulent flow for $Pr = 0.71$ and $\beta = \text{const.}$ [6]

β	$n_* = -2.6$	$n_* = -0.6$	$n_* = 0$	$n_* = 0.5$	$n_* = 1$	$n_* = 2$	$n_* = 3$	$n_* = 4$
0.0	0.0	0.0167	0.0185	0.0196	0.0218	0.0218	0.0227	0.0234
0.1	0.0	0.0157	0.0176	0.0188	0.0198	0.0213	0.0223	0.0232
0.3	0.0	0.0122	0.0141	0.0153	0.0164	0.0180	0.0193	0.0203
0.5	0.0	0.008	0.0095	0.0106	0.0116	0.0131	0.0144	0.0154
0.7	0.0	0.0039	0.354	0.00553	0.0062	0.0073	0.0082	0.009
0.9	0.0	0.00075	0.00096	0.00112	0.0013	0.0016	0.0019	0.0022
0.99	0.0	2.0×10^{-5}	2.6×10^{-5}	3.0×10^{-5}	3.5×10^{-5}	4.5×10^{-5}	4.4×10^{-5}	6.4×10^{-5}

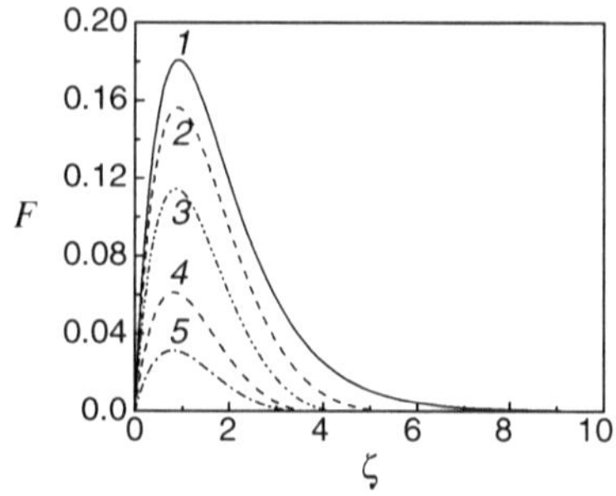

Fig. 4.3 Profiles of the radial velocity F [6]: *1*—$\beta = 0$; *2*—0.4; *3*—0.6; *4*—0.8; *5*—0.9

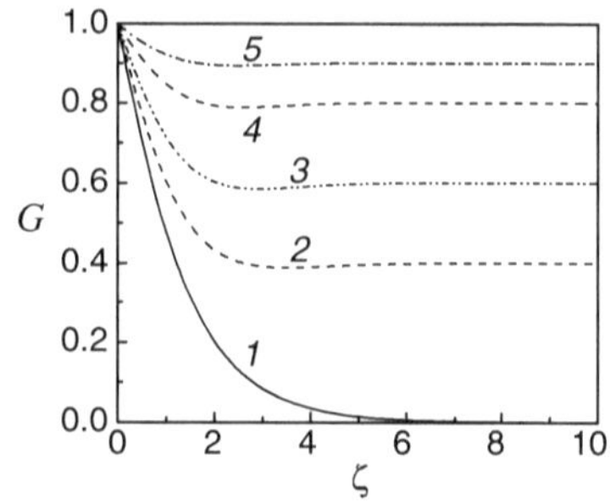

Fig. 4.4 Profiles of the tangential velocity G [6]: *1*—$\beta = 0$; *2*—0.4; *3*—0.6; *4*—0.8; *5*—0.9

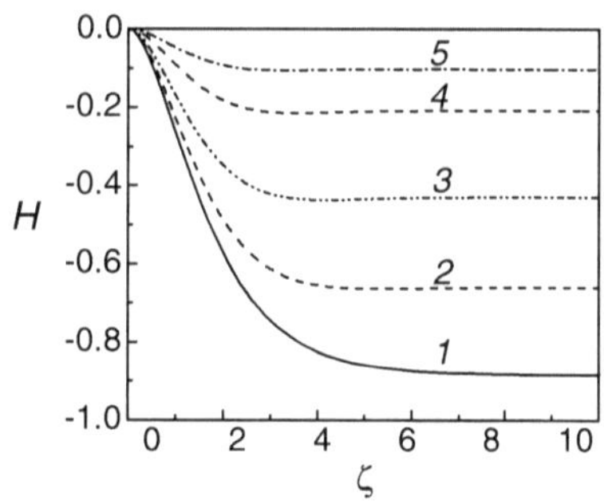

Fig. 4.5 Profiles of the axial velocity H [6]: *1*—$\beta = 0$; *2*—0.4; *3*—0.6; *4*—0.8; *5*—0.9

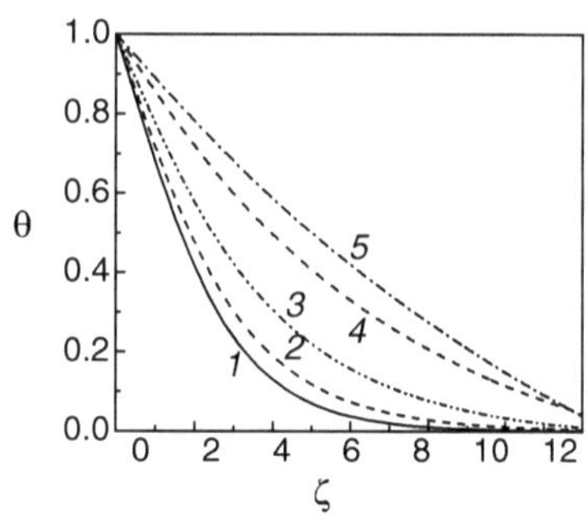

Fig. 4.6 Dimensionless temperature θ profiles [6]: *1*—$\beta = 0$; *2*—0.4; *3*—0.6; *4*—0.8; *5*—0.9

Table 4.4 Parameters for laminar flow for $N = 0$ and $\beta \leq 1$ [6]

β	F'_w	$-G'_w$	α	$\tau_{w\varphi}/\tau_{w\varphi 0}$	$-H_\infty$
0.0	0.5102	0.6159	0.8284	1.0	0.8845
0.1	0.5134	0.6016	0.8535	0.9767	0.9176
0.3	0.4769	0.5305	0.8989	0.8614	0.7687
0.5	0.3908	0.4177	0.9357	0.6782	0.5457
0.7	0.2622	0.2716	0.9654	0.4409	0.3178
0.9	0.09594	0.09695	0.9895	0.1574	0.1020
1.0	0.0	0.0	1.0	0.0	0.0

Table 4.5 Parameters for laminar flow for $N = 0$ and $\beta \geq 1$ [6]

β^{-1}	$-\beta^{-3/2}F'_w$	$\beta^{-3/2}G'_w$	α	$-\tau_{w\varphi}/\tau_{w\varphi 0}$	$\beta^{-3/2}H_\infty$
0.0	0.9420	0.7729	1.2188	1.2549	1.3696
0.1	0.8449	0.7184	1.1761	36.8861	0.1199
0.3	0.6602	0.5931	1.1132	4.8601	0.2633
0.5	0.4776	0.4469	1.0688	2.0522	0.2935
0.7	0.2915	0.2814	1.0359	0.7800	0.2307
0.9	0.09901	0.09798	1.0106	0.1863	0.07778
1.0	0.0	0.0	1.0	0.0	0.0

coefficients are made free parameters to be defined empirically via matching them with the exact solution [11]. For laminar and turbulent flow, the integral Eqs. (2.17) and (2.18) for the velocity boundary layer are completed with the models for the velocity profiles [11]

$$\bar{v}_\varphi = 1 - g, \quad \bar{v}_r = \frac{\alpha}{\alpha_0} f. \tag{4.12}$$

Here g and f are functions of the variable ξ; the parameter α is the tangent of the flow swirl angle at the wall: $\alpha_0 = 0.8284$ for a single rotating disk. Equation (4.12) yields

$$g = \frac{\beta - G}{\beta - 1}, \quad f = \frac{\alpha_0 F}{\alpha(1 - \beta)}. \tag{4.13}$$

In [2, 3], the functions g and f are independent of β and have been calculated using the profiles for a single rotating disk. This assumption entails noticeable deviations from the exact solution. Account for the effect of β on the functions g and f is a principal difference of the present integral method from the method of the authors [2, 3].

An important assumption of the present integral method is that the velocity profiles g and f plotted versus the coordinate $\xi = z/\delta$ depend only on β. Unknowns δ

and α are to be found by solving the integral equations. Equations (2.17) and (2.18), being integrated in view of Eq. (4.12), yield

$$\frac{\mathrm{d}}{\mathrm{d}r}\left\{\delta\omega^2 r^3(1-\beta)^2\left(-B_1\alpha^2\right)\right\}+\delta(\omega r)^2(\beta-1)C_3(\beta-C_5)=r\tau_{wr}/\rho, \quad (4.14)$$

$$\frac{\mathrm{d}}{\mathrm{d}r}\left\{\delta\omega^2 r^4(\beta-1)(1-\beta)\alpha D_1\right\}+\delta\omega r^2(\beta-1)\alpha A_1\frac{\mathrm{d}}{\mathrm{d}r}\left(\beta\omega r^2\right)=r^2\tau_{w\varphi}/\rho, \quad (4.15)$$

where

$$A_1=\frac{1}{\alpha_0}\int_0^1 f\,\mathrm{d}\xi, \quad A_3=\int_0^1 g\,\mathrm{d}\xi, \quad B_1=\frac{1}{\alpha_0^2}\int_0^1 f^2\,\mathrm{d}\xi, \quad (4.16)$$

$$C_1=\int_0^1 g^2\,\mathrm{d}\xi, \quad D_1=\frac{1}{\alpha_0}\int_0^1 fg\,\mathrm{d}\xi, \quad (4.17)$$

$$C_3=C_1-2A_3, \quad C_5=C_1/C_3. \quad (4.18)$$

Equations (4.14) and (4.15) hold for laminar and turbulent flows. In the latter case,

$$g=1-\xi^n, \quad f=\alpha_0\xi^n(1-\xi)^2 \quad (4.19)$$

are functions independent of r and β [the first of them coincides with Eq. (2.40), the second is a modified Eq. (2.42)]. For turbulent flow, coefficients and shear stresses in Eqs. (4.14) and (4.15) are described in Sect. 2.4.

For laminar flow, an integration of the solution of the self-similar Eqs. (2.32)–(2.35) and the further use of Eq. (4.13) gives the values of the constants in Eqs. (4.16)–(4.18)

$$\frac{a_1\delta}{\delta_0}=A_1\zeta_\delta=\int_0^{\zeta_\delta}\frac{F}{(1-\beta)\alpha}\mathrm{d}\zeta, \quad \frac{b_1\delta}{\delta_0}=B_1\zeta_\delta=\int_0^{\zeta_\delta}\frac{F^2}{(1-\beta)^2\alpha^2}\mathrm{d}\zeta,$$

$$\frac{a_3\delta}{\delta_0}=A_3\zeta_\delta=\int_0^{\zeta_\delta}\frac{\beta-G}{\beta-1}\mathrm{d}\zeta, \quad \frac{c_1\delta}{\delta_0}=C_1\zeta_\delta=\int_0^{\zeta_\delta}\frac{(\beta-G)^2}{(\beta-1)^2}\mathrm{d}\zeta, \quad (4.20)$$

$$\frac{d_1\delta}{\delta_0}=D_1\zeta_\delta=\int_0^{\zeta_\delta}\frac{F(G-\beta)}{(\beta-1)^2\alpha}\mathrm{d}\zeta,$$

where $\zeta_\delta = \delta\sqrt{\omega/\nu}$, $\zeta_{\delta 0} = \delta_0\sqrt{\omega/\nu}$. The integrals in Eq. (4.20) are constant for any β.

Shear stress components are given by the following relations:

$$\tau_{w\varphi} = \frac{\mu\omega r}{\delta}\cdot\left(\frac{\mathrm{d}g}{\mathrm{d}\xi}\right)_{\xi=0} = \mu\omega r\left(\frac{\omega}{\nu}\right)^{1/2}\cdot\left(\frac{\mathrm{d}G}{\mathrm{d}\zeta}\right)_{\zeta=0}, \tag{4.21}$$

$$\tau_{wr} = \frac{\mu\omega r}{\delta}\cdot\left(\frac{\mathrm{d}f}{\mathrm{d}\xi}\right)_{\xi=0} = \mu\omega r\left(\frac{\omega}{\nu}\right)^{1/2}\cdot\left(\frac{\mathrm{d}F}{\mathrm{d}\zeta}\right)_{\zeta=0}, \tag{4.22}$$

$$\frac{\tau_{wr}}{\tau_{w\varphi}} = \left(\frac{\mathrm{d}f}{\mathrm{d}\xi}\right)_{\xi=0}\Big/\left(\frac{\mathrm{d}g}{\mathrm{d}\xi}\right)_{\xi=0} = \left(\frac{\mathrm{d}F}{\mathrm{d}\zeta}\right)_{\zeta=0}\Big/\left(\frac{\mathrm{d}G}{\mathrm{d}\zeta}\right)_{\zeta=0} = -\alpha. \tag{4.23}$$

Here $(\mathrm{d}g/\mathrm{d}\xi)_{\xi=0} = \text{const.}$ is independent of β. For $\beta = 0$, Eq. (3.1) yields $(\mathrm{d}G/\mathrm{d}\zeta)_{\zeta=0} = -0.6159$ and $(\mathrm{d}F/\mathrm{d}\zeta)_{\zeta=0} = 0.5102$. Hence, Eqs. (4.21) and (4.22) result in

$$(\mathrm{d}F/\mathrm{d}\zeta)_{\zeta=0} = 0.5102\tau_{wr}/\tau_{wr0} = 0.5102(\alpha/\alpha_0)\left(\tau_{w\varphi}/\tau_{w\varphi 0}\right), \tag{4.24}$$

$$(\mathrm{d}G/\mathrm{d}\zeta)_{\zeta=0} = -0.6159\tau_{w\varphi}/\tau_{w\varphi 0} = -0.6159\delta_0/\delta, \tag{4.25}$$

Given that α = const., δ = const., one can solve Eqs. (4.14) and (4.15), which results in

$$\alpha = \left[-c_3(\beta - c_5)/(\beta h_7 + h_8)\right]^{1/2}, \tag{4.26}$$

$$h_7 = -3b_1 - 4d_1 + 2a_1, \quad h_8 = 3b_1 + 4d_1, \tag{4.27}$$

$$\delta/\delta_0 = \tau_{w\varphi 0}/\tau_{w\varphi}, \tag{4.28}$$

$$\frac{\tau_{w\varphi}}{\tau_{w\varphi 0}} = \frac{\delta_0}{\delta} = \left[(1-\beta)\frac{\alpha}{\alpha_0}\cdot\frac{d_1 + \beta(0.5a_1 - d_1)}{d_{1,0}}\right]^{1/2}. \tag{4.29}$$

The values of α and $\tau_{w\varphi}/\tau_{w\varphi 0}$ shown in Table 4.6 were found by using Eqs. (4.23), (4.25); coefficients a_1, a_3, b_1, c_1, and d_1 result from Eqs. (4.20) and (4.28).

Constants a_1, a_3, b_1, c_1, and d_1 were approximated by the fourth-order polynomials

$$a_1 = \sum_0^4 a_{1,i}\beta^i, \quad a_3 = \sum_0^4 a_{3,i}\beta^i, \quad b_1 = \sum_0^4 b_{1,i}\beta^i, \quad c_1 = \sum_0^4 c_{1,i}\beta^i,$$
$$d_1 = \sum_0^4 d_{1,i}\beta^i. \tag{4.30}$$

Table 4.6 Values α, $\tau_{w\varphi}/\tau_{w\varphi 0}$ a_1, a_3, b_1, c_1, and d_1: the exact solution and the integral method (*bold font*) [6]

β	α	$\tau_{w\varphi}/\tau_{w\varphi 0}$	a_1	a_3	b_1	c_1	d_1	d_2	K_V
0.0	0.8284	1	0.5338	1.2715	0.07887	0.6726	0.18587	0.34793	0.3482
	0.8284	**1**	**0.5338**	**1.2715**	**0.07887**	**0.6726**	**0.18587**	**0.34793**	**0.3482**
0.1	0.8535	0.9767	0.5834	1.0744	0.09580	0.5897	0.18004	0.40334	0.30862
	0.8570	**0.9745**	**0.5751**	**1.0744**	**0.09534**	**0.5902**	**0.17872**	**0.39639**	**0.31076**
0.3	0.8989	0.8614	0.5261	0.7469	0.10109	0.4433	0.14662	0.37949	0.27869
	0.8952	**0.8634**	**0.5323**	**0.7467**	**0.10157**	**0.4429**	**0.14761**	**0.38468**	**0.27732**
0.5	0.9357	0.6782	0.3955	0.4856	0.08429	0.3104	0.10497	0.29051	0.26543
	0.9318	**0.6797**	**0.3995**	**0.4854**	**0.08464**	**0.3101**	**0.10561**	**0.29386**	**0.26437**
0.7	0.9654	0.4409	0.2419	0.2691	0.05560	0.1841	0.06230	0.17961	0.25753
	0.9705	**0.4398**	**0.2393**	**0.2693**	**0.05533**	**0.1843**	**0.06187**	**0.17738**	**0.25860**
0.9	0.9895	0.1574	0.0811	0.08374	0.01978	0.0610	0.02046	0.06068	0.25212
	0.9975	**0.1569**	**0.0799**	**0.08385**	**0.01965**	**0.0611**	**0.02026**	**0.05966**	**0.25350**
0.999	0.9999	0.0016	8.1×10^{-4}	8.1×10^{-4}	2.0×10^{-3}	6.1×10^{-3}	2.03×10^{-3}	6.1×10^{-4}	0.25062
	1.0011	**0.0016**	$\mathbf{8.1 \times 10^{-4}}$	$\mathbf{8.1 \times 10^{-4}}$	$\mathbf{2.0 \times 10^{-3}}$	$\mathbf{6.1 \times 10^{-3}}$	$\mathbf{2.03 \times 10^{-3}}$	$\mathbf{6.1 \times 10^{-4}}$	**0.25062**

The polynomial coefficients are: $a_{1,i}$ = 0.5338, 0.6936, −3.0692, 2.7371, −0.8952; $a_{3,i}$ = 1.2715, −2.1025, 1.3781, −0.7233, 0.1761; $b_{1,i}$ = 0.07887, 0.2204, −0.5981, 0.4228, −0.1239; $c_{1,i}$ = 0.6726, −0.8612, 0.3994, −0.2969, 0.08607; $d_{1,i}$ = 0.1859, −0.03277, −0.4271, 0.4124, −0.1384 for i = 0, 1, 2, 3, 4, accordingly.

Constants predicted by Eq. (4.30), values α by Eq. (4.26) and $\tau_{w\varphi}/\tau_{w\varphi 0}$ by Eq. (4.29) given in Table 4.6 in boldface divert from the exact values by not more than 1 %.

Depicted in Fig. 4.7 is a zeroth-order solution for α and $\tau_{w\varphi}/\tau_{w\varphi 0}$, Eqs. (4.26), (4.29), whereas the profiles of g and f are independent of β, with coefficients a_1, a_3, b_1, c_1, and d_1 computed by Eqs. (4.30) using just the first terms of the power series (at i = 0). Deviation of this solution from the exact solution is high and reaches 34 % for α and 42 % for $\tau_{w\varphi}/\tau_{w\varphi 0}$ for $\beta \leq 0.5$; for $\beta \to 1$ this deviation is equal to 54.3 % for α and 3200 % for $\tau_{w\varphi}/\tau_{w\varphi 0}$. Thus the significant variation of the constants a_1, a_3, b_1, c_1, and d_1 with β must be taken into account in accurate modeling of the functions α, $\tau_{w\varphi}/\tau_{w\varphi 0}$, and others.

Integral method: heat transfer. The transformed Eq. (4.15) hat to be solved together with Eq. (2.70) of the thermal boundary layer [6]

$$(4+m)(1-\beta)\alpha D_1\delta/r + 2\beta\alpha A_1\delta/r = \frac{C_f}{2}(1-\beta)\left(1+\alpha^2\right)^{1/2}, \tag{4.31}$$

$$\frac{d}{dr}\left[r^2\delta\overline{\delta}_T^{**}(T_w - T_\infty)\right] + \frac{dT_\infty}{dr}\delta r^2\int_0^1 \frac{v_r}{\omega r}d\xi = \chi\frac{C_f}{2}r^2(1-\beta)\left(1+\alpha^2\right)^{1/2}(T_w - T_\infty). \tag{4.32}$$

For laminar flow, a generalization of the "turbulent" Eq. (4.5) brings [11]

$$\overline{\delta}_T^{**} = -(1-\beta)\alpha(-A_1/b_2 + \chi Pr^{n_p}D_2 b_1/b_2). \tag{4.33}$$

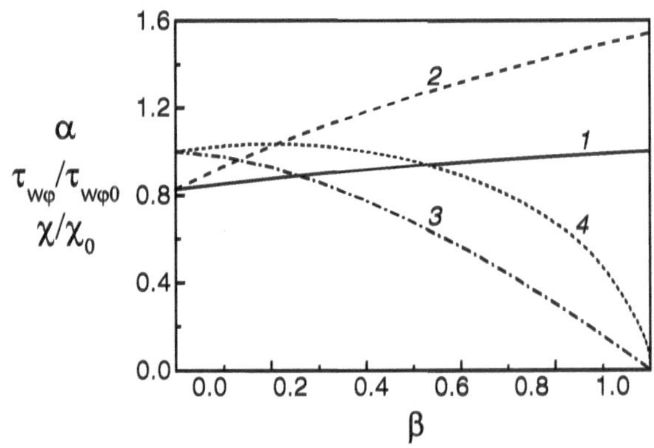

Fig. 4.7 Effect of the swirl parameter β on functions α (curves *1* and 2) and $\tau_{w\varphi}/\tau_{w\varphi 0}$ (curves *3* and *4*). Exact solution: curves *1* and *3*. Approximate zeroth-order solution: curves *2* and *4* [6]

For the boundary condition (2.31), Eqs. (4.31), (4.32), and (4.33) yield [11]

$$\chi = \frac{(2+m+n_*) + b_2 n_* \beta/(1-\beta)_*}{(2+m+n_*)Pr^{n_p}(1-K_V)b_1 + (4+m)K_V b_2 + b_2 2\beta/(1-\beta)}, \quad (4.34)$$

For the boundary condition (2.30), Eq. (4.34) reduces to

$$\chi = \frac{2+m+n_*}{(2+m+n_*)Pr^{n_p}(1-K_V)b_1 + (4+m)K_V b_2 + b_2 2\beta/(1-\beta)}. \quad (4.35)$$

For turbulent flow $b_1 = b_2 = 1$ and Eqs. (4.34), (4.35) reduce to Eqs. (4.8), (4.9), accordingly. For laminar flow: $m = 0$ and $K_V = 1 - D_2/A_1 = 1 - d_2/a_1$ (see Eq. (3.37)).

The constant b_1 is independent of the Prandtl number, while the functions $b_2(Pr)$ and $n_p(Pr)$ are identical to those for a single rotating disk [Sect. 3.4, Eqs. (3.54)–(3.56)].

The dependence of b_1 and b_2 on β is found in the following way. Based on the exact solution and on Eqs. (2.47), (3.5), and (4.21), the Reynolds analogy parameter χ is given as

$$\chi = -K_1(1-\beta)\left[Pr(\mathrm{d}G/\mathrm{d}\zeta)_{\zeta=0}\right]^{-1}. \quad (4.36)$$

Equating Eqs. (4.35) and (4.36) for any β at $Pr = 1$, $n_* = 0$, and $n_* = 2$, one can find the values of the coefficients b_1 and b_2. These values can be approximated as [11]

$$b_1 = \sum_0^4 b_{1,i}\beta^i, \quad b_2 = b_{2,0}\left(1 + \sum_1^4 b_{2,i}\beta^i\right), \quad (4.37)$$

where: $b_{1,0} = 0.6827$, $b_{1,1} = 0.1694$, $b_{1,2} = 0.1115$, $b_{1,3} = 0.7931$, and $b_{1,4} = -0.7569$. Equation (3.54) brings the value of $b_{2,0}$ for $\beta = 0$, while $b_{2,1} = 0.2866$, $b_{2,2} = -2.5420$, $b_{2,3} = 2.7414$, and $b_{2,4} = -0.8586$. At $\beta = 0.999$, the asymptotic values were $b_1 = 1$ and $b_2 = 1$.

As soon as the χ is known from Eq. (4.35), the coefficient K_1 can be calculated as

$$K_1 = 0.6159 \frac{\tau_{w\varphi}}{\tau_{w\varphi 0}} \chi \frac{Pr}{1-\beta}. \quad (4.38)$$

The function $\tau_{w\varphi}/\tau_{w\varphi 0}$ is given by Eq. (4.29). The parameters K_1 and χ for $Pr = 1$ and 0.71 computed using the integral method and the self-similar solution, Eqs. (4.38) and (4.35), are listed in Tables 4.7, 4.8, 4.9, and 4.10.

The dependence of the constant K_1 on the swirl parameter β has a maximum for $\beta = 0.1$. Authors [12] experimentally detected that at a stagnation point of a swirling jet impinging onto a stationary surface the *Nu* number also exhibits a maximum at small β.

For larger values of β, the constant K_1 decreases together with the absolute shear of the tangential velocity v_z. For $\beta \geq 0.8$, the exact solution is dominated by heat conduction with $K_1 = -\theta'_{\zeta=0} = 1/\zeta_\infty$ (independently of the values of Pr and n_*), where ζ_∞ is the coordinate at which $\theta = 0$. Accepted asymptotic values $b_1 = b_2 = 1$ for $\beta = 0.999$ are valid at $\zeta_\infty = \infty$. The exact solution (Tables 4.7, 4.8, 4.9, and 4.10) obtained for $\zeta_\infty = 7.05$ and $\beta = 0.999$, and for $\zeta_\infty = 9.93$ and $\beta = 0.99$ is in a good agreement with the integral method.

The exact and approximate solutions, Tables 4.7, 4.8, 4.9, and 4.10, are in a good agreement with each other for $\beta \leq 0.7$. For $\beta = 0.7$, inaccuracy of the approximate solution reduces with increasing n_* and (a) reach maximum 4.6 % for $Pr = 1$ and 13.2 % for $Pr = 0.71$ at $n_* = -1.5$ (hardly observed in practice); (b) does not go beyond 1.6 % for $Pr = 1$ and 2.5 % for $Pr = 0.71$ over the range $\beta = 0$ to 0.7 and $n_* = -1$ to 4. For $\beta = 0.8$, the approximate solution demonstrates 10.6 % inaccuracy for $n_* = -0.5$, which reduces to 1.5 % at $n_* = 4$. For $\beta = 0.9$, the integral method performs well only for $n_* = 1$ to 4, whereas for $\beta > 0.9$ convective heat transfer is negligible compared to heat conduction, which makes the integral method inapplicable [11].

Thus, in the present section a self-similar solution and an integral method for convective heat transfer over a disk rotating in a corotating fluid with $\beta < 1$ were obtained and validated.

Table 4.7 Constant K_1 for $Pr = 1.0$: exact solution and integral method (*bold font*) [6]

β	$n_* = -2$	$n_* = -1.5$	$n_* = -1$	$n_* = -0.5$	$n_* = 0$	$n_* = 0.5$	$n_* = 1$	$n_* = 2$	$n_* = 4$
0.0	0.0	0.1305	0.2352	0.3221	0.3962	0.4608	0.5180	0.6159	0.7693
	0.0	**0.1261**	**0.2311**	**0.3199**	**0.3961**	**0.4620**	**0.5197**	**0.6159**	**0.7557**
0.1	0.0	0.1333	0.2396	0.3275	0.4022	0.4672	0.5246	0.6228	0.7764
	0.0	**0.1287**	**0.2354**	**0.3254**	**0.4022**	**0.4686**	**0.5266**	**0.6228**	**0.7622**
0.3	0.0	0.1123	0.2243	0.3086	0.3809	0.4442	0.5003	0.5967	0.7479
	0.0	**0.1198**	**0.2205**	**0.3064**	**0.3804**	**0.445**	**0.5017**	**0.5969**	**0.7367**
0.5	0.0	0.1013	0.1868	0.2607	0.3255	0.3832	0.4351	0.5253	0.6688
	0.0	**0.0998**	**0.1858**	**0.2607**	**0.3264**	**0.3847**	**0.4366**	**0.5253**	**0.6591**
0.7	0.0	0.0721	0.1319	0.1861	0.2356	0.2811	0.323	0.3983	0.5229
	0.0	**0.0683**	**0.1297**	**0.1852**	**0.2356**	**0.2816**	**0.3237**	**0.3983**	**0.5174**
0.9	0.0	0.0643	0.0314	0.1065	0.1266	0.1461	0.1649	0.2009	0.2670
	0.0	**0.0247**	**0.0484**	**0.0712**	**0.0933**	**0.1145**	**0.1350**	**0.1739**	**0.2443**
0.999	0.0	0.1417	0.1418	0.1421	0.1423	0.1424	0.1426	0.1429	0.1437
	0.0	**0.0003**	**0.0005**	**0.0008**	**0.0010**	**0.0012**	**0.0015**	**0.0020**	**0.0030**

Table 4.8 Value χ for Pr = 1.0 from the exact solution and integral method (*bold font*) [6]

β	$n_* = -2$	$n_* = -1.5$	$n_* = -1$	$n_* = -0.5$	$n_* = 0$	$n_* = 0.5$	$n_* = 1$	$n_* = 2$	$n_* = 4$
0.0	0.0	0.2119	0.3818	0.5230	0.6433	0.7482	0.8410	1.0	1.1335
	0.0	**0.2047**	**0.3752**	**0.5195**	**0.6431**	**0.7502**	**0.8439**	**1.0**	**1.1249**
0.1	0.0	0.1995	0.3585	0.49	**0.6018**	0.699	**0.7849**	**0.9318**	1.055
	0.0	**0.193**	**0.353**	**0.4879**	**0.6031**	**0.7027**	**0.7895**	**0.9339**	**1.049**
0.3	0.0	0.1632	0.296	0.4072	0.5026	0.586	0.6601	0.7873	0.8943
	0.0	**0.1577**	**0.2903**	**0.4033**	**0.5008**	**0.5858**	**0.6605**	**0.7858**	**0.8867**
0.5	0.0	0.1213	0.2236	0.312	0.3897	0.4587	0.5208	0.6288	0.7206
	0.0	**0.1192**	**0.2219**	**0.3113**	**0.3899**	**0.4595**	**0.5215**	**0.6274**	**0.7144**
0.7	0.0	0.0796	0.1457	0.2056	0.2603	0.3105	0.3569	0.4400	0.5129
	0.0	**0.0756**	**0.1436**	**0.2051**	**0.2609**	**0.3119**	**0.3586**	**0.4411**	**0.5118**
0.9	0.0	0.0664	0.0885	0.1099	0.1306	0.1507	0.1701	0.2073	0.2423
	0.0	**0.0255**	**0.0501**	**0.0737**	**0.0965**	**0.1185**	**0.1397**	**0.1800**	**0.2176**
0.999	0.0	0.1417	0.1419	0.1421	0.1423	0.1424	0.1426	0.143	0.1434
	0.0	**0.00025**	**0.0005**	**0.00075**	**0.0010**	**0.0012**	**0.0015**	**0.0020**	**0.0025**

Table 4.9 Constant K_1 for Pr = 0.71: exact solution and integral method (*bold font*) [6]

β	$n_* = -2$	$n_* = -1.5$	$n_* = -1$	$n_* = -0.5$	$n_* = 0$	$n_* = 0.5$	$n_* = 1$	$n_* = 2$	$n_* = 4$
0.0	0.0	0.1035	0.1893	0.2624	0.3259	0.3818	0.4319	0.5185	0.6555
	0.0	**0.1009**	**0.187**	**0.2612**	**0.3259**	**0.3828**	**0.4332**	**0.5185**	**0.6457**
0.1	0.0	0.1060	0.1934	0.2675	0.3316	0.3881	0.4384	0.5254	0.6627
	0.0	**0.1031**	**0.1906**	**0.2658**	**0.3312**	**0.3885**	**0.4392**	**0.5248**	**0.6518**
0.3	0.0	0.0971	0.1789	0.2493	0.3108	0.3654	0.4145	0.4997	0.6350
	0.0	**0.0958**	**0.1782**	**0.2498**	**0.3125**	**0.3680**	**0.4174**	**0.5015**	**0.6281**
0.5	0.0	0.0795	0.1469	0.2067	0.2605	0.3091	0.3535	0.4320	0.5594
	0.0	**0.0797**	**0.1497**	**0.2117**	**0.2670**	**0.3165**	**0.3613**	**0.4387**	**0.5585**
0.7	0.0	0.0616	0.1065	0.1481	0.1869	0.2231	0.2572	0.3195	0.4257
	0.0	**0.0544**	**0.1039**	**0.1493**	**0.1910**	**0.2295**	**0.2651**	**0.3288**	**0.4328**
0.9	0.0	0.0668	0.0821	0.0970	0.1116	0.1258	0.1397	0.1667	0.2172
	0.0	**0.0195**	**0.0385**	**0.0568**	**0.0745**	**0.0916**	**0.1083**	**0.1401**	**0.1983**
0.999	0.0	0.1417	0.1418	0.1420	0.1421	0.1422	0.1424	0.1426	0.1432
	0.0	**0.0002**	**0.0004**	**0.0006**	**0.00079**	**0.00099**	**0.0012**	**0.0016**	**0.0024**

4.2 Flow Impingement onto an Orthogonal Disk

4.2.1 Experimental and Computational Data of Different Authors

One of the high-performance techniques used for heating/cooling of bodies, for instance, end wall surfaces of gas turbine rotors, is impingement cooling [1, 4, 13–17].

Table 4.10 Value χ for $Pr = 0.71$: exact solution and integral method (*bold font*) [6]

β	$n_* = -2$	$n_* = -1.5$	$n_* = -1$	$n_* = -0.5$	$n_* = 0$	$n_* = 0.5$	$n_* = 1$	$n_* = 2$	$n_* = 4$
0.0	0.0	0.2366	0.4330	0.6001	0.7452	0.8732	0.9876	1.1856	1.3533
	0.0	**0.2308**	**0.4276**	**0.5973**	**0.7453**	**0.8753**	**0.9906**	**1.1857**	**1.3446**
0.1	0.0	0.2233	0.4076	0.5637	0.6988	0.8177	0.9238	1.1071	1.2619
	0.0	**0.2177**	**0.4025**	**0.5614**	**0.6995**	**0.8206**	**0.9276**	**1.1084**	**1.2551**
0.3	0.0	0.1805	0.3325	0.4632	0.5776	0.6791	0.7702	0.9286	1.0630
	0.0	**0.1777**	**0.3304**	**0.4631**	**0.5794**	**0.6823**	**0.7738**	**0.9298**	**1.0577**
0.5	0.0	0.1340	0.2476	0.3486	0.4391	0.5212	0.596	0.7284	0.8426
	0.0	**0.1341**	**0.2519**	**0.3561**	**0.4491**	**0.5325**	**0.6077**	**0.7381**	**0.8471**
0.7	0.0	0.0959	0.1656	0.2304	0.2908	0.3472	0.4002	0.4971	0.5839
	0.0	**0.0848**	**0.1621**	**0.2329**	**0.2980**	**0.3580**	**0.4135**	**0.5129**	**0.5993**
0.9	0.0	0.0970	0.1192	0.1409	0.1621	0.1828	0.2030	0.2421	0.2796
	0.0	**0.0285**	**0.0561**	**0.0827**	**0.1086**	**0.1336**	**0.1579**	**0.2042**	**0.2479**
0.999	0.0	0.1997	0.1998	0.2000	0.2002	0.2004	0.2006	0.2009	0.2013
	0.0	**0.00029**	**0.00056**	**0.00084**	**0.0011**	**0.0014**	**0.0017**	**0.0022**	**0.0028**

Main approaches to simulations of axisymmetric laminar flows orthogonally impinging onto a stationary flat surface or a disk (Fig. 4.8), e.g., a well-known theoretical solution for this case, are well described in the works [5, 18–20].

Authors of the works [5, 13, 21–23] investigated single axisymmetric jets impinging onto a stationary surface/disk of a diameter D. Detailed investigations of axisymmetric air jets impinging onto rotating disks were performed in the works [12, 19, 24–40]. These studies demonstrated that a constant-thickness boundary layer develops in the immediate neighborhood of the stagnation point, whereas Eq. (2.27) describe the behavior of the radial $v_{r,\infty}$ and axial $v_{z,\infty}$ velocity components in potential flow.

If a disk is placed perpendicular to a potential uniform flow, whose velocity at infinity is V_j, the coefficient a in Eq. (2.27) is given by the expression [19]

$$a = V_j A/D_j = 2V_j/(\pi b), \quad A = 4/\pi. \tag{4.39}$$

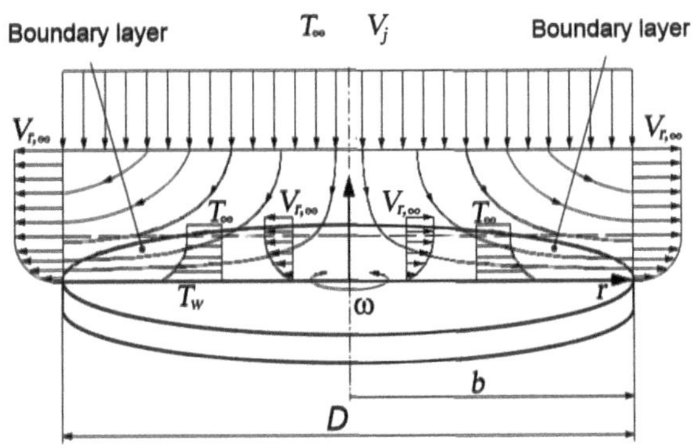

Fig. 4.8 Fluid flow and heat transfer between a rotating disk and a uniform flow perpendicular to it [6]

Once the flow impinges onto an entire disk, $D_j = D$ in the definition of $A = aD_j/V_j$. On a disk, the local stagnation pressure p_s can be written in two different ways

$$p_s = p(r) + (1/2)\rho v_{r,\infty}^2, \tag{4.40}$$

$$p_s = p_\infty + (1/2)\rho V_j^2. \tag{4.41}$$

The potential flow velocity $v_{r,\infty}$ is given by Eqs. (2.27), (4.39). Consequently, [41]

$$\frac{p - p_\infty}{(1/2)\rho V_j^2} = 1 - \left(\frac{2}{\pi}\right)^2 \left(\frac{r}{b}\right)^2. \tag{4.42}$$

Experiments [41] for the case where flow impinges onto the entire disk agree well with the theoretical Eq. (4.42) up to a radial location $r/b = 0.85$. This confirms the validity of Eqs. (2.27) and (4.39) modeling the radial velocity $v_{r,\infty}$. Flow separation at the outer rim of the disk is a reason of the deviation of Eq. (4.42) from experiments [41] at $r/b > 0.85$.

If a disk is subject to the impingement of a single perpendicular jet, the coefficient a in Eq. (2.27) is practically constant in the neighborhood of the stagnation point. Further, as the velocity $v_{r,\infty}$ tends to its maximum for increasing r, the parameter a decreases. Having passed the point of maximum, velocity $v_{r,\infty}$ diminishes. Here starts a region of a turbulent wall jet not investigated in the present work. In [13], one can find empirical equations predicting all parameters of the point of maximum of $v_{r,\infty}$.

The jet-to-disk diameter ratio D_j/D and dimensionless jet-to-disk distance h_j/D_j are important parameters of real impinging jets. Here, the value of A deviates from $4/\pi$ (see Eq. (4.39)) and becomes a function of h_j/D_j (as well as of the turbulence level and the shape of the velocity profile at nozzle outlet). Different experimental investigations suggest different dependences of A on h_j/D_j. We will follow the findings of [21], where the parameter A together with the stagnation Nusselt number diminish at high values of Re_j, while the parameter h_j/D_j increases over the range $2 \le h_j/D_j \le 6$

$$A = 1.5\left(h_j/D_j\right)^{-0.22}, \tag{4.43}$$

valid for laminar flow with a uniform velocity profile at nozzle outlet.

Experimental investigations [12, 24, 26, 28, 32–34, 42] outline results for heat and mass transfer of jets impinging onto a perpendicular wall, whereas [19, 27, 29, 35, 38, 39] represent results of theoretical studies. For heat transfer at uniform flow impingement onto an orthogonal disk, authors [29] obtained a self-similar solution for $Pr = 0.7$, 1, 10 and $T_w = \text{const}$. For mass transfer in the same geometrical configuration, a self-similar solution was investigated at $Sc = 0.2$, 3.0 in the paper [35].

The self-similar Eqs. (2.32)–(2.36) for $\beta = 0$ and boundary conditions (2.27)–(2.30) describe the problem considered here. As a result, for a stationary disk, the Nusselt numbers may be described by the following relations [13]:

$$Nu_D = K_1 \cdot Re_a^{1/2}, \quad Nu_D = 0.763 \cdot Re_{\rm j}^{1/2} \cdot Pr^{0.4} \cdot A^{1/2}. \tag{4.44}$$

For a disk placed in the uniform flow, the second of Eq. (4.44) holds, provided that $Re_a = Re_{\rm j}A$, $D_{\rm j} = D$, and T_w = const.

Laminar flow and heat transfer at a uniform perpendicular air impingement onto a single rotating disk and T_w = const. were investigated experimentally and theoretically in [41, 43–45] for a wide range of the parameter $0 \leq a/\omega \leq \infty$ (ratio of the impingement velocity and angular velocity of rotation). The Nu numbers obtained from the self-similar solution agree well with the experiments [41, 43] (see Sect. 3.2.3).

A self-similar solution for mass transfer of a rotating disk subject to an orthogonal impingement at C_w = const. was approximated by the following relation [27]:

$$\left[\frac{Sh}{\left(Re_a^2 + Re_\omega^2\right)^{1/4}}\right]^m = \left[\beta^{1/4}\frac{Sh_a}{Re_a^{1/2}}\right]^m + \left[(1+\beta)^{1/4}\frac{Sh_\omega}{Re_\omega^{1/2}}\right]^m, \tag{4.45}$$

which is valid for the Schmidt numbers $Sc = 0\ldots\infty$. Here $m = 2.65$ for $Sc = 0.1$–1.0 or $m = 3.85$ for $Sc = 1$–$10{,}000$. For a single rotating disk, Sherwood numbers Sh_ω are represented by Eq. (3.4) (for $n_R = 1/2$) and Eq. (3.7), where Nu is replaced by Sh, and Pr by Sc, respectively. For a stationary disk placed in a uniform flow, the Sherwood number Sh_a is described by the following relation [27]:

$$Sh_a/Re_a^{1/2} = K_1 = 0.65693Sc^{1/2}/\left(0.3098 + 1.015Sc^{1/2} + Sc\right)^{1/6}. \tag{4.46}$$

In [27], it is said that deviation of Eq. (4.45) from the self-similar solution is less than 7.55 % for $Sc = 0.1$–1.0 and 3.85 % for $Sc = 1$–$10{,}000$. Our analysis indicated, however, the deviation of Eq. (4.45) from the self-similar solution reaches 12 % for $Sc = 1$.

Experimental data [12, 33, 34] demonstrate that both jet impingement and disk rotation engender heat transfer enhancement. A comparison of the data [12, 33, 34] with the self-similar solution is outlined below. Average heat transfer and average mass transfer in naphthalene sublimation in air for an *entire* disk subject to jet impingement were measured in experiments [24, 32] and [26, 28], respectively.

For turbulent flow, [46] obtained a solution of Eqs. (2.17), (2.18) together with Eqs. (2.37)–(2.41) for $\beta = 0$, N = const., $n = 1/7$ and boundary conditions (2.27). The authors of [2] translated this solution for arbitrary values of n in the following equations:

$$\alpha^2 + R\alpha N = S + TN^2, \quad \gamma = K^{\frac{n+1}{3n+1}} \left[\frac{(3n+1)(1+\alpha^2)^{0.5(1-n)/(1+n)}}{(5+11n)(G\alpha + HN)} \right]^{(n+1)/(3n+1)}, \tag{4.47}$$

$$C_{\mathrm{M}} = \gamma(G\alpha + HN)Re_{\varphi}^{-2n/(3n+1)}. \tag{4.48}$$

The authors of [2] tabulated the functions R, S, T, G, H, and K. The most important parameters of this solution are presented in Table 4.11 [2]. These data are compared below with the results predicted by the present integral method (see Chap. 2).

4.2.2 Turbulent Flow

A study of turbulent flow at $N = \text{const.}$ is applicable to cases of converging channels [2, 19, 46]. The constants α, γ, ε_m, and ε_{M} calculated by Eqs. (2.83)–(2.86) with the help of the present integral method for $n = 1/7$, as well as by Eqs. (4.47), (4.48) [2, 46] are listed in Table 4.11. Both solutions approach each other with increasing N; in fact, values α and ε_m by both models merge already for $N > 0.4$.

The thermal boundary layer Eq. (2.89) for $\beta = 0$ can be reduced to the following form:

$$\frac{\overline{\delta}_{\mathrm{T}}^{**}}{\alpha A_1} = -\left[\left(-1 + \chi Pr^{n_p}(1 - K_{V0})\right) + \frac{\kappa}{\alpha}\left(-\frac{A_2}{A_1} + \chi Pr^{n_p}\frac{D_4}{A_1} \right) \right] - \frac{\kappa}{\alpha}\frac{1}{A_1}\left(\frac{(\chi Pr^{n_p})^{-1/n} - \chi Pr^{n_p}}{n+1} - (\chi Pr^{n_p})^{-1/n} + 1 \right). \tag{4.49}$$

Table 4.11 Constants in Eqs. (2.77)–(2.81) for $N = \text{const.}$, $\beta = 0$ [6]

N	Solution (2.83)–(2.86) [10], $n = 1/7$				Solution (4.47), (4.48) [2, 46], $n = 1/7$			
	α	γ	ε_m	ε_{M}	α	γ	ε_m	ε_{M}
0	0.2087	0.5299	0.1806	0.1466	0.162	0.526	0.219	0.146
0.2	0.2818	0.2892	0.3566	0.1727	0.2587	0.2918	0.3648	0.172
0.6	0.6408	0.1484	0.4994	0.2255	0.6288	0.1486	0.5010	0.224
1	1.037	0.1122	0.6236	0.2794	1.026	0.1122	0.6242	0.278
2	2.046	0.0853	0.9446	0.422	2.033	0.0853	0.9448	0.42
4	4.073	0.0708	1.567	0.699	4.056	0.0708	1.568	0.696

This yields a transcendental equation for χ

$$-1+\chi Pr^{n_p}(1-K_{V0})+\frac{\kappa}{\alpha}\left(-\frac{A_2}{A_1}+\chi Pr^{n_p}\frac{D_4}{A_1}\right)$$
$$-\frac{\kappa}{\alpha}\frac{1}{A_1}\left(\frac{-n(\chi Pr^{n_p})^{-1/n}-\chi Pr^{n_p}}{n+1}+1\right)=\chi\frac{4+m}{2+m+n_*}K_{V0}\left[1+\frac{\kappa}{\alpha}\left(\frac{B_5}{D_1}-1\right)\right]. \tag{4.50}$$

Here $K_{V0}=1-D_2/A_1$, $B_5=D_1+D_3$; Eq. (4.7) connects the parameters χ and Δ.

Equation (4.50) is transcendental, and its analytical solution can be only approximate. For Pr = 0.5–1, one can assume $\chi Pr^{n_p}\approx 1$ and $(\chi Pr^{n_p})^{-1/n}\neq 1$, which gives

$$\chi=Pr^{-n_p}\left[1-\frac{n+1}{n}\left(\frac{\alpha}{\kappa}D_1+D_3\right)\left(1-Pr^{-n_p}\frac{4+m}{2+m+n_*}\right)\right]^{-n}. \tag{4.51}$$

In case of small values of Pr, one can derive another approximate solution of Eq. (4.50). Setting $\chi Pr^{n_p}\to 0$ and $(\chi Pr^{n_p})^{-1/n}\to\infty$, one can deduce

$$\chi=Pr^{-n_p}\left[\frac{n+1}{n}Pr^{-n_p}\left(\frac{\alpha}{\kappa}D_1+D_3\right)\frac{4+m}{2+m+n_*}\right]^{\frac{-n}{n+1}}. \tag{4.52}$$

Numerical data for the parameters K_1 and χ for Pr = 0.71 (air) are listed in Tables 4.12 and 4.13 and enable making the following important conclusions.

Table 4.12 Values of the constant K_1 according to a numerical solution of Eq. (4.50) and its approximate solutions (4.51) and (4.52) for turbulent flow and Pr = 0.71 [6]

$\kappa = a/\omega$	κ/α	$n_* = -1.5$	$n_* = -1$	$n_* = -0.6$	$n_* = 0$	$n_* = 1$	$n_* = 2$	$n_* = 3$	$n_* = 4$
0.0 (4.50)		0.0125	0.0152	0.0167	0.0185	0.0205	0.0218	0.0227	0.0234
0.1 (4.50)	0.4491	0.0197	0.0207	0.0213	0.0221	0.0231	0.0239	0.0246	0.0251
0.1 (4.51)		0.0189	0.0200	0.0207	0.0216	0.0227	0.0237	0.0246	0.0254
0.1 (4.52)		0.0194	0.0203	0.0209	0.0216	0.0225	0.0232	0.0238	0.0243
0.2 (4.50)	0.7098	0.0219	0.0229	0.0236	0.0243	0.0253	0.0261	0.0267	0.0273
0.2 (4.51)		0.0213	0.0225	0.0232	0.0241	0.0252	0.0260	0.0267	0.0273
0.2 (4.52)		0.0220	0.0230	0.0237	0.0244	0.0255	0.0262	0.0269	0.0275
1.0 (4.50)	0.9649	0.0359	0.0375	0.0385	0.0397	0.0412	0.0423	0.0433	0.0441
1.0 (4.51)		0.0352	0.0370	0.0381	0.0394	0.0411	0.0423	0.0433	0.0441
1.0 (4.52)		0.0363	0.0381	0.0392	0.0405	0.0422	0.0435	0.0445	0.0455
4.0 (4.50)	0.9813	0.1068	0.1116	0.1145	0.1180	0.1224	0.1258	0.1286	0.1310
4.0 (4.51)		0.1048	0.1102	0.1134	0.1173	0.1221	0.1257	0.1286	0.1309
4.0 (4.52)		0.1082	0.1134	0.1166	0.1204	0.1255	0.1294	0.1326	0.1353

Table 4.13 Values of the Reynolds analogy parameter χ according to a numerical solution of Eq. (4.50) and its approximate Eqs. (4.51) and (4.52) for turbulent flow and $Pr = 0.71$ [6]

$\kappa = a/\omega$	κ/α	$n_* = -1.5$	$n_* = -1$	$n_* = -0.6$	$n_* = 0$	$n_* = 1$	$n_* = 2$	$n_* = 3$	$n_* = 4$
0.0 (4.50)	0	0.6580	0.7971	0.8788	0.9707	1.0747	1.144	1.193	1.231
0.1 (4.50)	0.4491	0.954	1.004	1.035	1.072	1.122	1.160	1.192	1.218
0.1 (4.51)		0.916	0.970	1.004	1.046	1.104	1.151	1.193	1.231
0.1 (4.52)		0.941	0.986	1.014	1.048	1.091	1.125	1.153	1.177
0.2 (4.50)	0.7098	0.976	1.022	1.051	1.085	1.129	1.163	1.191	1.215
0.2 (4.51)		0.950	1.002	1.033	1.072	1.122	1.160	1.192	1.218
0.2 (4.52)		0.978	1.025	1.054	1.089	1.134	1.170	1.197	1.224
1.0 (4.50)	0.9649	0.989	1.033	1.060	1.092	1.133	1.165	1.191	1.213
1.0 (4.51)		0.969	1.019	1.050	1.086	1.131	1.164	1.191	1.213
1.0 (4.52)		1.001	1.048	1.078	1.114	1.160	1.196	1.226	1.252
4.0 (4.50)	0.9813	0.989	1.033	1.060	1.093	1.133	1.165	1.191	1.213
4.0 (4.51)		0.970	1.020	1.050	1.086	1.131	1.164	1.191	1.212
4.0 (4.52)		1.002	1.050	1.079	1.115	1.162	1.198	1.228	1.253

First, an increase in the parameter κ results in the growing values of the constant K_1 caused by the increase in the coefficient $A_c(1+\alpha^2)^{1/2}$ in accordance with Eq. (4.7).

Second, within the range $\kappa = 0{-}\infty$, the Reynolds analogy parameter χ preserves the same order of magnitude. Variation of χ is most significant (33.5 %) for $n_* = -1.5$, and smallest (1.5 %) for $n_* = 4$.

Third, parameter χ remains constant to within four meaningful digits for $\kappa > 5$; within 2 % deviation, χ = const. already for $\kappa = 0.2$. Thus, a turbulent thermal boundary layer adapts itself very fast to a negative pressure gradient.

Fourth, both variants of approximate solutions, Eqs. (4.51) and (4.52), deviate by not more than 2 % from the exact numerical solution of the transcendental Eq. (4.50).

4.2.2.1 Laminar Flow

Exact solution. The self-similar solution of Eqs. (2.32)–(2.36) at N = const. and $\beta = 0$ was obtained in [47–50]. The following boundary conditions have been used:

$$\begin{aligned} &\zeta \to 0: \quad F = H = 0, G = 1/(1+\kappa), \quad \theta = 1, \\ &\zeta \to \infty: \quad G = 0, F = \kappa/(1+\kappa), \qquad \theta = 0, \end{aligned} \tag{4.53}$$

where $\kappa = N = a/\omega$. The Nusselt number can be calculated by the relations

$$Nu_D = K_1(Re_{\omega D} + Re_a)^{1/2}, \quad K_1 = -\theta'_{\zeta=0}. \tag{4.54}$$

In fact, Eqs. (3.4) and (4.44) are the limiting cases of Eq. (4.54).

One can transform Eqs. (2.32)–(2.36) to the case of axisymmetric orthogonal flow onto a nonrotating wall [5, 18] by setting $\varphi = -H/2$ and limiting transition $a/\omega \to \infty$.

Equations (2.32)–(2.36) were solved using a Mathcad code. Figures 4.9, 4.10, 4.11, and 4.12 show profiles of the velocity components F, G, H and temperature θ.

With increasing values of κ (in fact, already for $\kappa > 0.5$), the radial velocity profiles reshape themselves from distributions typical for flow over a rotating disk to boundary layer profiles (Fig. 4.9). This reshaping engenders significant changes in the heat transfer rate described below. Contrary to that, the shapes of the axial and tangential velocity, as well as the temperature profiles, remain qualitatively unchanged (see Figs. 4.10, 4.11, and 4.12). For larger values of κ, the boundary layer becomes thinner, whereas tangential shear stresses increase.

Listed in Table 4.14 are computed values of the ζ-derivatives of the radial and tangential velocity components and relative tangential shear stresses at the wall (subscript "0" denotes a single disk).

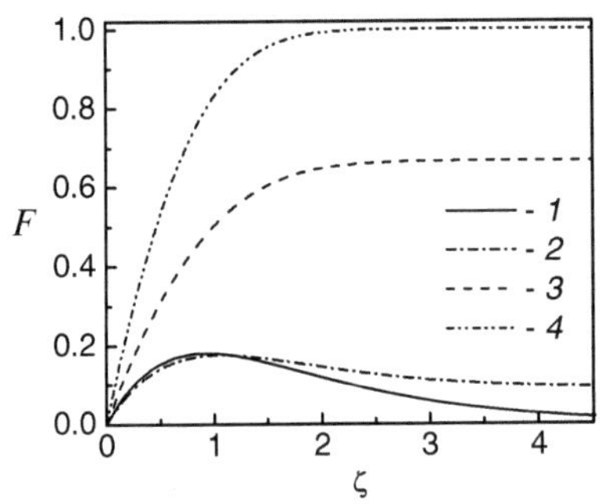

Fig. 4.9 Radial velocity profiles for laminar coaxial uniform orthogonal flow impingement onto a rotating disk [6]. *1*—$\kappa = 0$; *2*—$\kappa = 0.1$; *3*—$\kappa = 2$; *4*—$\kappa = 4000$

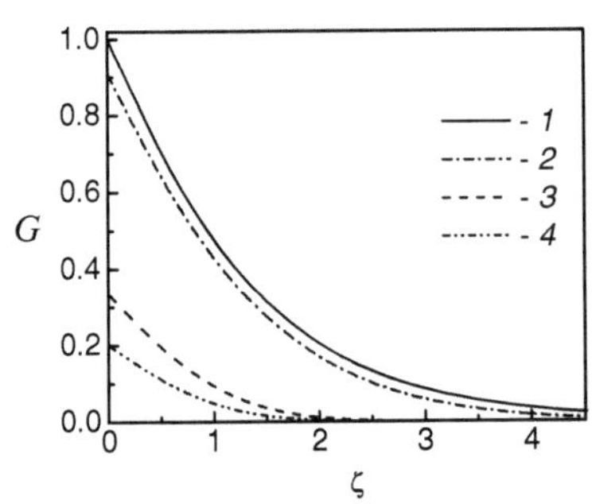

Fig. 4.10 Tangential velocity profiles for laminar coaxial uniform orthogonal flow impingement onto a rotating disk [6]. *1*—$\kappa = 0$; *2*—$\kappa = 0.1$; *3*—$\kappa = 2$; *4*—$\kappa = 4000$

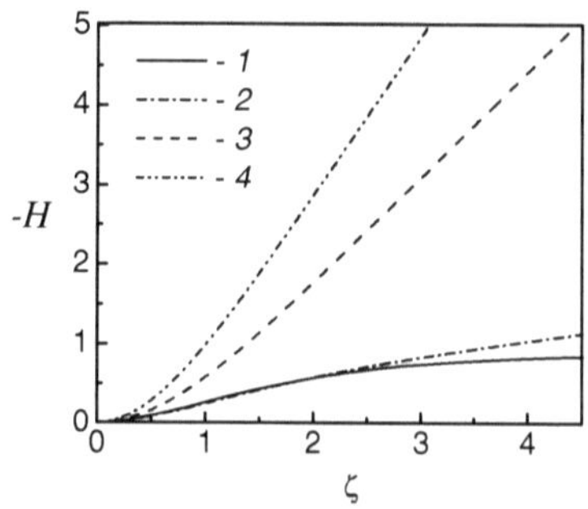

Fig. 4.11 Axial velocity profiles for laminar coaxial uniform orthogonal flow impingement onto a rotating disk [6]. *1*—$\kappa = 0$; *2*—$\kappa = 0.1$; *3*—$\kappa = 2$; *4*—$\kappa = 4000$

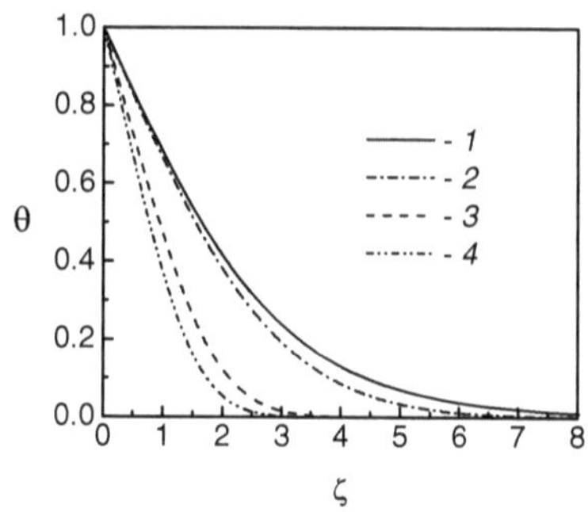

Fig. 4.12 Temperature profiles for laminar uniform orthogonal flow impingement onto a rotating disk, $Pr = 0.71$ (air) and $n_* = 1$ [6]. *1*—$\kappa = 0$; *2*—$\kappa = 0.1$; *3*—$\kappa = 2$; *4*—$\kappa = 4000$

Table 4.14 Values of constants of the exact self-similar solution of the Navier–Stokes equations for uniform flow impingement onto a rotating disk [6]

$\kappa = a/\omega$	$\alpha = -F'_w/G'_w$	$\tau_{w\varphi}/\tau_{w\varphi 0}$,	F'_w	$-G'_w$
0.0	0.8284	1.0	0.5102	0.6159
0.1	0.8127	1.0438	0.4529	0.5573
0.4	0.8939	1.2834	0.4266	0.4772
1.0	1.4180	1.8022	0.5565	0.3924
2.0	2.5440	2.4888	0.7505	0.2950
4.0	6.1454	3.9070	1.0062	0.1637
10.0	12.229	4.5195	1.1395	0.0932
50	61.043	12.338	1.2736	0.0209
100	122.08	17.448	1.2925	0.0106
4×10^3	4883.1	110.35	1.3114	0.00027

Tables 4.15 and 4.16 contain numerical data for the constant K_1 in Eq. (4.54) for the Nusselt number depending on the parameters $\kappa = a/\omega$ and n_* for $Pr = 0.71$ and 1.0. In the asymptotical case of $\kappa \to \infty$ at $n_* = 0$ (i.e., $T_w = \text{const.}$), the constant K_1 coincides with Eq. (4.44) [13]. For another asymptotic case of $\kappa = a/\omega = 0$, the

Table 4.15 Values of K_1 according to the exact self-similar solution for $Pr = 0.71$ [6]

$\kappa = a/\omega$	$n_* = -2$	$n_* = -1.5$	$n_* = -1$	$n_* = 0$	$n_* = 0.5$	$n_* = 1$	$n_* = 2$	$n_* = 3$	$n_* = 4$
0.0	0.0	0.1035	0.1893	0.3259	0.3818	0.4319	0.5185	0.5918	0.6555
0.1	0.0	0.1174	0.2078	0.3429	0.3964	0.4436	0.5246	0.5927	0.6519
0.3	0.0	0.1424	0.2425	0.3826	0.4360	0.4825	0.5612	0.6269	0.6836
0.5	0.0	0.1612	0.2705	0.4194	0.4752	0.5236	0.6049	0.6723	0.7305
0.7	0.0	0.1752	0.2922	0.4498	0.5084	0.5590	0.6439	0.7141	0.7745
1.0	0.0	0.1905	0.3165	0.4847	0.547	0.6007	0.6905	0.7646	0.8283
3.0	0.0	0.2304	0.3812	0.5811	0.6546	0.7179	0.8235	0.9106	0.9853
4.0	0.0	0.2426	0.4013	0.6114	0.6887	0.7551	0.8661	0.9575	1.0359
10	0.0	0.2533	0.4189	0.6381	0.7187	0.788	0.9037	0.9991	1.0808
50	0.0	0.2630	0.4349	0.6624	0.7462	0.8181	0.9382	1.0372	1.122
1000	0.0	0.2654	0.4390	0.6687	0.7532	0.8258	0.9471	1.0469	1.1331

Table 4.16 Values of K_1 according to the exact self-similar solution for $Pr = 1$ [6]

$\kappa = a/\omega$	$n_* = -2$	$n_* = -1.5$	$n_* = -1$	$n_* = 0$	$n_* = 0.5$	$n_* = 1$	$n_* = 2$	$n_* = 3$	$n_* = 4$
0.0	0.0	0.1305	0.2352	0.3963	0.4608	0.5180	0.6159	0.6982	0.7693
0.1	0.0	0.1414	0.2483	0.4055	0.4669	0.5209	0.6130	0.6901	0.7568
0.3	0.0	0.1663	0.2821	0.4426	0.5034	0.5562	0.6453	0.7195	0.7836
0.5	0.0	0.1867	0.3121	0.4815	0.5447	0.5993	0.6910	0.7670	0.8324
0.7	0.0	0.2023	0.3361	0.5147	0.5809	0.6379	0.7334	0.8123	0.8801
1.0	0.0	0.2196	0.3633	0.5536	0.6237	0.6840	0.7849	0.8681	0.9395
3.0	0.0	0.2652	0.4368	0.6622	0.7449	0.8159	0.9343	1.0318	1.1154
4.0	0.0	0.2792	0.4597	0.6966	0.7835	0.8580	0.9824	1.0848	1.1726
10	0.0	0.2915	0.4798	0.7270	0.8176	0.8953	1.0250	1.1318	1.2233
50	0.0	0.3026	0.4981	0.7547	0.8488	0.9295	1.0641	1.1749	1.2699
1000	0.0	0.3055	0.5028	0.7619	0.8568	0.9383	1.0741	1.1860	1.2819

coefficient K_1 from Tables 4.15 and 4.16 fully agrees with data for a free rotating disk (see Chap. 3).

In [41, 43, 51] the authors compared the original experimental data with the exact solution for a single disk rotating in quiescent air at $T_w = \text{const.}$ within a wide range of definition of the parameter $a/\omega = 0 \ldots \infty$. Equation (4.54) can be rewritten as $Nu_D/Nu_{D,a=0} = K_1(1+\kappa)^{1/2}/K_{1,a=0}$ or $Nu_D/Nu_{D,\omega=0} = K_1(1+\kappa^{-1})^{1/2}/K_{1,\omega=0}$.

Figures 4.13 and 4.14 demonstrate that the self-similar solution agrees well with the experiments. The experimental data were averaged over the radial span of a disk $r/b \leq 0.8$, where the boundary conditions for the radial velocity $v_{r,\infty}$, Eqs. (2.27) and (4.39), held with a reasonable level of accuracy [see also comments to Eqs. (4.40)–(4.42)].

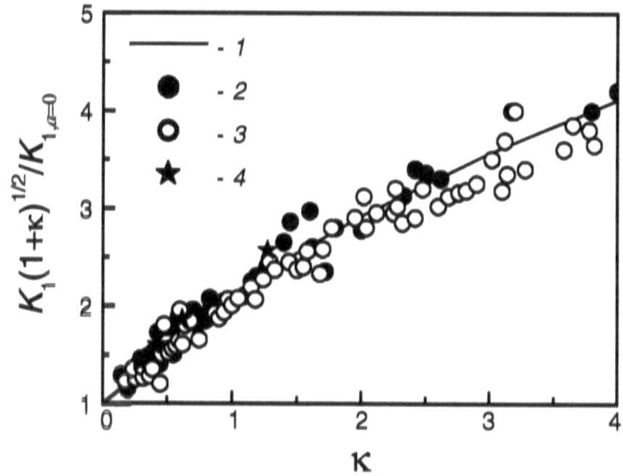

Fig. 4.13 Effect of the parameter $\kappa = a/\omega$ on the heat transfer rate at orthogonal uniform air flow impingement onto an isothermal rotating disk ($Pr = 0.71$, $T_w = \text{const.}$) [6]. *1*—Self-similar solution, Table 4.15. Experiments: *2*—$b = 74.3$ mm [41]; *3*—$b = 25$ mm [41]; *4*—[43]

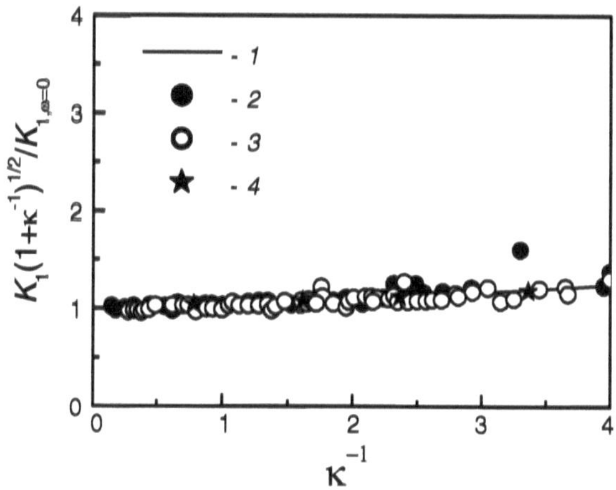

Fig. 4.14 Effect of the parameter $\kappa^{-1} = \omega/a$ on the heat transfer rate at orthogonal uniform air flow impingement onto an isothermal rotating disk ($Pr = 0.71$, $T_w = \text{const.}$) [6]. *1*—Self-similar solution, Table 4.15. Experiments: *2*—$b = 74.3$ mm [41]; *3*—$b = 25$ mm [41]; *4*—[43]

The expression (4.54) for the Nusselt number can be transformed as follows:

$$Nu_D = K_1(1 + \omega/a)^{1/2} Re_a^{1/2}. \tag{4.55}$$

The product $K_{1*} = K_1(1 + \omega/a)^{1/2}$ presented in Table 4.17 for $Pr = 0.71$ is almost constant for $\kappa = 1.5$–∞. Thus, when $\kappa = a/\omega$ trespasses a threshold value, the heat transfer of a rotating disk is determined only by flow impingement. In other words, under conditions of flow impingement, rotation causes an increase in heat

Table 4.17 Values of $K_{1*} = K_1(1 + \omega/a)^{1/2}$, self-similar solution for Pr = 0.71 [6, 48, 49]

$\kappa = a/\omega$	$n_* = -2$	$n_* = -1.5$	$n_* = -1$	$n_* = 0$	$n_* = 0.5$	$n_* = 1$	$n_* = 2$	$n_* = 3$	$n_* = 4$
0.0	0.0	∞	∞	∞	∞	∞	∞	∞	∞
0.1	0.0	0.3896	0.6891	1.1373	1.3147	1.4714	1.7398	1.9657	2.1620
0.2	0.0	0.3199	0.5531	0.8881	1.0176	1.1310	1.3240	1.4856	1.6257
0.3	0.0	0.2965	0.5048	0.7965	0.9076	1.0044	1.1683	1.3049	1.4231
0.5	0.0	0.2793	0.4685	0.7265	0.8231	0.9069	1.0477	1.1645	1.2652
1.0	0.0	0.2694	0.4476	0.6855	0.7736	0.8495	0.9765	1.0813	1.1714
1.5	0.0	0.2674	0.4431	0.6766	0.7628	0.8369	0.9608	1.063	1.1508
2.0	0.0	0.2666	0.4414	0.6733	0.7588	0.8323	0.9551	1.0563	1.1432
10	0.0	0.2657	0.4393	0.6692	0.7538	0.8265	0.9478	1.0478	1.1336
100.	0.0	0.2656	0.4392	0.669	0.7536	0.8262	0.9475	1.0475	1.1332

transfer for the same values Re_a only if the values of κ are smaller than a threshold point. The constant K_{1*} is useful to compare computations with experiments (excluding the point $\kappa = 0$).

Computed values of the Reynolds analogy parameter χ, Eq. (4.7), for $Pr = 1$ and 0.71 at $\beta = 0$ are listed in Tables 4.18 and 4.19. The function $\chi(\kappa)$ varies much weaker than the function $K_1(\kappa)$. For $Pr = 0.71$, the most significant (32 %) increase in χ occurs at $n_* = -1.5$, whereas a weak decrease of χ (0.9 %) takes place at $n_* = 4$. Also variation of the function $\chi(\kappa)$ in laminar flow is weaker than that in turbulent flow; however, this variation stretches over a much larger span of the parameter κ. For practical purposes, χ may be considered constant already for κ = 1.5–∞ for all values of n_* and both Prandtl numbers (see Tables 4.18 and 4.19).

Comparisons with experiments for coaxial orthogonal single jet impingement onto a rotating disk. As known from the literature [5, 18–20], the stagnation region of a single jet (see Fig. 4.15) is similar to a uniform orthogonal flow over a single disk studied above. *Non-coaxial* jet impingement studied experimentally in [25] dominated over the disk rotation effects under the following conditions:

Table 4.18 Values of χ, self-similar solution and the present integral method for Pr = 1 [6]

$\kappa = a/\omega$	$n_* = -2$	$n_* = -1.5$	$n_* = -1$	$n_* = 0$	$n_* = 0.5$	$n_* = 1$	$n_* = 2$	$n_* = 3$	$n_* = 4$
0.0	0.0	0.2119	0.3818	0.6433	0.7482	0.8410	1.0000	1.1335	1.2490
0.0	**0.0**	**0.2047**	**0.3752**	**0.6431**	**0.7502**	**0.8439**	**1.0000**	**1.1249**	**1.2270**
0.5	0.0	0.2701	0.4516	0.6968	0.7882	0.8673	1.0000	1.1099	1.2046
0.5	**0.0**	**0.2704**	**0.4454**	**0.6975**	**0.7920**	**0.8719**	**1.0000**	**1.0981**	**1.1756**
1.5	0.0	0.2823	0.4656	0.7074	0.7962	0.8726	1.0000	1.1050	1.1951
1.5	**0.0**	**0.2822**	**0.4584**	**0.7075**	**0.7995**	**0.8770**	**1.0000**	**1.0934**	**1.1668**
50.	0.0	0.2844	0.4681	0.7093	0.7976	0.8735	1.0000	1.1041	1.1934
50.	**0.0**	**0.2844**	**0.4608**	**0.7093**	**0.8009**	**0.8779**	**1.0000**	**1.0926**	**1.1653**

Table 4.19 Parameter χ, self-similar solution and present integral method, Pr = 0.71 [6, 50]

$\kappa = a/\omega$	$n_* = -2$	$n_* = -1.5$	$n_* = -1$	$n_* = 0$	$n_* = 0.5$	$n_* = 1$	$n_* = 2$	$n_* = 3$	$n_* = 4$
0.0	0.0	0.2366	0.4330	0.7452	0.8732	0.9876	1.1856	1.3533	1.4990
0.0	**0.0**	**0.2308**	**0.4276**	**0.7453**	**0.8753**	**0.9906**	**1.1857**	**1.3446**	**1.4766**
0.5	0.0	0.3286	0.5513	0.8549	0.9687	1.0672	1.2329	1.3704	1.4889
2 0.5	**0.0**	**0.3276**	**0.5421**	**0.8540**	**0.9717**	**1.0717**	**1.2326**	**1.3563**	**1.4546**
1.5	0.0	0.3453	0.5721	0.8736	0.9848	1.0805	1.2405	1.3725	1.4857
1.5	**0.0**	**0.3449**	**0.5630**	**0.8733**	**0.9885**	**1.0856**	**1.2404**	**1.3583**	**1.4510**
50.	0.0	0.3481	0.5757	0.8768	0.9876	1.0829	1.2418	1.3728	1.4852
50.	**0.0**	**0.3481**	**0.5668**	**0.8768**	**0.9916**	**1.0882**	**1.2418**	**1.3586**	**1.4504**

$$Re_{\mathrm{j}}/Re_{\mathrm{ri}} = V_{\mathrm{j}}D_{\mathrm{j}}/\omega r_{\mathrm{i}}^2 > 0.125, \tag{4.56}$$

$$a/\omega > 0.125A\left(r_{\mathrm{i}}/D_{\mathrm{j}}\right)^2, \tag{4.57}$$

where the impingement radius r_{i} is a characteristic length in $Re_{\mathrm{ri}} = \omega r_{\mathrm{i}}^2/\nu$.

The jet domination regime arises also at coaxial impingement. However, Eqs. (4.56) and (4.57) are inapplicable for r_{i} = 0. Alternatively, replacing r_{i} with b (because the jets in [25] impinged near the outer radius of the disk), one can replace Eq. (4.57) with a relation $a/\omega > 0.03125A(D/D_{\mathrm{j}})^2$. Using Eq. (4.43), one can estimate the parameter A in the experiments [12, 34] (where D_{j}/D = 0.09): A = 1.0–1.29. The onset of the regime of impinging jet domination occurred at $a/\omega > 3.86$–4. Thus, Eq. (4.56) predicts the beginning of this regime fairly close to our suggestion ($a/\omega > 1.5$).

Fig. 4.15 Geometric arrangement, fluid flow and heat transfer of a single impinging round jet perpendicular to and coaxial with a rotating disk [6, 48]

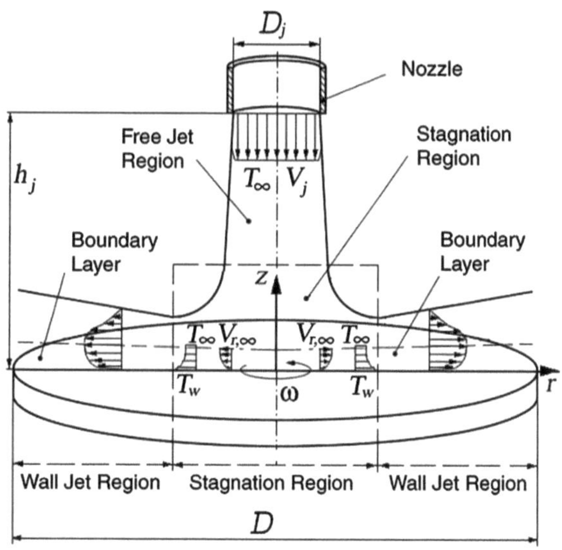

If the diameter D_j of the coaxial jet is significantly smaller than the disk diameter D, the exact solution holds only near the stagnation point for $r \le D_j/2$. For these conditions, Eq. (4.55) should be replaced with

$$Nu_{Dj} = K_{1*} Re_j^{1/2} A^{1/2}, \quad Nu_b = Nu_{Dj}(b/D_j). \tag{4.58}$$

In experiments [12, 34], the ratio $D_j/D = 0.09$ was constant, and the disk surface in the stagnation region was practically isothermal ($n_* = 0$). Using the definition of the parameter A, experimental data [12, 34] can be recalculated to find out the parameter κ

$$\kappa = \frac{AV_j}{\omega D_j} = A\frac{Re_j}{Re_{\omega j}}, \tag{4.59}$$

where $Re_{\omega j} = Re_\varphi \cdot b^2/D_j^2 = 0.25 Re_\varphi (D/D_j)^2$.

Equation (4.43) predicts the values of the parameter A equal to 1.29, 1.14, and 1.01 for h_j/D_j = 2, 4, and 6, accordingly. As seen from Fig. 4.16, this provides a good agreement of the simulations with experiments [12, 34] for the parameter K_{1*} at Re_j = 24,700. For Re_j = 6800, experiments contradict to the tendency of decrease in K_{1*} at larger values of h_j/D_j. Value A = 1.12 independent of h_j/D_j enables attaining the best coincidence with the experimental data for Re_j = 6800.

Integral method: fluid flow. A further application of the integral method for the considered problem was performed in [47, 48, 50]. Integral Eqs. (2.21)–(2.25) for $v_{\varphi,\infty} = 0$ (or $\beta = 0$) were used. Velocity profiles were approximated as

$$v_\varphi/(\omega r) = g, \quad v_r/(\omega r) = (1 - g_*)\tan\varphi, \tag{4.60}$$

$$(\tan\varphi - \kappa)/(\alpha - \kappa) = T. \tag{4.61}$$

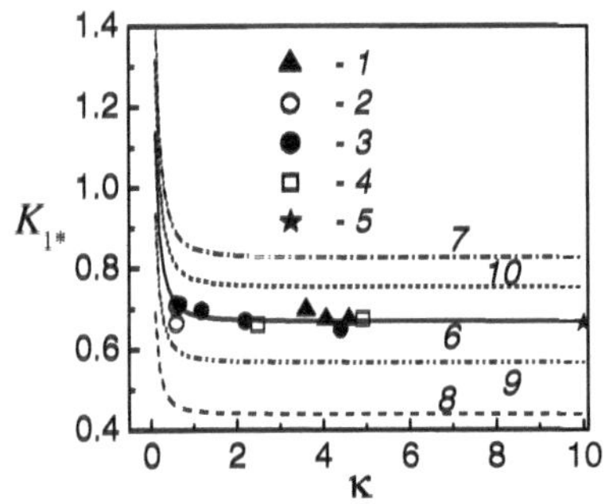

Fig. 4.16 Effect of the parameter $\kappa = a/\omega$ on the constant K_{1*} for Pr = 0.71 [6]. Experiments [12, 34]: *1*—Re_j = 24,700, Re_b = 214,000, h_j/D_j = 2, 4, and 6; *2*—Re_j = 6800, Re_b = 396,000, h_j/D_j = 2, 4, and 6; *3*—Re_j = 6800, Re_b = 53,500, 107,000, 200,000, and 353,000, h_j/D_j = 2; *4*—Re_j = 24,700, Re_b = 2000 and 396,000, h_j/D_j = 2. Experiments [13, 52]: *5*—n_* = 0, various Re_j and h_j/D_j. Predictions: *6*—n_* = 0; *7*—n_* = 1; *8*—n_* = −1; *9*—n_* = −0.5; *10*—n_* = 0.4

Given $K = 0$, these velocity profiles reduce to

$$[v_r/(\omega r)]_{\kappa=0} = \alpha_0(1 - g_*)\Pi = f, \tag{4.62}$$

$$v_r/v_{r,\infty} = (1 - g_*)\kappa + (\alpha - \kappa)f/\alpha_0. \tag{4.63}$$

Here profiles of g, g_*, f, $\tan\varphi$, and Π depend only on the variable ξ. This yields the expressions for the integral parameters

$$\overline{\delta}_r^* = A_4 - A_1 X, \quad \overline{\delta}_r^{**} = B_7 + B_8 X - B_1 X^2, \tag{4.64}$$

$$\overline{\delta}_{\varphi r}^{**} = \kappa(B_6 + D_1 X) = \alpha D_1 + \kappa(B_6 - D_1), \quad \int_0^1 v_\varphi^2 \mathrm{d}\xi = (\omega r)^2 C_1, \tag{4.65}$$

$$A_4 = \int_0^1 g_*\,\mathrm{d}\xi, \quad B_6 = \int_0^1 (1 - g_*)g\,\mathrm{d}\xi, \quad B_7 = \int_0^1 (1 - g_*)g_*\,\mathrm{d}\xi, \tag{4.66}$$

$$B_8 = -A_1 + 2D_5, \quad D_5 = \frac{1}{\alpha_0}\int_0^1 f g_*\,\mathrm{d}\xi, \quad X = \alpha/\kappa - 1. \tag{4.67}$$

whereas constants A_1, B_1, C_1, D_1 are described by Eqs. (4.16)–(4.18). For both flow regimes, Eqs. (4.60)–(4.67) hold. For laminar flow, g_*, $\tan\ \varphi$, and Π are unknown; functions $g = G_0 \equiv G_{a=0}$ and $f = F_0 = F_{a=0}$ are taken from the exact solution for a single disk (subscript "0", $\kappa = 0$); $\alpha = \tan\varphi_w$; $\alpha_0 = 0.8284$. For turbulent flow, Eq. (4.19) for velocity components with $g = g_*$ are valid; constants A_2, $A_3 = 1 - A_1 - A_2$, B_2, B_3, $B_4 = -A_1 + 2D_1$, $B_5 = \int_0^1 (1 - g)g\mathrm{d}\xi$, D_2, D_3, and D_4 are seen in Sects. 2.4 and 2.5; $B_8 \equiv B_4$, $B_6 \equiv B_7 \equiv B_4$.

The constants (4.66), (4.67) are the same for a particular flow regime. The integral parameters $\overline{\delta}_r^*$, $\overline{\delta}_r^{**}$, and $\overline{\delta}_{\varphi r}^{**}$ depend solely on the variable X.

For laminar flow, functions f and g (and g_*), like in Sect. 4.1.2, were integrated numerically with respect to the self-similar coordinate ζ with an accuracy of four significant digits. Thus, Eq. (4.20) for laminar flow are valid accompanied with relations for the other constants, which do not depend on κ

$$A_4 = a_4/\zeta_{\delta 0}, \quad B_6 = b_6/\zeta_{\delta 0}, \quad B_7 = b_7/\zeta_{\delta 0}, \quad B_8 = b_8/\zeta_{\delta 0}, \quad D_5 = d_5/\zeta_{\delta 0}, \tag{4.68}$$

$$a_4 = \int_0^{\zeta_{\delta 0}} g_*\,\mathrm{d}\zeta_0, \quad b_6 = \int_0^{\zeta_{\delta 0}} (1 - g_*)g\,\mathrm{d}\zeta_0, \quad b_7 = \int_0^{\zeta_{\delta 0}} (1 - g_*)g_*\mathrm{d}\zeta_0. \tag{4.69}$$

$$b_8 = -a_1 + 2d_5, \quad d_5 = \frac{1}{\alpha_0}\int_0^{\zeta_{\delta 0}} f g_* \mathrm{d}\zeta_0. \tag{4.70}$$

In accordance with Table 4.6 for $\beta = 0$, the constants a_1, b_1, c_1, and d_1 take the values

$$a_1 = 0.5338, \quad b_1 = 0.07887, \quad c_1 = 0.6726, \quad d_1 = 0.18587. \tag{4.71}$$

The constants (4.71) conform with earlier calculations [2]. The shear stresses are given by Eqs. (4.21)–(4.23) accompanied with the following relations:

$$\tau_{wr} = \frac{\mu v_{r,\infty}}{\delta}\left(\frac{\mathrm{d}\bar{v}_r}{\mathrm{d}\xi}\right)_{\xi=0}, \quad \left(\frac{\mathrm{d}\bar{v}_r}{\mathrm{d}\xi}\right)_{\xi=0} = -\left(\frac{\mathrm{d}g_*}{\mathrm{d}\xi}\right)_{\xi=0} + \frac{X}{\alpha}\left(\frac{\mathrm{d}f}{\mathrm{d}\xi}\right)_{\xi=0},$$
$$\left(\frac{\mathrm{d}f}{\mathrm{d}\xi}\right)_{\xi=0} = \left(\frac{\mathrm{d}f}{\mathrm{d}\xi}\right)_{\xi=0}\cdot\zeta_{\delta 0}, \quad \left(\frac{\mathrm{d}g_*}{\mathrm{d}\xi}\right)_{\xi=0} = \left(\frac{\mathrm{d}g_*}{\mathrm{d}\xi_0}\right)_{\xi=0}\cdot\zeta_{\delta 0}.$$

Values $(\mathrm{d}g_*/\mathrm{d}\zeta_0)_{\zeta=0}$ and $(\mathrm{d}f/\mathrm{d}\zeta_0)_{\zeta=0}$ are given by Eqs. (3.1) (single disk).

Assuming α = const., δ = const., and X = const., Equations (2.21) and (2.22) can be reduced to

$$\begin{aligned} &a^2\delta\left[3\left(B_7 + B_8X - B_1X^2\right) + A_4 - A_1X\right] + \delta\omega^2C_1 \\ &\quad = \frac{\nu a}{\delta}\left[-\left(\frac{\mathrm{d}g_*}{\mathrm{d}}\xi\right)_{\xi=0} + \frac{X}{\alpha_0}\left(\frac{\mathrm{d}f}{\mathrm{d}\xi}\right)_{\xi=0}\right], \\ &4a\delta[B_6 + D_1X] = -\frac{\nu}{\delta}\left(\frac{\mathrm{d}g}{\mathrm{d}\xi}\right)_{\xi=0}. \end{aligned} \tag{4.72}$$

The analysis of the exact solution shows that $(\mathrm{d}g_*/\mathrm{d}\xi)_{\xi=0}= (\mathrm{d}g/\mathrm{d}\xi)_{\xi=0}$. In view of this, Eq. (4.72) transform to a quadratic equation, whose solution is

$$X = X_{s1} + \left(X_{s2} + X_{s3}/\kappa^2\right)^{1/2}, \tag{4.73}$$

$$\alpha = (X_{s1} + 1)\kappa + \left(X_{s2}\kappa^2 + X_{s3}\right)^{1/2}, \tag{4.74}$$

$$X_{s1} = -h_2/(2h_3), \quad X_{s2} = X_{s1}^2 - h_4/h_3, \quad X_{s3} = -c_1/h_3. \tag{4.75}$$

$$h_3 = -3b_1 - 4d_1, \quad h_2 = 3b_8 - a_1 - 4b_6 - 4d_1, \tag{4.76}$$

$$h_1 = h_4 + c_1/\kappa^2, \quad h_4 = 3b_7 + a_4 - 4b_6. \tag{4.77}$$

Matching Eqs. (4.73)–(4.75) with the exact solution gives the unknown constants. As constants X_{s3} and h_3 result from the free disk solution, Eq. (4.74) automatically agrees with the conditions for $\kappa = 0$. An assumption that Eqs. (4.73)–(4.75) must merge with the self-similar solution at $\kappa = 4000$ (a nonrotating disk) and $\kappa = 1.65$ (empirical value), yields

$$h_2 = -2.79818, \quad h_3 = -0.98009, \quad h_4 = 0.66556, \tag{4.78}$$

$$X_{s1} = -1.42751, \quad X_{s2} = 2.71687, \quad X_{s3} = 0.68625. \tag{4.79}$$

Rearranging Eq. (4.72) for the function g and splicing it with the exact solution gives

$$\frac{\tau_{w\varphi}}{\tau_{w\varphi 0}} = \frac{\delta_0}{\delta} = \left[\frac{\alpha + \kappa(b_6/d_1 - 1)}{\alpha_0}\right]^{1/2}. \tag{4.80}$$

$$b_6 = 0.42776, \quad b_6/d_1 = 2.30137. \tag{4.81}$$

Equation (4.64) for $\overline{\delta}_r^*$ at $\kappa = 4000$ enables finding the constant a_4. Equations (4.69), (4.70), and (4.77) yield the rest of the constants b_7, b_8, and d_5

$$a_4 = \overline{\delta}_{r\zeta}^*(\delta_0/\delta)(1+\kappa)^{-1/2}, \quad \text{where} \quad \overline{\delta}_{r\zeta}^* = \int_0^\infty \left(1 - F\frac{1+\kappa}{\kappa}\right) d\zeta. \tag{4.82}$$

$$a_4 = 1.11052, \quad b_7 = 0.42202, \quad b_8 = 0.063373, \quad d_5 = 0.29859. \tag{4.83}$$

Table 4.20 contains numerical values of fluid flow parameters computed by the self-similar solution and the integral method, which agree with each other very well.

Table 4.20 Hydrodynamic parameters of the problem by the data of the exact solution and the integral method [6]

$\kappa = a/\omega$	$\overline{\delta}_{r\zeta}^*$, exact	$\overline{\delta}_{r\zeta}^*$	$\overline{\delta}_{r\zeta}^{**}$, exact	$\overline{\delta}_{r\zeta}^{**}$	α, exact	α	$\tau_{w\varphi}/\tau_{w\varphi 0}$, exact	$\tau_{w\varphi}/\tau_{w\varphi 0}$
0.2	0.0290	−0.4896	−0.1604	−0.1064	0.8172	0.8061	1.1124	1.1346
0.3	0.4849	0.1461	0.2281	0.2652	0.8452	0.8365	1.1947	1.2170
0.4	0.6491	0.4171	0.3207	0.3467	0.8939	0.8877	1.2834	1.3038
0.5	0.7150	0.5499	0.3459	0.3646	0.9591	0.9548	1.3741	1.3921
1.0	0.7399	0.6931	0.3340	0.3394	1.4180	1.4172	1.8022	1.8116
2.0	0.6821	0.6708	0.2999	0.3012	2.5440	2.5441	2.4888	2.4926
6.0	0.6130	0.6119	0.2672	0.2673	7.3593	7.3593	4.2781	4.2790
10	0.5962	0.5958	0.2597	0.2597	12.229	12.229	5.5197	5.5201
50	0.5746	0.5745	0.2501	0.2501	61.043	61.033	12.338	12.338
100.	0.5717	0.5717	0.2489	0.2489	122.08	122.06	17.449	17.449
1000	0.5692	0.5692	0.2478	0.2478	1220.8	1220.8	54.178	54.178
4000	0.5690	0.5690	0.2477	0.2477	4883.1	4883.1	110.36	110.36

The maximal deviation between both solutions is 1.7 % and diminishes with increasing κ.

Integral method: heat transfer. The thermal boundary layer Eq. (2.23) together with the rewritten Eq. (2.22) can be transformed similar to Eqs. (3.42), (3.43). This results in

$$\frac{\mathrm{d}}{\mathrm{d}r}\left[r^4\delta\overline{\delta}_{\varphi r}^{**}\right]=\frac{C_\mathrm{f}}{2}r^4(1+\alpha^2)^{1/2}, \tag{4.84}$$

$$\frac{\mathrm{d}}{\mathrm{d}r}\left[r^2\delta\overline{\delta}_\mathrm{T}^{**}(T_w-T_\infty)\right]=\chi\frac{C_\mathrm{f}}{2}r^2(1+\alpha^2)^{1/2}(T_w-T_\infty). \tag{4.85}$$

Equations (4.84) and (4.85), subject to the condition (2.30), can be rearranged as

$$(2+m+n_*)\frac{\overline{\delta}_\mathrm{T}^{**}}{\alpha A_1}=\chi(4+m)\frac{\overline{\delta}_{\varphi r}^{**}}{\alpha A_1}, \tag{4.86}$$

$$(2+m+n_*)\frac{\overline{\delta}_\mathrm{T}^{**}}{\alpha A_1}=\chi(4+m)K_{V0}\left[1+\frac{\kappa}{\alpha}\left(\frac{b_6}{d_1}-1\right)\right]. \tag{4.87}$$

In doing this, Eqs. (2.77)–(2.82) and (4.64)–(4.72) were involved. Here $K_{V0}=0.3482$ (Eq. (3.44)) for the laminar regime. In the turbulent regime, $K_{V0}=1-D_2/A_1$ (Eq. (3.37)), while B_5/D_1 (where as $B_5=C_2/2=D_1+D_3$) is employed in place of b_6/d_1.

The model relation for $\overline{\delta}_\mathrm{T}^{**}$ in laminar flow (where $n=1$) is based on Eq. (4.49) developed for turbulent flow in Sect. 4.2.2

$$\frac{\overline{\delta}_\mathrm{T}^{**}}{\alpha A_1}=\frac{1}{b_2}-\chi Pr^{n_p}(1-K_{V0})\frac{b_1}{b_2}+\frac{\kappa}{\alpha}\left(e_1\chi^{-1}+e_2\chi+e_3\right). \tag{4.88}$$

Here the coefficient b_1 (not to confuse with Eq. (4.71)), b_2, and n_p are presented by Eqs. (3.54)–(3.56). Hence Eq. (4.87) for the laminar flow regime results in

$$\begin{aligned}&\frac{1}{b_2}-\chi Pr^{n_p}(1-K_{V0})\frac{b_1}{b_2}+\frac{\kappa}{\alpha}\left(e_1\chi^{-1}+e_2\chi+e_3\right)\\&=\chi\frac{4}{2+n_*}K_{V0}\left[1+\frac{\kappa}{\alpha}\left(\frac{b_6}{d_1}-1\right)\right].\end{aligned} \tag{4.89}$$

The constants e_1, e_2, and e_3 are depending on the Prandtl number only. Splicing the parameter χ in Eq. (4.89) and the self-similar solution at the points $n_*=-1.5$; 0 and 2 yields values e_1, e_2, and e_3 listed in Table 4.21. Parameter χ by the self-similar solution is given as

Table 4.21 Values of the coefficients e_1, e_2, and e_3 [6, 47, 48, 50]

Pr	e_1	e_2	e_3
1	0.1426	−0.7227	1.0333
0.72	0.2007	−0.7495	1.3556
0.71	0.2042	−0.7497	1.3696
0.5	0.2957	−0.7733	1.8104
0.1	1.4528	−0.8052	4.7291

$$\chi = -K_1\left[(1+\kappa)Pr(\mathrm{d}G/\mathrm{d}\zeta)_{\zeta=0}\right]^{-1}. \tag{4.90}$$

Having obtained the coefficients e_1, e_2, and e_3, one can rewrite Eq. (4.89) as a quadratic equation, whose solution has the following form:

$$\chi = (-f_2 - D^{1/2})/(2f_1), \quad D = f_2^2 - 4f_1f_3, \tag{4.91}$$

where

$$f_1 = \frac{\kappa}{\alpha}e_2 - \frac{4}{2+n_*}K_{V0}\left[1+\frac{\kappa}{\alpha}\left(\frac{b_6}{d_1}-1\right)\right] - Pr^{n_p}(1-K_{V0})\frac{b_1}{b_2},$$
$$f_2 = \frac{1}{b_2} + \frac{\kappa}{\alpha}e_3, \quad f_3 = \frac{\kappa}{\alpha}e_1.$$

Having calculated the parameter χ, one can estimate the constants K_1 and K_{1*} with the help of the expressions

$$K_1 = -\chi\frac{\tau_{w\varphi}}{\tau_{w\varphi0}}\left(\frac{\mathrm{d}g}{\mathrm{d}\zeta_0}\right)_{\zeta=0}(1+\kappa)^{-1/2}Pr, \tag{4.92}$$

$$K_{1*} = -\chi\frac{\tau_{w\varphi}}{\tau_{w\varphi0}}\left(\frac{\mathrm{d}g}{\mathrm{d}\zeta_0}\right)_{\zeta=0}\kappa^{-1/2}Pr, \tag{4.93}$$

which take into account Eqs. (3.1) and (4.80).

As seen from Tables 4.18 and 4.19, the maximal deviation of the parameter χ, predicted by Eq. (4.91), as compared to the self-similar solution is less than 2.4 %. This estimate is valid also for K_{1*} for $Pr = 0.1 - 1$ (see Tables 4.17 and 4.22).

Values of the parameter K_{1*} from the self-similar solution and the integral method for a nonrotating disk ($\kappa \geq 50$) are listed in Table 4.23. In particular, for $Pr = 0.71$

$$K_{1*} = K_1 = 0.763\chi, \quad \chi = \frac{-1.8205 - [3.8863 + 1.9253/(n_* + 2)]^{1/2}}{-1.7102 + 5.7551/(n_* + 2)}. \tag{4.94}$$

Table 4.22 Values of K_{1*}, integral method for $Pr = 0.71$ (the first line represents K_1) [6]

$\kappa = a/\omega$	$n_* = -2$	$n_* = -1.5$	$n_* = -1$	$n_* = 0$	$n_* = 1$	$n_* = 2$	$n_* = 3$	$n_* = 4$
0.0	0.0	0.1009	0.187	0.3259	0.4332	0.5185	0.5880	0.6457
0.1	0.0	0.3957	0.6933	1.1562	1.5007	1.7671	1.9793	2.1524
0.2	0.0	0.3266	0.5569	0.9055	1.1582	1.3499	1.5004	1.6217
0.3	0.0	0.3015	0.5063	0.8106	1.0273	1.1897	1.3159	1.4169
0.5	0.0	0.2821	0.4667	0.7352	0.9226	1.0611	1.1677	1.2523
1.0	0.0	0.2704	0.4426	0.6887	0.8578	0.9814	1.0757	1.1500
1.5	0.0	0.2678	0.4371	0.6781	0.8430	0.9632	1.0547	1.1268
2.0	0.0	0.2669	0.4351	0.6742	0.8376	0.9565	1.0470	1.1181
10.	0.0	0.2657	0.4326	0.6692	0.8306	0.9479	1.0371	1.1071
50.	0.0	0.2656	0.4325	0.6690	0.8303	0.9475	1.0366	1.1067

Table 4.23 Values of K_{1*} the data of the exact solution (*boldface*) and the integral method for different values of Pr and $\kappa \geq 50$ for a stationary disk [6]

Pr	$n_* = -2$	$n_* = -1.5$	$n_* = -1$	$n_* = 0$	$n_* = 1$	$n_* = 2$	$n_* = 3$	$n_* = 4$
1.0	**0.0**	**0.3057**	**0.5031**	**0.7622**	**0.9387**	**1.0747**	**1.1866**	**1.2825**
	0.0	0.3056	0.4952	0.7623	0.9434	1.0747	1.1742	1.2523
0.72	**0.0**	**0.2671**	**0.4417**	**0.6726**	**0.8306**	**0.9525**	**1.0529**	**1.139**
	0.0	0.2671	0.4349	0.6726	0.8347	0.9524	1.0419	1.1123
0.5	**0.0**	**0.2295**	**0.3812**	**0.5839**	**0.7235**	**0.8314**	**0.9203**	**0.9967**
	0.0	0.2295	0.3755	0.5839	0.7269	0.8314	0.9110	0.9737
0.1	**0.0**	**0.1124**	**0.1911**	**0.3015**	**0.3802**	**0.4421**	**0.4936**	**0.5381**
	0.0	0.1124	0.1886	0.3015	0.3819	0.4420	0.4888	0.5262

Apparently, the deviations of the data in Table 4.23 predicted by the integral method are less than 2.4 % for any value of Pr.

To conclude, in this section an approximate analytical solution by means of the present integral method was derived, whose maximal deviation with respect to the self-similar solution are less than 2.4 % for values of the Prandtl number in the range $Pr = 0.1$–1.

4.3 Forced Outward Flow Between Corotating Disks

4.3.1 Ekman Layers

If fluid rotates with a velocity close to that of the disk, i.e., $\beta \to 1$ and $v_{r,\infty} = 0$, Ekman layers emerge on the disk, where the tangential velocity ωr is by an order of magnitude larger than the velocities v_r, v_z, and $(v_\varphi - \omega r)$ [2, 3]. The boundary layer equations are simplified to the following form:

$$-2\rho\omega(v_\varphi - v_{\varphi,\infty}) = \frac{\partial \tau_r}{\partial z}, \quad 2\rho\omega v_r = \frac{\partial \tau_\varphi}{\partial z}, \tag{4.95}$$

$$2\omega\rho\delta \int_0^1 (v_\varphi - v_{\varphi,\infty})\mathrm{d}\xi = \tau_{wr}, \quad 2\omega\rho\delta \int_0^1 v_r \mathrm{d}\xi = -\tau_{w\varphi}. \tag{4.96}$$

A solution of Eqs. (4.95) and (4.96) for laminar flow is [3]

$$v_r = \omega r(1-\beta)\exp(-\zeta)\sin(\zeta), \quad v_\varphi - v_{\varphi,\infty} = \omega r(1-\beta)\exp(-\zeta)\cos(\zeta), \tag{4.97}$$

$$\dot{m}_{\mathrm{d}}/(\mu r) = \pi(1-\beta)Re_\omega^{1/2}, \quad \delta/r = \pi Re_\omega^{1/2}, \tag{4.98}$$

$$\tau_{wr}/(\rho\omega^2 r^2) = (1-\beta)Re_\omega^{-1/2}, \quad \alpha_E = 0, \quad \zeta = z\sqrt{\omega/\nu}. \tag{4.99}$$

For turbulent Ekman layers [3], obtained with the help of an integral method a solution that looks similar to Eqs. (2.77)–(2.81), (2.85), and (5.2), whose coefficients $\alpha_{\mathrm{E}} = 0.553$, $\gamma_{*\mathrm{E}} = 0.0983$, and $\varepsilon^*_{m\mathrm{E}} = 0.1395$ coincide with the data obtained by Eqs. (5.1) at $N = 0$ and $\beta = 1$.

The author of [10] obtained an analogous solution for the turbulent Ekman layers with the help of the present integral method, which looks as

$$\alpha_{\mathrm{E}} = \left[A_1^{-1} n/(n+1)\right]^{1/2}, \tag{4.100}$$

$$\tau_{wr}/\rho = [2n/(n+1)](1-\beta)\omega^2 r\delta, \tag{4.101}$$

$$\tau_{w\varphi}/\rho = 2A_1\alpha_E(1-\beta)\omega^2 r\delta. \tag{4.102}$$

The functions δ/r, $\dot{m}_{\mathrm{d}}/(\rho\omega r^3)$ and C_{M} for the turbulent Ekman layers obtained in [10] also follow Eqs. (2.77)–(2.81), (2.84)–(2.86) for $N = 0$, where H_9 is replaced by $2A_1\alpha_{\mathrm{E}}$.

The constants α_{E}, γ_{E}, and $\varepsilon^*_{m\mathrm{E}}$ by Eqs. (4.100)–(4.102) for the turbulent Ekman layers (see Table 4.24) have the same values as those for $N = 0$ and $\beta = 1$ (Sect. 4.1).

Table 4.24 Constants α_{E}, γ_{E}, and $\varepsilon^*_{m\mathrm{E}}$ for Ekman layers [6, 10]

Coefficient	$n = 1/7$	$n = 1/8$	$n = 1/9$	$n = 1/10$
α_{E}	0.6936	0.6442	0.604	0.5705
γ_{E}	0.1224	0.0978	0.0807	0.0677
$\varepsilon^*_{m\mathrm{E}}$	0.1386	0.106	0.084	0.0678

The value $\varepsilon^*_{m\mathrm{E}} = 0.1386$ at $n = 1/7$ listed in Table 4.24 agrees well with $\varepsilon^*_{m\mathrm{E}} = 0.1395$ reported in [2], although the constants α_E from both sources are different by 25 %.

The Ekman layer model yields a noticeably simpler solution than that resulting from the complete boundary layer model at $N = 0$ and $\beta = \mathrm{const}$. In spite of this, the Ekman layer model enables predicting fluid flow parameters over the range $0.5 \le \beta \le 2$, whose accuracy suffices practical needs. It is important that the radial variation $\beta(r)$ can be arbitrary.

4.3.2 *Flow Structure in Forced Outward Flow Between Corotating Disks*

End walls of gas turbine rotors are often cooled using secondary air [1–4, 31, 53–69]. Compressed cooling air is delivered in a gap between two rotating disks, moves in the direction of larger radii, and afterwards leaves the gap (Figs. 4.17 and 4.18).

We consider here cases of the radial/axial inlet into the gap with nonmerging boundary layers over the disks, and arbitrary distributions of the surface temperature T_w.

For the purely radial inlet into the cavity and swirl parameters less than unity ($\beta_\mathrm{i} \le 1$), the flow pattern according to experimental results of [3] looks as shown in Fig. 4.17. Starting at the radial position r_i and up to a location r_e, centrifugal radial flow occupies the entire gap, where boundary layers emerge on the walls and entrain air from the core flow (so-called source region).

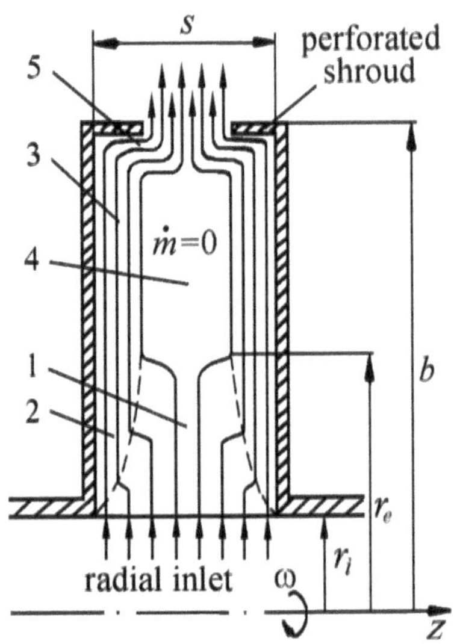

Fig. 4.17 Structure of outward flow in a rotating cavity for $\beta \le 1$ and a uniform radial inlet [3]. *1*—Source region; *2*—entraining boundary layer; *3*—Ekman-type layer; *4*—internal core (no radial flow); *5*—region of the outlet. Subscript "i": inlet into the cavity

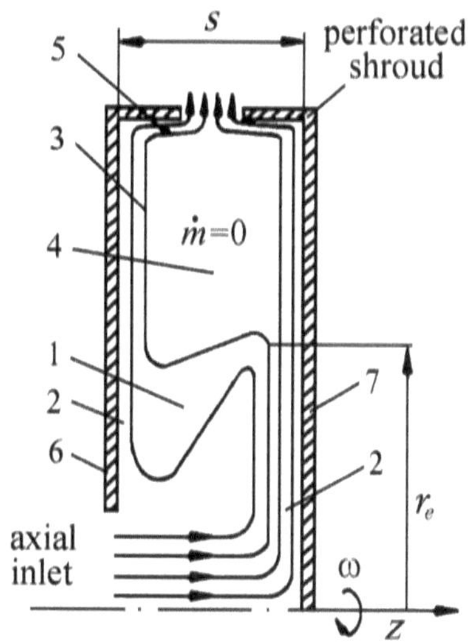

Fig. 4.18 Schematic of streamlines in a rotating cavity for $\beta \leq 1$ and axial inlet [3]. *1*—Source region; *2*—entraining boundary layer; *3*—Ekman-type layer; *4*—internal core (no radial flow); *5*—region of the outlet; *6*—upstream disk; *7*—downstream disk

Flow in the source region develops as a free vortex

$$v_{\phi,\infty} = \left(v_{\phi,\infty} r\right)_{\rm i} / r. \qquad (4.103)$$

Flow fed into the cavity axially (Fig. 4.18) impinges first onto a downstream disk and moves close to it toward the larger radii. However, at the radial coordinate $r = r_e$, a part of air is released to the source region and entrained by the upstream disk's boundary layer.

At $r > r_e$, air is fully entrained into the Ekman-type layers [3].

Here the mass flow rate $\dot{m}_{\rm d} = 0.5\dot{m} =$ const. both for the radial and axial inlets into the cavity [3]. Rotation of disks causes the swirl parameter to increase in the core flow.

4.3.3 *Radial Variation of the Swirl Parameter*

The numerical version of the *present integral method* was involved in simulations described below. Equation (2.72) and/or Eq. (2.74) were solved using the Runge–Kutta method [6, 70, 71]. The *local swirl parameter* $\beta(r)$ was validated in Fig. 4.19.

In the unshrouded cavity (Fig. 4.19), experiments [55] revealed air ingress from the atmosphere near the center plane of the gap ($z = s/2$), which causes the mass flow rate in the boundary layers near the periphery to increase with the coordinate r (instead of being constant as in the Ekman-type layers). This entails a decrease in the swirl parameter β not modeled by the present integral method. Curve 4 for $C_w = \dot{m}/(\mu b) = 2963$ and 6173 correlates fairly well with the experiments [55] up to the point of the maximum of β. Lines *A* for C_w = 1111 exhibit a reduced

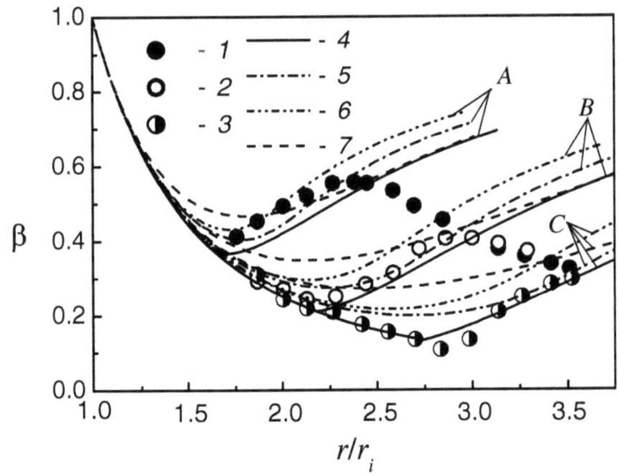

Fig. 4.19 Local swirl parameter β in the cavity open to atmosphere at $Re_\varphi = 4.97 \times 10^5$, $\beta_i = 1$, $s/b = 0.068$, $r_i/b = 0.285$, and $b = 0.1$ m [6]. *1–3*—Experiments [55]; *4*—present integral method [71], Eqs. (2.72) and (2.74), $n = 1/7$; *5*—"erroneous" Eq. (4.106) [55]; *6*—corrected Eq. (4.106); *7*—Eq. (4.105) [4]. Lines *A* and experiments *1*—$C_w = 1111$; *B* and *2*—2963; *C* and *3*—6173

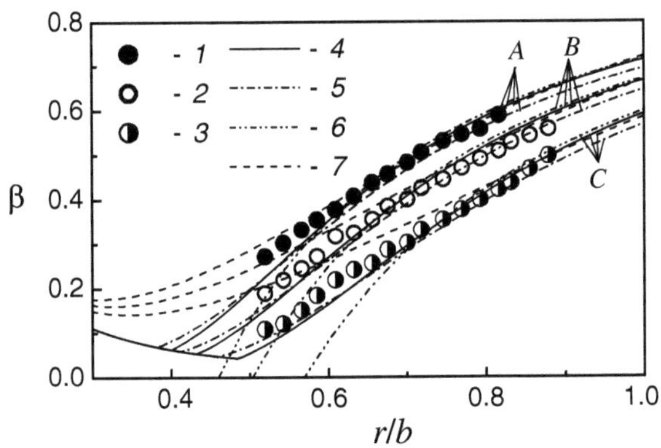

Fig. 4.20 Local swirl parameter β in a cavity with a perforated shroud at $C_w = 2500$, $\beta_i = 1$, $s/b = 0.1$, and $r_i/b = 0.1$ [6]. *1–3*—Experiments [3]. Present integral method: *4*—$n = 1/7$ [71]; *5*—1/9. Lines *6*—Ekman-type layer model (4.104), $n = 1/7$; lines *7*—Eq. (4.105) [4]. Lines *A* and experiments *1*—$Re_\varphi = 1.1 \times 10^6$; *B* and *2*—$6.177 \times 10^5$; *C* and *3*—$5.47 \times 10^5$

accuracy, because the flow near the radial location $r/r_i = 2.1$ is transitional, which is not properly captured by the present integral method [55].

A perforated shroud at the periphery prevents the air ingress from the atmosphere. Curves 4 and 5 in Fig. 4.20 obtained with the present integral method agree well with the experiments [2]. With an increase in the Reynolds number Re_{V_*} or r/b, turbulence becomes fully developed, which improves the agreement with the experiments. Curves for $n = 1/7$ and $n = 1/9$ lie close to each other.

The Ekman layer model yields a relation for the mass flow rate through the boundary layer $\dot{m}_\mathrm{d}$ and, given $\dot{m}_\mathrm{d}/\dot{m} = 0.5$, for the swirl parameter β

$$\begin{aligned} \dot{m}_\mathrm{d}/(\mu r) &= \varepsilon_m^* \,\mathrm{sgn}(1-\beta)|1-\beta|^{2(n+1)/(3n+1)} Re_\omega^{(n+1)/(3n+1)}, \\ \beta &= 1 - \varepsilon_\beta Y^{-0.5(5n+1)/(n+1)}, \end{aligned} \tag{4.104}$$

where $Y = x\lambda_\mathrm{T}^{-(3n+1)/(5n+3)}$, $\lambda_\mathrm{T} = C_w Re_\varphi^{-(n+1)/(3n+1)}$. The coefficient $\varepsilon_\beta = (0.5/\varepsilon_m^*)^{0.5(2n+1)/(n+1)}$ for n = 1/7; 1/8; 1/9; 1/10 takes the values ε_β = 2.23; 2.581; 2.916; 3.256, respectively. The authors [2] obtained that ε_β = 2.21 for n = 1/7. Curves of β calculated by Eq. (4.104) for n = 1/7 and depicted in Fig. 4.20 (curve 6) noticeably deviate from the experimental data at the boundary between the source region and the Ekman-type layers. However, curve 6 for larger x-coordinates merges with the curves predicted by the integral method.

In [4], the differential equation for β has been derived

$$\frac{\mathrm{d}\beta}{\mathrm{d}x} = -\frac{2\beta}{x} + 2Ax^{1.6}(1-\beta)^{1.2}, \tag{4.105}$$

where $A = 0.0274 \cdot 2\pi \cdot Re_\varphi^{0.8}/C_w$. In [55], another differential equation has been proposed

$$\frac{\mathrm{d}\beta}{\mathrm{d}x} = -\frac{2\beta}{x} + \frac{0.6c_1(1-\beta)^{c_2}[c_3(1-\beta)+\beta]Re_\varphi^{(n+1)/(3n+1)}x^{2(n+1)/(3n+1)}}{0.5C_w b/r_\mathrm{i} - c_1(1-\beta)^{c_2}Re_\varphi^{(n+1)/(3n+1)}x^{(5n+3)/(3n+1)}}, \tag{4.106}$$

where $c_1 = 0.2054\pi \mathrm{e}^{-0.175/n}$, $c_2 = 0.753n^{-0.424}$, $c_3 = 1.543n^{0.625}$, n = 1/5. Erroneously, the first term $0.5C_w b/r$ on the right-hand side of Eq. (4.106) in [55] was multiplied by ρ.

Close to the end of the Ekman-type layers region (curve 7 in Fig. 4.20), and in the initial part of the source region (curve 7 in Fig. 4.19), Eq. (4.105) agrees well with the experimental data. Between these two regions, deviations of Eq. (4.105) are significant.

Equation (4.106) agrees well with the original experiments [55] (curves 5, 6 in Fig. 4.19). However, as shown in [6], Eq. (4.106) diverts noticeably from the data [3] for the well-developed Ekman-type layers. Thus, both Eqs. (4.105) and (4.106) are far less accurate and universal than the present integral method [6].

4.3.4 *Local Nusselt Numbers*

In the experiments [59] for $\beta_\mathrm{i} = 1$, $T_w = \mathrm{const.}$ and an axial inlet into the cavity, solely the source region existed. Measured *local* Nusselt numbers (data *1* and empirical curve 2 in Figs. 4.11 and 6.9) were approximated by the empirical equations

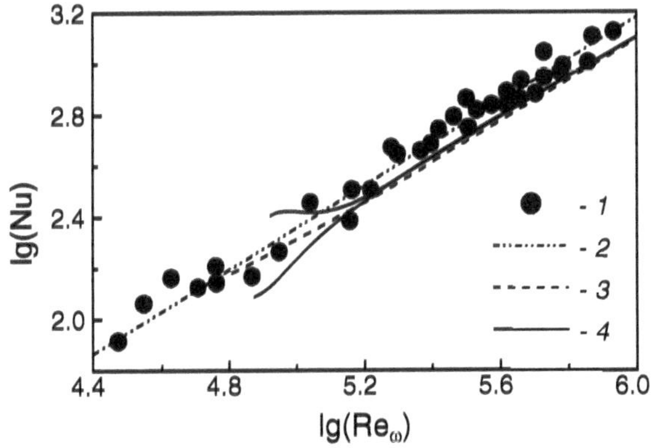

Fig. 4.21 Variation of the local Nusselt numbers versus Re_ω for $\beta_i = 1$, $N_i = 0.3333$, $r_i/b = 0.245$, $s/r_i = 0.18$, T_w = const., and $Re_{\omega i} = 6 \times 10^4$ [6]. *1*, *2*—Experiments and their approximation [59]; *3*—Eq. (4.107) [59]; *4*—present integral method, *upper curve* for $\alpha_i = 6.5$, *lower curve* for $\alpha_i = 4.4$

$$Nu = 0.024 Re_\omega^{0.8} Pr^{0.6} V_M, \tag{4.107}$$

$$\begin{aligned} V_M = 2.461 + 0.128\bar{r}^2 + 0.122\bar{r}/N_i + 0.02N_i^{-2} \\ - 0.858\bar{r} - 0.392/N_i, \quad \bar{r} = 1.2\text{–}2 \end{aligned} \tag{4.108}$$

$$V_M = 1.08 N_i^{0.07}, \quad \bar{r} = 2\text{–}3. \tag{4.109}$$

Here $N_i = v_{r,i}/(\omega r_i) = 0.1667$–$0.5$, $\bar{r} = r/r_i$, $Re_\omega = 6 \times 10^4 - 10^6$, T_w = const.

Experiments [59] surpass Eq. (4.107) (see curve 3 in Fig. 4.21) by 10–20 %, because heaters glued over the disk surface caused a turbulization effect [59].

For $Re_\omega \geq 1.6 \times 10^5$, computations by the present integral method practically merge with curve 3 by Eq. (4.107) in Fig. 4.21. A proper selection of the initial value α_i, as a boundary condition of the present integral method, enables to account for the heat transfer enhancement near the axial inlet into the cavity (see Fig. 4.21), while Eqs. (4.108) and (4.109) do not provide such an opportunity and also fail to perform for $N_i = 0.1667$ [6].

In the experimental investigations for $\beta_i = 0$ [65, 66], the outer core temperature is $T_\infty = T_i$ = const. in the region of the entraining boundary layers. The cold air from the flow's core is not entrained into the Ekman-type layers, which start to release heat back to the flow's core. As a result, the temperature T_∞ increases and becomes itself an unknown.

Therefore, the Nusselt number is redefined as $Nu = q_w r/[\lambda(T_w - T_i)]$ using a known temperature T_i instead of the unknown T_∞, which brings [3]

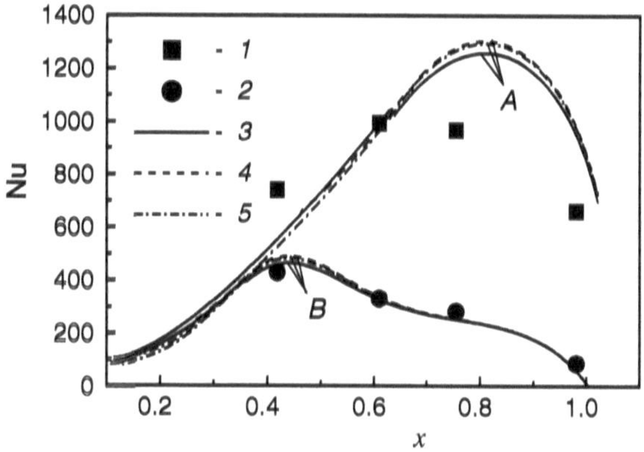

Fig. 4.22 Radial variation of the Nusselt numbers for $Re_\varphi = 1.9 \times 10^6$, $\beta_\mathrm{i} = 0$, $r_\mathrm{i}/b = 0.103$, $s/b = 0.138$, $b = 0.428$ m, $\mathrm{d}T_w/\mathrm{d}r > 0$ [6, 70]. *1*, *2*—Experiments [65, 66]. Present integral method, $n = 1/7$: *3*—$k = 0.5$; *4*—0.25; *5*—0.03. Lines *A* and experiments *1*—$C_w = 13{,}000$; *B* and *2*—2800

$$Nu = St\,Re_{V*r}Pr\frac{T_w - T_\infty}{T_w - T_\mathrm{i}}, \qquad (4.110)$$

where $Re_{V*r} = V_* r/\nu$. Experiments [65, 66] and simulations using the present integral method (see Fig. 4.22) demonstrate that in the source region, given $T_\infty = T_\mathrm{i}$, the Nu values grow up like in Fig. 4.21. In the Ekman-type layers, the temperature T_∞ (predicted by Eq. (2.75)) increases, while the last factor in Eq. (4.110) decreases together with the Nusselt number. Predictions at $C_w = 2800$ are consistent with experiments [65, 66], while at $C_w = 13{,}000$, when the near-inlet region is dominated by an impinging jet, noticeable differences are observed (like in [2, 65]).

In Fig. 4.22, a relation $v_{r,\infty} = (v_{r,\infty}r)_\mathrm{i}/r$ was used in the beginning of the source region followed with a relation $v_{r,\infty} = 0$ over the remaining length. Splicing took place at the coordinate, where a condition of $\dot{m}_\mathrm{d} = k\dot{m}$ was attained with alternatively $k = 0.03$, 0.25, or 0.5. Curves for the Nusselt number in Fig. 4.22 predicted for different values of the coefficient k are nevertheless close to each other for the same values of C_w.

4.3.5 *Effect of the Radial Distribution of the Disk Temperature*

To employ experimental distributions of the wall temperature $T_w(r)$ obtained in [3, 65] for an upstream disk, they were approximated by [72] by a seventh-order polynomial (Fig. 4.23). Experiments [3, 65] were performed for $\beta_\mathrm{i} = 0$, $Re_\varphi = (3.2\text{–}3.3) \times 10^6$, $r_\mathrm{i}/b = 0.103$, $s/b = 0.138$, $b = 0.428$ m, $C_w = 7000$.

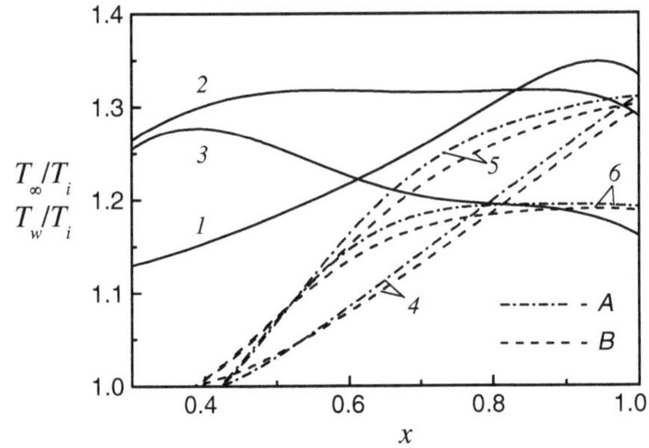

Fig. 4.23 Wall temperature T_w/T_i: *solid curves 1–3*—are experiments [65] approximated in [70]. Center plane temperature T_∞/T_i: curves *4–6*—computed by the present integral method, $n_T = 1/7$ and 1/10 [6, 70]. *1, 4*—$dT_w/dr > 0$; *2, 5*—$dT_w/dr \approx 0$; *3, 6*—$dT_w/dr < 0$. *A*—$n = 1/7$; *B*—$n = 1/10$

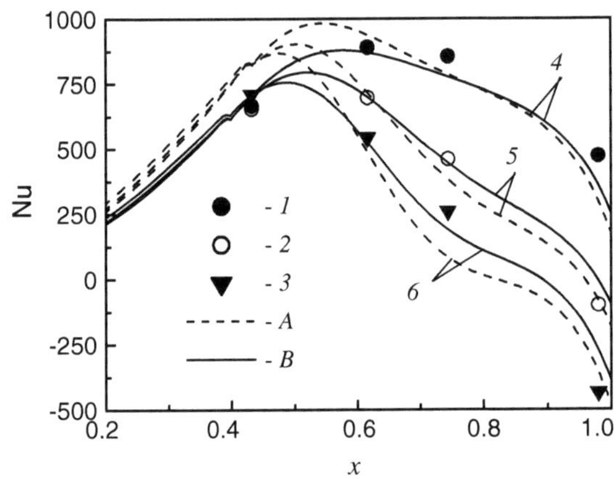

Fig. 4.24 Nusselt numbers in a cavity ($Re_\varphi = 3.2 \times 10^6$ for cases *1, 4*; $Re_\varphi = 3.3 \times 10^6$ for cases *2, 3, 5, 6*): *1–3*—experiments [3, 65]; *4–6*—present integral method, $n_T = 1/7$ [6, 70]. *1, 4*—$dT_w/dr > 0$; *2, 5*—$dT_w/dr \approx 0$; *3, 6*—$dT_w/dr < 0$. *A*—$n = 1/7$; *B*—$n = 1/10$

Center plane temperature T_∞ and Nusselt numbers for positive, approximately constant and negative distributions of dT_w/dr are depicted in Figs. 4.23 and 4.24.

Most of the simulations were performed for $n = 1/10$ in agreement with high Re_φ numbers measured in experiments. A larger value of $n = 1/7$ causes a larger deviation from the experiments (see Fig. 4.24). Temperature profiles are less sensitive to the effect of Re_φ; this entailed selection of the value $n_T = 1/7$ [72].

Both experiments and computations indicate that at $dT_w/dr \approx 0$ and $dT_w/dr < 0$ the Nusselt numbers are negative in the area of larger radii. This arises from the fast drop of the wall temperature T_w and the simultaneous rise of the center plane temperature T_∞, which leads to a region where T_∞ is higher than T_w (Fig. 4.23). As a result, the sign of the heat flux on the disk surface changes: air heats up the disk rather than the reverse.

To conclude, it was demonstrated in this section that the present integral method provides a good match of the simulations with known experimental data [3, 59, 65, 66] for rotation cavities. In comparison with the integral method developed in [3], the present integral method possesses higher accuracy at the expense of more accurate approximation of the radial velocity and temperature profiles in the boundary layer, as well as the possibility to vary the power exponents in these profiles.

References

1. Lokai VI, Bodunov MN, Zhuikov VV, Shchukin AV (1985) Heat transfer in cooled parts of aircraft gas-turbine engines. Izdatel'stvo Mashinostroenie, Moscow, USSR (in Russian)
2. Owen JM, Rogers RH (1989) Flow and heat transfer in rotating-disc systems. Volume 1: Rotor-stator systems. Research Studies Press Ltd., Taunton (Somerset, England)
3. Owen JM, Rogers RH (1995) Flow and heat transfer in rotating-disc systems. Volume 2: Rotating cavities. Research Studies Press Ltd., Taunton (Somerset, England)
4. Shvets IT, Dyban EP (1974) Air cooling of gas turbine parts. Naukova Dumka, Kiev, Ukraine (in Russian)
5. Schlichting G (1968) Boundary-layer theory. McGraw-Hill Book Company, New York
6. Shevchuk IV (2009) Convective heat and mass transfer in rotating disk systems. Springer, Berlin
7. Shevchuk IV, Delas NI (2005) Aerodynamics and turbulent flow heat exchange in the rotary disk air cleaner. Heat Transfer Res 36(1–2):104–113
8. Karman Th. von. (1921) Über laminare und turbulente Reibung. Z Angew Math Mech 1 (4):233–252
9. Newman BG (1983) Flow and heat transfer on a disk rotating beneath a forced vortex. AIAA J 21(8):1066–1070
10. Shevchuk IV, Khalatov AA (1997) Integral method for calculating the characteristics of a turbulent boundary layer on a rotating disk: quadratic approximation of the tangent of the flow swirl angle. Heat Transfer Res 28(4–6):402–413
11. Shevchuk IV, Buschmann MH (2005) Rotating disk heat transfer in a fluid swirling as a forced vortex. Heat Mass Transfer 41(12):1112–1121
12. Saniei N, Yan XT, Schooley WW (1998) Local heat transfer characteristics of a rotating disk under jet impingement cooling. In: Proceedings of 11th IHTC heat transfer 1998 (Kyongju, Korea), vol 5, pp 445–450
13. Dyban YP, Mazur AI (1982) Convective heat transfer in jet flows past bodies. Naukova Dumka, Kiev, USSR (in Russian)
14. Lallave JC, Rahman MM, Kumar A (2007) Numerical analysis of heat transfer on a rotating disk surface under confined liquid jet impingement. Int J Heat Fluid Flow 28(4):720–734
15. Rahman MM, Lallave JC (2007) A comprehensive study of conjugate heat transfer during free liquid jet impingement on a rotating disk. Numer Heat Transfer A Appl 51(11):1041–1064
16. Rahman MM, Lallave JC (2008) Thermal transport during liquid jet impingement from a confined spinning nozzle. AIAA J Thermophys Heat Transfer 22(2):210–218
17. Şara ON, Erkmen J, Yapici S, Çopur M (2008) Electrochemical mass transfer between an impinging jet and a rotating disk in a confined system. Int Commun Heat Mass Transfer 35 (3):289–298

18. Cebeci T, Bradshaw P (1984) Physical and computational aspects of convective heat transfer. Springer, Berlin
19. Dorfman LA (1963) Hydrodynamic resistance and the heat loss of rotating solids. Oliver and Boyd, Edinburgh, UK
20. Kochin NE, Kibel IA, Roze NV (1964) Theoretical hydrodynamics. Interscience, New York
21. Brdlik PM, Savin VK (1966) Heat transfer in the vicinity of the stagnation point in an axisymmetric jet flowing over flat surfaces normal to the flow. J Eng Phys Thermophys 10 (4):241–245
22. Han B, Goldstein RE (2001) Jet-impingement heat transfer in gas turbine systems. Ann N Y Acad Sci 234:147–161
23. Viskanta R (1993) Heat transfer to impinging isothermal gas and flame jets. Exp Thermal Fluid Sci 6(2):111–134
24. Bogdan Z (1982) Cooling of a rotating disk by means of an impinging jet. In: Proceedings of 7th IHTC heat transfer 1982 (Munich, Germany), vol 3, pp 333–336
25. Brodersen S, Metzger DE, Fernando HJS (1996) Flows generated by the impingement of a jet on a rotating surface. Part I. Basic flow patterns. Trans ASME J Fluids Eng 118 (1): 61–67
26. Chen Y-M, Lee W-T, Wu S-J (1998) Heat (mass) transfer between an impinging jet and a rotating disk. Heat Mass Transfer 34(2–3):101–108
27. Cheng W-T, Lin H-T (1994) Unsteady and steady mass transfer by laminar forced flow against a rotating disk. Wärme und Stoffübertragung 30(2):101–108
28. He Y, Ma LX, Huang S (2005) Convection heat and mass transfer from a disk. Heat Mass Transfer 41(8):766–772
29. Koh JCY, Price JF (1967) Non-similar boundary layer heat transfer of a rotating cone in forced flow. Trans ASME J Heat Transfer 89:139–145
30. Metzger DE, Bunker RS, Bosch G (1991) Transient liquid crystal measurements of local heat transfer on a rotating disk with jet impingement. Trans ASME J Turbomach 113(1):52–59
31. Minagawa Y, Obi S (2004) Development of turbulent impinging jet on a rotating disk. Int J Heat Fluid Flow 25(5):759–766
32. Popiel CO, Boguslawski L (1986) Local heat transfer from a rotating disk in an impinging round jet. Trans ASME J Heat Transfer 108(2):357–364
33. Saniei N, Yan X (1998) Effect of jet location on impingement cooling of a rotating disk. Proc ASME HTD 361–1:203–210
34. Saniei N, Yan XT (2000) An experimental study of heat transfer from a disk rotating in an infinite environment including heat transfer enhancement by jet impingement cooling. J Enhanced Heat Transfer 7:231–245
35. Soong CY, Chyuan CH (1998) Similarity solutions of mixed convection heat and mass transfer in combined stagnation and rotation-induced flows over a rotating disk. Heat Mass Transfer 34(2–3):171–180
36. Theodorsen Th, Regier A (1944) Experiments on drag of revolving discs, cylinders and streamline rods at high speeds. NASA Report No 793
37. Tien CL, Campbell DT (1963) Heat and mass transfer from rotating cones. J Fluid Mech 17:105–112
38. Tien CL, Tsuji IJ (1964) Heat transfer by laminar forced flow against a non-isothermal rotating disk. Int J Heat Mass Transfer 7(2):247–252
39. Tien CL, Tsuji IJ (1965) A theoretical analysis of laminar forced flow and heat transfer about a rotating cone. Trans ASME J Heat Transfer 87(2):184–190
40. Tifford AN, Chu ST (1954) On the flow and temperature fields in forced flow against a rotating disc. In: Proceedings of 2nd US national congress of applied mechanics (Ann Arbor, Michigan, USA), pp 793–800
41. Mabuchi I, Tanaka T, Sakakibara Y (1971) Studies of convective heat transfer from a rotating disk (5th Report, Experiment on the laminar heat transfer from a rotating isothermal disk in a uniformed forced stream). Bull JSME 14(72):581–589
42. Astarita T, Cardone G (2008) Convective heat transfer on a rotating disk with a centred impinging round jet. Int J Heat Mass Transfer 51(7–8):1562–1572

43. Helcig C, aus der Wiesche S (2013) The effect of the incidence angle on the flow over a rotating disk subjected to forced air streams. In: Proceedings of 2013 FEDSM (Incline Village, Nevada, USA), Paper FEDSM2013-16360
44. Trinkl CM, Bardas U, Weyck A, aus der Wiesche S (2011) Experimental study of the convective heat transfer from a rotating disc subjected to forced air streams. Int J Thermal Sci 50 (1):73–80
45. aus der Wiesche S (2010) Heat transfer from rotating discs with finite thickness subjected to outer air streams. In: Proceedings of IHTC14 (Washington, DC, USA), Paper IHTC14-22558
46. Truckenbrodt E (1954) Die turbulente Strömung an einer angeblasenen rotierenden Scheibe. Z Angew Math Mech 34:150–162
47. Shevchuk IV (2002) Laminar heat transfer in a rotating disk under conditions of forced air impingement cooling: approximate analytical solution. High Temp 40(5):684–692
48. Shevchuk IV, Saniei N, Yan XT (2002) Impinging jet heat transfer over a rotating disk: exact solution and experiments. In: Proceedings of 8th AIAA/ASME joint thermophys heat transfer conference (St. Louis, Missouri, USA), Paper AIAA 2002–3015
49. Shevchuk IV (2003) Exact solution of the heat transfer problem for a rotating disk under uniform jet impingement. Fluid Dyn 38(1):18–27
50. Shevchuk IV, Saniei N, Yan XT (2003) Impingement heat transfer over a rotating disk: integral method. AIAA J Thermophys Heat Transfer 17(2):291–293
51. Helcig C, aus der Wiesche S, Shevchuk IV (2014) Internal symmetries, fundamental invariants and convective heat transfer from a rotating disc. In: Proceedings of IHTC-15, (Kyoto, Japan), Paper IHTC15-8483
52. Sparrow EM, Geiger GT (1985) Local and average heat transfer characteristics for a disk situated perpendicular to a uniform flow. Trans ASME J Heat Transfer 107(2):321–326
53. Abrahamson S, Lonnes S (1994) An improved model for radial injection between corotating disks. Trans ASME J Fluids Eng 116(2):255–257
54. Chen JX, Gan X, Owen JM (2001) Heat transfer from air-cooled contra-rotating disks. Trans ASME J Turbomach 119(1):61–67
55. Dyban YP, Kabkov VI (1979) Experimental study of flow of air in a gap formed by two rotating disks. Fluid Mech-Soviet Res 8(4):99–104
56. Kapinos VM (1964) Hydraulic resistance and heat transfer of a free disc with a cob. Izvestiya vuzov. Energetika 11:85–92 (in Russian)
57. Kapinos VM (1965) Heat transfer of a disk rotating in a housing. Izvestiya vuzov. Aviatsyonnaya Tekhnika 2:76–86 (in Russian)
58. Kapinos VM (1965) Effect of a radial gradient of a relative circumferential velocity component on heat transfer in forced flow between two rotating disks. Izvestiya vuzov. Energetika 3:111–120 (in Russian)
59. Kapinos VM, Pustovalov VN, Rud'ko AP (1971) Heat transfer in a coolant flow from a center to a periphery between two rotating disks. Izvestiya vuzov. Energetika 6:116–124 (in Russian)
60. Karabay H, Chen J-X, Pilbrow R, Wilson M, Owen JM (1999) Flow in a "cover-plate" preswirl rotor-stator system. Trans ASME J Turbomach 121(1):161–166
61. Karabay H, Wilson M, Owen JM (2001) Predictions of effect of swirl on flow and heat transfer in a rotating cavity. Int J Heat Fluid Flow 22:143–155
62. Kilic M, Owen JM (2003) Computation of flow between two disks rotating at different speeds. Trans ASME J Turbomach 125(2):394–400
63. Morse AP (1988) Numerical prediction of turbulent flow in rotating cavities. Trans ASME J Turbomach 110(2):202–215
64. Morse AP (1991) Assessment of laminar-turbulent transition in closed disc geometries. Trans ASME J Turbomach 113(2):131–138
65. Northrop A, Owen JM (1988) Heat transfer measurements in rotating-disc systems. Part 2: The rotating cavity with a radial outflow of cooling air. Int J Heat Fluid Flow 9(1):27–36
66. Northrop A (1984) Heat transfer in a cylindrical rotating cavity. D. Phil. thesis. University of Sussex, Brighton, UK

67. Pilbrow R, Karabay H, Wilson M, Owen JM (1999) Heat transfer in a 'cover-plate' pre-swirl rotating-disc system. Trans ASME J Turbomach 121(2):249–256
68. Prata AT, Pilichi CDM, Ferreira RTS (1995) Local heat transfer in axially feeding radial flow between parallel disks. Trans ASME J Heat Transfer 117(1):47–53
69. Yan Y, Gord MF, Lock GD, Owen JM (2003) Fluid dynamics of a pre-swirl rotor-stator system. Trans ASME J Turbomach 125(4):641–647
70. Shevchuk IV (1998) The effect of distribution of radial velocity in the flow core on heat transfer under conditions of centrifugal flow in a gap between parallel rotating disks. High Temp 36(6):972–974
71. Shevchuk IV (1999) Integral method of calculation of a turbulent centrifugal underswirl flow in a gap between parallel rotating disks. Heat Transfer Res 30(4–6):238–248
72. Shevchuk IV (1999) Effect of wall-temperature distribution on heat transfer in centrifugal flow in the gap between parallel rotating disks. J Eng Phys Thermophys 72(5):896–899

Chapter 5
Heat and Mass Transfer in Rotating Cone-and-Disk Systems for Laminar Flows

5.1 General Characterization of the Problem

In the past, non-rotating conical diffusers (Fig. 5.1) were modeled using simplified Navier–Stokes equations without flow pre-swirl at the inlet [1]. Flow pre-swirl effects on the heat transfer were for the first time studied by the author of this work [2].

Cone-and-plate devices, where flow develops in a gap with small angles $\gamma = 1\ldots0.5°$ between a rotating cone and a stationary plate (Fig. 5.2), are used in viscosimetry [3–5]. Medicine employs such devices for nurturing endothelial cells that grow as a monolayer on the non-rotating plate, whereas a cone rotates slowly to renew the feeding fluid and simultaneously not to damage the cells [6–8].

Flow regimes in cone-and-plate devices were studied experimentally [8], simulated using CFD codes [3, 8] and using perturbation techniques [5–7]. Self-similar Navier–Stokes and energy equations were derived and solved by the author of this work [2, 9–11].

Convective heat transfer in cone-and-disk configurations, with one of them rotating or both co-rotating/contra-rotating, along with a stationary conical diffuser, depends strongly on the radial temperature distribution on the disk [2, 9, 10]. Simulations were done mostly for air ($Pr = 0.71$); new phenomena in heat and mass transfer for other values of the Prandtl and Schmidt numbers were first investigated by the author in [11].

For the small angle γ, the Navier–Stokes Eqs. (2.1)–(2.3) can be simplified [5–7] for the considered laminar flow

$$-\frac{v_\varphi^2}{r} = -\frac{1}{\rho}\frac{\partial p}{\partial r} + \nu\frac{\partial^2 v_r}{\partial z^2}, \tag{5.1}$$

$$v_r\frac{\partial v_\varphi}{\partial r} + v_z\frac{\partial v_\varphi}{\partial z} + \frac{v_r v_\varphi}{r} = \nu\frac{\partial^2 v_\varphi}{\partial z^2}, \tag{5.2}$$

I.V. Shevchuk, *Modelling of Convective Heat and Mass Transfer in Rotating Flows*, Mathematical Engineering, DOI 10.1007/978-3-319-20961-6_5

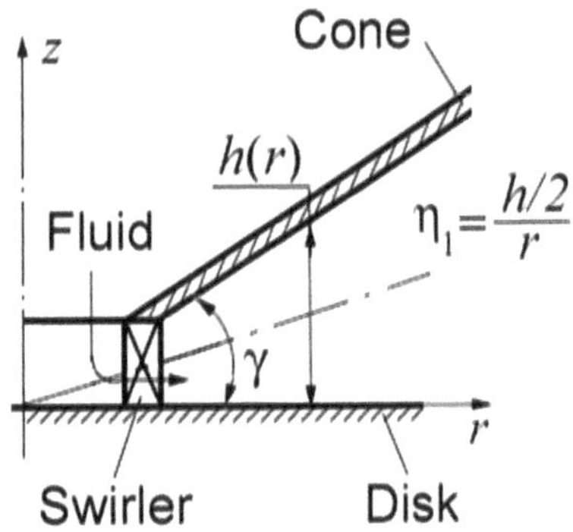

Fig. 5.1 Schematic of swirling flow in a stationary conical diffuser [9]

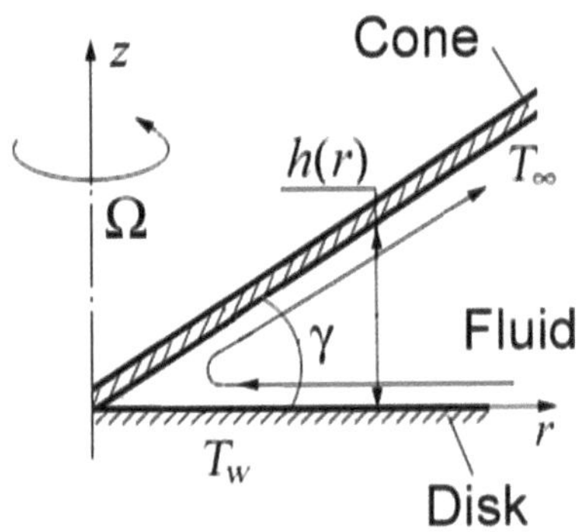

Fig. 5.2 Schematic of fluid flow in a gap with a rotating cone and a stationary disk [9]

$$-\frac{1}{\rho}\frac{\partial p}{\partial z}+\nu\frac{\partial^2 v_z}{\partial z^2}=0. \tag{5.3}$$

A perturbation solution of Eqs. (5.1)–(5.3) by the method of expansion in the small parameter $Re = Re_\Omega \eta_1^2/12$ (where $\eta_1 = h/r$) yields [5]

$$v_r/(\Omega r) = Re(1.8\tilde{z}^2 - \tilde{z}^4 - 0.8\tilde{z}), \tag{5.4}$$

$$v_\varphi/(\Omega r) = \tilde{z} + Re^2(-83\tilde{z} + 70\tilde{z}^4 + 63\tilde{z}^5 - 50\tilde{z}^7)/175, \tag{5.5}$$

$$v_z/(\Omega r) = Re\eta_1(\tilde{z}^2 - \tilde{z}^3). \tag{5.6}$$

Based on Eqs. (5.4) and (5.5), one can derive expressions for the flow swirl angle on the surface of a stationary disk φ_w, whereas a cone is rotating

$$\varphi_w = \arctan[0.8Re/(1 - 83Re^2/175)] \quad \text{for } Re = 0 - 1.452, \tag{5.7}$$

$$\varphi_w = \arctan(0.8Re) \quad \text{for } Re \ll 1. \tag{5.8}$$

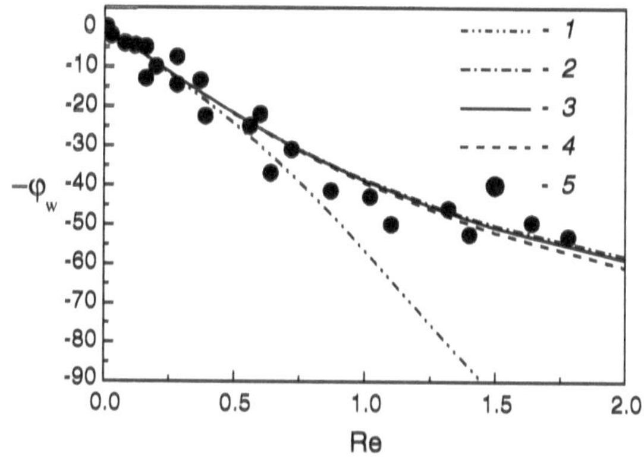

Fig. 5.3 Flow swirl angle on the surface of a stationary disk with a rotating cone [9]. Computations: *1*—Eq. (5.7); *2*—Eq. (5.8) [5]; *3*—self-similar Eqs. (5.19)–(5.22) [10]; *4*—self-similar Eqs. (5.24)–(5.26) [10]. Data *5*—experiments [5]

Equation (5.7) agrees well with measurements [5] and Eq. (5.8) just for $Re \leq 0.5$ (Fig. 5.3). Equation (5.8) that formally holds just for $Re \ll 1$ correlates nevertheless with the measurements up to $Re = 2$. Authors [5] deduced only Eq. (5.8), whereas Eq. (5.7) automatically stemming from Eqs. (5.4) and (5.5) was ignored in [5]. At $Re = 1.452$, Eq. (5.7) predicts the value $\varphi_w = 90°$, whereas the expression in brackets of the function arctan tends effectively to infinity; this contradicts to the physics of the problem.

A series expansion in the small parameter Re, with up to 70 terms, were used also in the works [6, 7] to solve Eqs. (5.1)–(5.3). However, the parameter φ_w predicted by the authors [6, 7] deviated from experiments [5] at $Re = 0.5$–1 more noticeably than Eq. (5.8). In addition, the parameter φ_w predicted in [6, 7] at $Re = 1.2928$ exhibits an asymptotical trend to infinity contradictive to the physics of the problem.

This chapter summarizes results of simulations of convective heat transfer in the geometries "stationary conical diffuser" (Fig. 5.1) and "rotating cone-and-disk" without initial flow swirl (Fig. 5.2). Such pioneering studies based on full self-similar forms of the Navier–Stokes equations together with the thermal boundary layer equation have been for the first time performed by the author of the present work [2, 9–11].

5.2 Self-similar Navier–Stokes and Energy Equations

Considering a steady-state axisymmetric laminar flow and heat transfer, we will solve the Navier–Stokes Eqs. (2.1)–(2.3) and the reduced Eqs. (5.1)–(5.3) together with the energy Eq. (2.12) for laminar flow. For the configurations "rotating cone-and-disk" without initial swirl, the boundary conditions are given by

$$z = 0:\quad v_r = 0,\quad v_z = 0,\quad v_\varphi = \omega r,\quad T_w - T_\infty = c_0 r^{n_*},\tag{5.9}$$

$$z = h:\quad v_r = 0,\quad v_z = 0,\quad v_\varphi = \Omega r,\quad T = T_\infty.\tag{5.10}$$

For the geometry "stationary conical diffuser," the boundary conditions are given by

$$z = 0:\quad v_r = v_\varphi = v_z = 0,\quad T_w - T_\infty = c_0 r^{n_*},\tag{5.11}$$

$$z = z_1 = h/2:\quad v_r = v_{r1},\quad v_\varphi = v_{\varphi 1},\quad \mathrm{d}v_r/\mathrm{d}z = 0,\quad T = T_\infty.\tag{5.12}$$

Here, c_0 and n_* are the constants, while the conditions at $z = z_1 = h/2$ are denoted with a subscript "1." We will study convective heat transfer of a *disk* (but not a cone) under the wall boundary conditions (5.9) and (5.11) that match with Eq. (2.30) for a single disk.

The exponent n_* in Eqs. (5.9) and (5.11) takes negative, zero, or positive values $-2 \le n_* \le 4$. Cone heat/mass transfer is unimportant for the current study; therefore, the temperature T_∞ and the concentration C_∞ on the surface of the cone are assumed to be constant and equal to those of the fluid at infinity. In case of convective diffusion in bioengineering applications, the boundary concentration on the plate/disk C_w is lower than that on the cone/infinity C_∞, because endothelial cells digest feeding culture from the fluid.

The boundary layer equation for the temperature is used instead of the full energy equation; this model assumption is justified above in Chaps. 2–4.

Self-similar variables and functions enable simplifying partial differential Eqs. (2.1)–(2.3), (5.1)–(5.3) and (2.12) and translating them to ordinary non-linear differential equations to be solved numerically with the help of the software like Mathcad [1, 12–17].

Self-similar variables and functions can be derived with the help of group theory [2, 9, 10]. Let us enter a linear transformation of differential equations

$$\begin{gathered} r = A^{\alpha_1}\overline{r},\quad z = A^{\alpha_2}\overline{z},\quad v_r = A^{\alpha_3}\overline{v}_r,\\ v_\varphi = A^{\alpha_4}\overline{v}_\varphi,\quad v_z = A^{\alpha_5}\overline{v}_z,\quad p = A^{\alpha_6}\overline{p}, \end{gathered}\tag{5.13}$$

with α_k ($k = 1, \ldots, 6$) and the parameter of transformation A being constants [12]. Transformations (5.13) are substituted into Eqs. (2.1)–(2.3), (5.1)–(5.3) and (2.12). If the exponents at the constant A are identical for every summand, this means that the non-transformed and transformed equations are invariant, which yields

$$\alpha_1 = \alpha_2 = \alpha,\quad \alpha_3 = \alpha_4 = \alpha_5 = -\alpha,\quad \alpha_6 = -2\alpha,\tag{5.14}$$

$$\begin{aligned} A &= \left(\frac{r}{\overline{r}}\right)^{1/\alpha} = \left(\frac{z}{\overline{z}}\right)^{1/\alpha} = \left(\frac{\overline{v}_r}{v_r}\right)^{1/\alpha}\\ &= \left(\frac{\overline{v}_\varphi}{v_\varphi}\right)^{1/\alpha} = \left(\frac{\overline{v}_z}{v_z}\right)^{1/\alpha} = \left(\frac{\overline{p}}{p}\right)^{1/(2\alpha)}, \end{aligned}\tag{5.15}$$

$$\overline{z}/z = \overline{r}/r, \quad \overline{v}_r\overline{r} = v_r r, \quad \overline{v}_\varphi\overline{r} = v_\varphi r, \quad \overline{v}_z\overline{r} = v_z r, \quad \overline{p}\cdot\overline{r}^2 = p\cdot r^2. \tag{5.16}$$

Morgan's theorem states [12] that Eq. (5.16) serve as similarity variables, if the boundary conditions are independent of the coordinate r.

The self-similar variables and functions were formulated using Eqs. (5.15) and (5.16)

$$\eta = z/r, \quad F = v_r r/v, \quad G = v_\varphi r/v, \quad H = v_z r/v, \tag{5.17}$$

$$P = pr^2/(\rho v^2), \theta = (T - T_\infty)/(T_w - T_\infty). \tag{5.18}$$

Function θ does not change its form because of the linearity of the energy equation. Substituting Eqs. (5.17) and (5.18) into Eqs. (2.1)–(2.3) and (2.12), one can deduce

$$F^2 + G^2 + 2P + F'L + \eta P' + F''M = 0, \tag{5.19}$$

$$G'L + G''M = 0, \tag{5.20}$$

$$P' - H(1 + F) - H'L - H''M = 0, \tag{5.21}$$

$$H' - \eta F'' = 0, \tag{5.22}$$

$$\theta'' = \Pr[n_* F\theta + \theta'(H - \eta F)]. \tag{5.23}$$

Here, $M = 1 + \eta^2$ and $L = 3\eta + \eta F - H$. In ordinary differential Eqs. (5.19)–(5.23), primes denote derivatives with respect to the η-coordinate.

A substitution of Eqs. (5.17) and (5.18) into Eqs. (5.1)–(5.3) gives

$$G^2 + 2P + \eta P' + F'' = 0, \tag{5.24}$$

$$G'(\eta F - H) + G'' = 0, \tag{5.25}$$

$$P' = H''. \tag{5.26}$$

Boundary conditions (5.9) and (5.10) can be rewritten as

$$\eta = 0: \quad F = H = 0, \quad G = G_0, \quad \theta = 1, \tag{5.27}$$

$$\eta = \eta_1: \quad F = H = 0, \quad G = G_1, \quad \theta = 0, \tag{5.28}$$

with $\eta_1 = h/r$, $G_0 = Re_\omega = \omega r^2/v$, $G_1 = Re_\Omega = \Omega r^2/v$.

Boundary conditions (5.11) and (5.12) can be rearranged as

$$\eta = 0: \quad F = G = H = 0, \quad \theta = 1, \tag{5.29}$$

$$\eta = \eta_1: \quad G = G_1, \quad F = F_1, \quad F' = 0, \quad \theta = 0. \tag{5.30}$$

Here, $\eta_1 = 0.5h/r$, and subscripts "0" and "1" denote functions at η = 0 and η = η_1, accordingly.

Boundary conditions (5.27) and (5.28) for the functions $G_0 = Re_\omega$ and $G_1 = Re_\Omega$ are *r*-dependent and do not comply with the self-similarity requirements that the self-similar functions must be constant at the boundaries. Self-similar fuctions are Eq. (5.29) (G_0 = 0) for a stationary disk and Eq. (5.30) with G_1 = const., F_1 = const., which imply the free vortex laws for the velocity components v_r and v_φ in the middle of the stationary conical diffuser

$$(v_\varphi)_{\eta=\eta_1} = G_1 v/r, \quad (v_r)_{\eta=\eta_1} = F_1 v/r. \tag{5.31}$$

We treat here Eqs. (5.27) and (5.28) as locally self-similar, with G_0 and G_1 being parameters at each specific *r*-coordinate [2, 10]. As demonstrated beneath, this model yields the results that are in good agreement with experiments and theoretical predictions.

5.3 Rotating Disk and/or Cone

5.3.1 Numerical Values of Parameters in the Computations

The Mathcad software has been used to obtain a numerical solution of Eqs. (5.19)–(5.26). Angles of conicity for the simulations were γ = 4° (small η_1 = 0.0698) and γ = 45° (relatively large η_1 = 1). The value of η_1 varying over the span of γ = 1–5° was shown to have no influence on the results of simulations.

Values of the Prandtl and Schmidt numbers were *Pr* = *Sc* = 0.1–100 for a configuration with a rotating cone and a stationary disk, *Pr* = *Sc* = 0.1–800 for a stationary cone and a rotating disk, and *Pr* = *Sc* = 0.71 for the rest of the geometries. In the simulations, the value of $Re = Re_\omega \eta_1^2/12$ (or $Re = Re_\Omega \eta_1^2/12$) was set to be unity, which yields Re_Ω = 12, Re_ω = 12 at η_1 = 1, and Re_ω = 2463, Re_Ω = 2463 at η_1 = 0.0698. The exponent n_* in Eq. (5.23) took negative, zero, or positive values $-2 \le n_* \le 4$ that enable modeling different radially decreasing, constant, or increasing distributions of T_w on the disk surface.

5.3.2 Rotating Cone and Stationary Disk

Figures 5.4, 5.5, and 5.6 depict velocity profiles predicted in [9] for *Re* = 1 (Re_ω = 2463) by Eqs. (5.19)–(5.22), (5.24)–(5.26) and those computed in the work [5] with a help of the method of expansion in the small parameter *Re*.

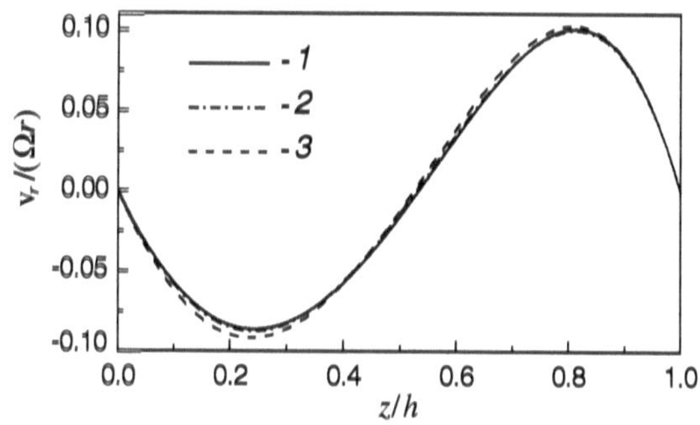

Fig. 5.4 Profiles of the radial velocity component in a gap between a rotating cone and a stationary disk [9]. *1*—Eqs. (5.19)–(5.22); *2*—Eqs. (5.24)–(5.26); *3*—Eq. (5.4) [5]

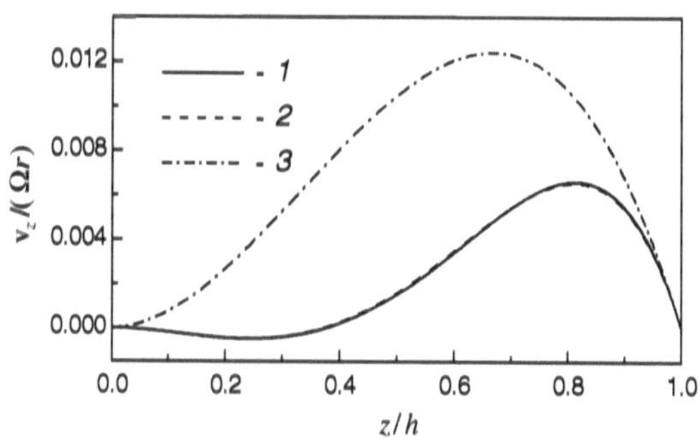

Fig. 5.5 Profiles of the axial velocity component in a gap between a rotating cone and a stationary disk [9]. *1*—Eqs. (5.19)–(5.22); *2*—Eqs. (5.24)–(5.26); *3*—Eq. (5.6) [5]

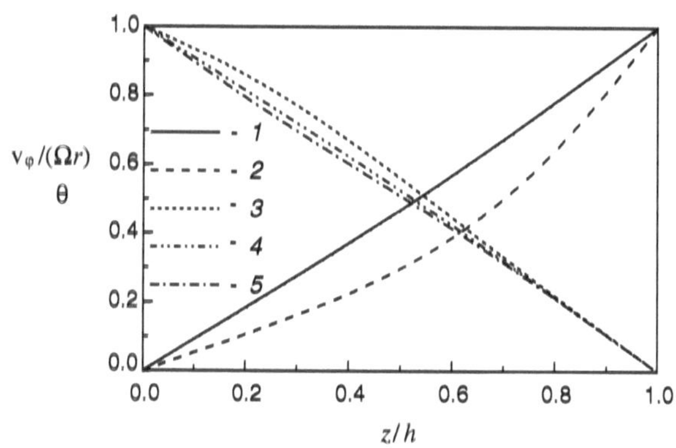

Fig. 5.6 Profiles of the tangential velocity component (*1, 2*) and temperature θ (*3–5*) in a gap between a rotating cone and a stationary disk [9]. *1*—Eqs. (5.19)–(5.22) and (5.24)–(5.26); *2*—Eq. (5.5) [5]; *3*—Eqs. (5.19)–(5.23) at $n_* = 2$; *4*—(5.19)–(5.23) at $n_* = 0$; *5*—(5.19)–(5.23) at $n_* = -1$

The axial velocity component (Fig. 5.5) is an order of magnitude smaller than the radial velocity component (Fig. 5.4), which, in turn, is an order of magnitude smaller than the tangential velocity component (Fig. 5.6). Curves predicted by Eqs. (5.19)–(5.22) and (5.24)–(5.26) practically merge, which certifies validity of the simplified Eqs. (5.24)–(5.26) for small angles of conicity γ. Data of [5] for the radial velocity v_r agree well with the simulations in Fig. 5.4; however, discrepancies between the data of [5] and the simulations for components v_z and v_φ are more distinct (Figs. 5.5 and 5.6).

To validate the accuracy of the simulations of the tangential velocity, experimental data [5] and predictions [9] for the flow swirl angle on the disk surface $\varphi_w = \arctan[v_r/(\Omega r - v_\varphi)]_{z=0} = \arctan(-F'_w/G'_w)$ were compared. Predictions and experiments correlate for the Reynolds number depicted in Fig. 5.3. It can be also concluded that the velocity profiles in Figs. 5.4, 5.5, and 5.6 predicted by Eqs. (5.19)–(5.22) and (5.24)–(5.28) model the flow in the gap more realistically than those by Eqs. (5.4)–(5.8) [5].

Figure 5.6 shows temperature profiles in the gap computed at for $Pr = 0.71$. The temperature curves demonstrate the decreasing trend from unity at the disk surface to zero at the cone wall. The form of the curves is affected by the value of n_*. Near the disk, derivatives of the θ profiles diminish with increasing n_*.

To compute the local Nusselt number, the following expression was used:

$$Nu = \frac{q_w r}{\lambda(T_w - T_\infty)} = -\theta'_{\eta=0}. \quad (5.32)$$

To enable comparisons with Eqs. (3.4) and (3.5) for the rotating disks, the Nusselt number may be rearranged using a derivative with respect to the variable $\zeta = z\sqrt{\Omega/v}$

$$Nu = K_1 Re_\Omega^{1/2} \quad (5.33)$$

$$K_1 = \frac{-(\mathrm{d}\theta/\mathrm{d}\eta)_{\eta=0}}{Re_\Omega^{1/2}} = -\left(\frac{\mathrm{d}\theta}{\mathrm{d}\zeta}\right)_{\zeta=0}. \quad (5.34)$$

Based on these expressions, it was calculated at $\eta_1 = 0.0698$ (or $Re_\Omega = 2463$) that $Nu = 15.28, 13.40, 9.35$ and $K_1 = 0.308, 0.270, 0.188$ at $n_* = -1, 0, 2$, accordingly. These values of the coefficient K_1 match fairly well with those for a single rotating disk (see Table 3.8). For larger values of n_*, the coefficient K_1 diminishes, which is observed in centripetal flow over a stationary disk imposed by a rotating cone (to compare, an increase in the coefficient K_1 together with n_* occurs in centrifugal flow over a rotating disk, see Chaps. 3 and 4).

Given $\eta_1 = 1$ and $Re = 1$ (or $Re_\Omega = 12$), the Nusselt numbers are $Nu = 1.047$, 0.954, 0.760 with $K_1 = 0.302, 0.275, 0.219$ for the same exponents n_*. One can conclude that the coefficient K_1 is conservative and only weakly dependent on the conicity angle γ.

5.3.3 Rotating Disk and Stationary Cone

Radial flow pattern here is opposite to that considered above: the flow is centripetal over the cone, and centrifugal over the disk (Fig. 5.7).

The tangential velocity v_φ demonstrates a trend linearly subsiding from a disk toward a cone, while the profile of axial velocity component v_z looks mirror-symmetrical as compared to the v_z profile in Fig. 5.5. The temperature profile θ in Fig. 5.7 for $Pr = 0.71$ almost merges with the $v_\varphi/(\omega r)$ curve.

To compute the Nusselt number at the disk, Eqs. (5.32)–(5.34) are again employed; as $\Omega = 0$, it must be replaced with ω while defining the Re number and coordinate ζ. Based on this, the calculated Nusselt numbers for $\eta_1 = 0.0698$, $Re = 1$ ($Re_\omega = 2463$) are $Nu = 13.33, 15.35, 19.13$ and $K_1 = 0.269, 0.309, 0.386$ at $Pr = 0.71$, and $n_* = -1, 0, 2$, accordingly. It is evident that the coefficient K_1 is an increasing function of n_*. However, the rate of increase is lower than that for a single rotating disk, where $K_1 = 0.189, 0.326, 0.519$ for the identical values n_* (Table 3.1). Given $\eta_1 = 1$ and $Re = 1$ ($Re_\omega = 12$), the computed Nusselt numbers are $Nu = 0.96, 1.041, 1.197$ and $K_1 = 0.277, 0.301, 0.345$ for identical exponents n_* and $Pr = 0.71$. Thus, the coefficient K_1 is again very weakly dependent on the conicity angle γ.

5.3.4 Effects of Prandtl and Schmidt Numbers

Effects of the Prandtl or Schmidt numbers are considered for the geometries with a rotating disk and stationary cone or a stationary disk and a rotating cone [11].

Rotating disk and stationary cone. As the Pr numbers increase, curves of the temperature profiles θ for $n_* = 0$ and $n_* = -1$ shift downward exhibiting a non-linear

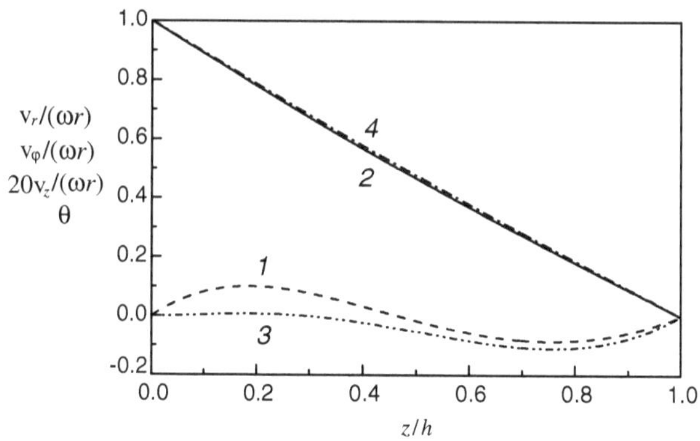

Fig. 5.7 Velocity and temperature profiles in the gap between a rotating disk and a stationary cone at $Re = 1$ ($Re_\omega = 2463$) and $\eta_1 = 0.0698$ [9]. *1*—$v_r/(\omega r)$; *2*—$v_\varphi/(\omega r)$; *3*—$20v_z/(\omega r)$; *4*—$\theta(Pr = 0.71, n_* = 0)$

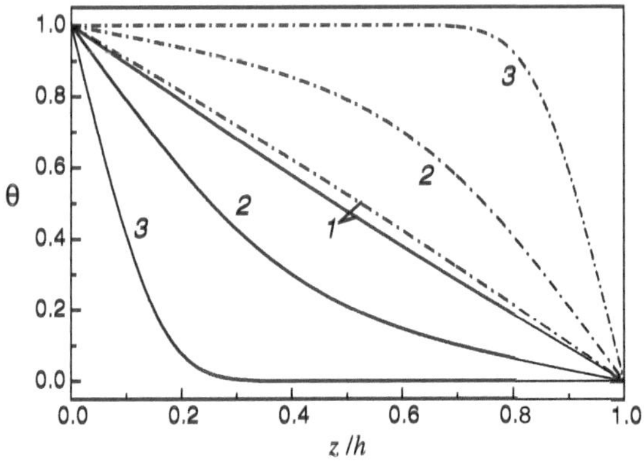

Fig. 5.8 Temperature profiles θ in the gap for $n_* = 0$ [11]. *Solid lines* rotating disk and stationary cone. *Dash-dot lines* stationary disk and rotating cone. *1*—$Pr = 0.71$; *2*—$Pr = 10$; *3*—$Pr = 100$

trend of variation due to diminished heat conduction, whereas the function θ at $Pr \geq 100$ becomes zero inside the gap between the cone and the disk (Figs. 5.8 and 5.9). For $n_* \geq 0$, curves of θ demonstrate qualitatively analogous trend. With respect to the profiles of θ for $n_* = -1$, the condition $\mathrm{d}\theta/\mathrm{d}\eta \to 0$ in the vicinity of the cone is attained already at $Pr \geq 20$ (Fig. 5.9).

Over the range $0 \leq n_* \leq 4$, the constant K_1 increases with the Prandtl number (Table 5.1); the trend persists also at $n_* = -0.5$, i.e., when negative gradient $\mathrm{d}T_w/\mathrm{d}r$ is weak.

As seen from Table 5.1, signs of v_r and $\mathrm{d}T_w/\mathrm{d}r$ become different and coefficient K_1 diminishes for larger Prandtl numbers, when the wall temperature gradient $\mathrm{d}T_w/\mathrm{d}r$ is strongly negative ($n_* = -1$). Let us write the coefficient K_1 at $n_* = 0$ as follows:

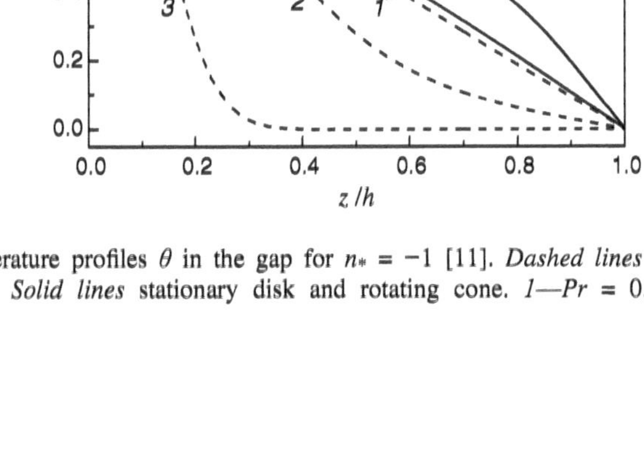

Fig. 5.9 Temperature profiles θ in the gap for $n_* = -1$ [11]. *Dashed lines* rotating disk and stationary cone. *Solid lines* stationary disk and rotating cone. *1*—$Pr = 0.71$; 2—$Pr = 10$; *3*—$Pr = 100$

Table 5.1 Coefficient K_1, for rotating disk and stationary cone [11]

$Pr(Sc)$	$n_* = -1$	$n_* = -0.5$	$n_* = 0$	$n_* = 1$	$n_* = 2$	$n_* = 4$
0.1	0.2858	0.2887	0.2915	0.2972	0.3029	0.3141
0.5	0.2745	0.2890	0.3032	0.3309	0.3579	0.4094
0.71	0.2687	0.2892	0.3094	0.3483	0.3855	0.4556
1.0	0.2607	0.2898	0.3180	0.3718	0.4225	0.5155
5.0	0.1650	0.3155	0.4447	0.6552	0.8172	0.9048
10.0	0.0842	0.3829	0.6082	0.9265	0.9860	1.5737
50	0.00008	0.8744	1.3572	1.9492	2.3432	2.8986
100	0.0	1.1745	1.7882	2.5295	3.0225	3.7184
400	0.0	2.0005	2.9876	4.1627	4.9419	6.0423
800	0.0	2.5742	3.8435	5.3051	6.2862	7.6717

$$K_1 = K_{1,Pr=1} Pr^{m_p}, \tag{5.35}$$

where $K_{1,Pr=1} = 0.318$. This enables determining a function for the exponent $m_p(Pr)$ presented in Table 5.2, whose asymptotic limit is $m_p = 0.372$ for high Prandtl numbers. To compare, this limit for a single rotating disk is $m_p = 1/3$ at $Pr \to \infty$ [9] (see Chap. 6).

Rotating cone and stationary disk. Profiles of θ for $n_* = 0$ and $n_* = -1$ at $Pr = 0.71$ span practically linear between unity on the disk surface and zero on the cone wall; further, for larger Prandtl numbers, profiles of θ shift upward demonstrating a non-linear trend of variation owing to reduced heat conduction (Figs. 5.8 and 5.9). For $Pr \geq 100$, the derivative $d\theta/d\eta$ in the vicinity of the disk exhibits zero values.

For $n_* > 0$, $Pr \leq 1$ and $n_* = -1$, $Pr \leq 1$, curves of θ practically merge with the profile predicted for $Pr = 0.71$, $n_* = 0$. Curves of θ demonstrate a *S*-shape at $n_* = -1$ and $Pr = 1$–10 (Fig. 5.9). Profiles of θ become non-physical at $Pr > 1$, $n_* > 0$ and $Pr > 10$, $n_* = -1$.

For larger Pr numbers and $n_* < 0$, the constant K_1 increases (because signs of v_r and dT_w/dr are the same); at $n_* \geq 0$, the coefficient K_1 diminishes due to the opposite signs of v_r and dT_w/dr (Table 5.3).

Application to the cone-and-plate devices. The results described in Sect. 5.3.4 become applicable to mass transfer upon replacement of *T*, *Pr*, *Nu* with *C*, *Sc*, *Sh*,

Table 5.2 Exponent m_p in Eq. (5.35) at T_w = const., rotating disk, and stationary cone [11]

$Pr(Sc)$	0.71	0.9	1.1	2	2.28	2.4	2.5	10
m_p	0.0805	0.0901	0.0978	0.1319	0.1409	0.1446	0.1476	0.2817
$Pr(Sc)$	100	200	400	600	800			
m_p	0.375	0.3752	0.3739	0.3729	0.3724			

Table 5.3 Coefficient K_1, rotating cone, and stationary disk [11]

$Pr(Sc)$	$n_* = -1$	$n_* = -0.5$	$n_* = 0$	$n_* = 1$	$n_* = 2$	$n_* = 4$
0.1	0.2913	0.2887	0.2860	0.2807	0.2752	0.2643
0.5	0.3021	0.2889	0.2755	0.2479	0.2193	0.1587
0.71	0.3078	0.2892	0.2700	0.2302	0.1883	0.0971
0.9	0.3130	0.2894	0.2651	0.2140	0.1593	0.0373
1.0	0.3158	0.2896	0.2625	0.2054	0.1436	0.0040
2.0	0.3436	0.2924	0.2375			
10.0	0.5712	0.3679	0.0895			
50			0.0001			
100			0.0			

respectively. Here, data for K_1 at $n_* = 0$ from Tables 5.1, 5.2, and 5.3 are to be used, since the wall boundary condition for mass transfer is C_w = const. [9]. It is evident that the coefficients K_1 for a stationary disk and a rotating cone are always smaller than the K_1 values for a rotating disk and a stationary cone. This difference becomes more pronounced at larger Schmidt numbers and is equal to 14.6 % at $Sc = 0.71$; 2.6 times at $Sc = 5$; 46.1 times at $Sc = 20$, and asymptotically tends to infinity in the limit at infinite Schmidt numbers.

Thus, one can enhance efficiency of a cone-and-plate device used in bioengineering for nurturing endothelial cells spread on the plate via assigning the disk to rotate and fixing the cone instead of the currently used devices "rotating cone—stationary plate."

5.3.5 *Co-rotating Disk and Cone*

Here, the ratio between the Re_Ω and Re_ω numbers makes a crucial influence on the flow pattern. If $Re_\Omega > Re_\omega$ (cone revolves faster), fluid flow over the cone is centrifugal, and centripetal over the disk. If $Re_\Omega < Re_\omega$ (disk revolves faster), a reverse flow pattern emerges. Equations (5.32)–(5.34) (reference angular speed Ω) were used to compute the Nusselt number for $Pr = 0.71$. A situation with approximately the same angular speeds of a disk and a cone was considered. Given $Re_\omega = 1.01 Re_\Omega$, Eqs. (5.32)–(5.34) yield $Nu = 14.31$, 14.35, 14.43 and $K_1 = 0.288$, 0.289, 0.291; given $Re_\omega = 0.99 Re_\omega$, one can obtain $Nu = 14.35$, 14.31, 14.23 and $K_1 = 0.289$, 0.288, 0.287 at $n_* = -1$, 0, 2, accordingly. In both cases, we pre-set $Re = 1$ and $\eta_1 = 0.0698$ ($Re_\Omega = 2463$). The computed Nusselt numbers were $Nu = 0.999$, 1.001, 1.004 and $K_1 = 0.288$, 0.289, 0.290 at $Re_\omega = 1.01 Re_\Omega$, $Re = 1$ and $\eta_1 = 1$ ($Re_\Omega = 12$).

The coefficient K_1 is practically the same for all considered cases. For larger values of n_*, Nusselt numbers increase in centrifugal flow over the disk and decrease in centripetal flow. Variation of the conical spacing practically does not affect the coefficient K_1.

5.3.6 Counter-Rotating Disk and Cone

The most complex flow pattern emerges here with centrifugal flow over a disk and a cone and centripetal flow in the center of the conical cavity (Fig. 5.10). The axial velocity v_z is negative in the vicinity of the walls and positive in the center of the gap; the tangential velocity $v_\varphi/(\Omega r)$ behaves as a linear function increasing between -1 and 1, whereas the temperature function θ at $Pr = 0.71$ monotonically diminishes from unity to zero.

Given $\eta_1 = 0.0698$, $Re = 1$, and $Re_\omega = -Re_\Omega = 2463$, profile of v_r is symmetrical relative to the center of the gap (curve 2, Fig. 5.10).

Equations (5.32)–(5.34) (reference velocity Ω) were used to compute the Nusselt number for $Pr = 0.71$. Nusselt number increases with n_* over the disk surface: $Nu = 14.21$, 14.44, 14.85 and $K_1 = 0.286$, 0.201, 0.299 at $n_* = -1$, 0, 2. The conditions $\eta_1 = 1$, $Re = 1$, $Re_\omega = 12$, and $Re_\Omega = -12$ yield a non-symmetrical radial velocity profile v_r, since the radial flow is stronger near the cone (curve 4, Fig. 5.8). As a result, increasing n_* is accompanied with a decreasing Nusselt number on the disk: $Nu = 1.011$, 0.989, 0.942 and $K_1 = 0.292$, 0.285, 0.272, given the same set of the n_* values as that used above.

Thus, here heat transfer is almost insensitive to the value and sign of dT_w/dr on the disk surface. Variation of the conical gap spacing and revolution speeds of a cone and a disk influence on the v_r profile and qualitative trend of the dependence of Nu on n_*.

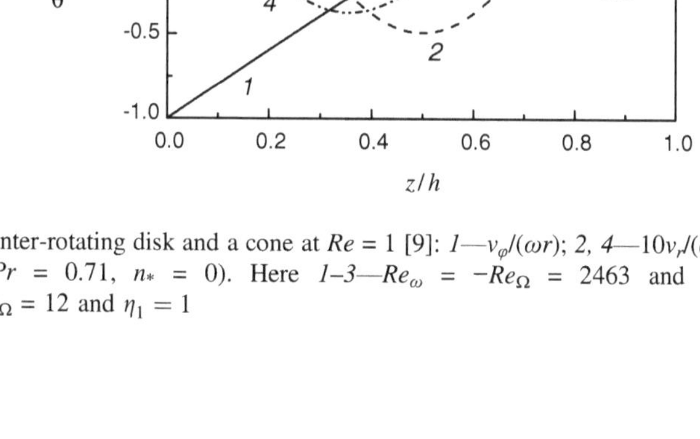

Fig. 5.10 Counter-rotating disk and a cone at $Re = 1$ [9]: *1*—$v_\varphi/(\omega r)$; *2, 4*—$10v_r/(\omega r)$; *3*—$100v_z/(\omega r)$; *5*—θ ($Pr = 0.71$, $n_* = 0$). Here *1–3*—$Re_\omega = -Re_\Omega = 2463$ and $\eta_1 = 0.0698$; *4*—$Re_\omega = -Re_\Omega = 12$ and $\eta_1 = 1$

5.4 Radially Outward Swirling Flow in a Stationary Conical Diffuser

A non-rotating diffuser with conicity of $\gamma = 35°$ or $\eta_1 = 0.35$ was studied here (Fig. 5.2). The physical interpretation of Eqs. (5.30) and (5.31) is that they describe a free vortex expanding along the centerline of the gap. In practice, potential flow in the form of a free vortex spans over a significant height of the conical gap pushing the boundary layers toward the walls. Hence, Eqs. (5.30) and (5.31) describe a somewhat idealized vortex flow pattern.

Simulations demonstrated that non-swirling purely radial flow ($G_1 = 0$) does not undergo separation from the walls at $F_1 < 63$. Separation starts at $F_1 \approx 63$, whereas at $F_1 > 63$, a pronounced recirculation flow region is visible over the disk (Fig. 5.11).

Responsible for the onset of separation is the large conicity of the diffuser: a reduced conicity $\eta_1 = 0.035$ shifts the separation value of F_1 to about 7500. Flow swirl ($G_1 = 97.96$, $Re = G_1\eta_1^2/12 = 1$) causes accentuated recirculation region over the disk (Fig. 5.11).

Given a zero radial velocity $F_1 = 0$ at the inlet to the diffuser, the radial velocity v_r becomes negative over the entire gap height excluding the point $\eta = \eta_1$ (curve 6 in Fig. 5.11; $|F|_{max}$ relates to the minimum point of the plot of $F/|F|_{max}$ at $\eta/\eta_1 \approx 0.4$, i.e., $F_{max} = -24.28$). For larger values of F_1, the recirculation area reduces, whereas the centrifugal flow area near the center of the conical gap grows up. The tangential velocity G/G_1 shows a linear distribution between 0 at $\eta = 0$ (disk) and 1 (cone) for $\eta = \eta_1$ (curve 1 in Fig. 5.12).

The diffuser is used to restore the static pressure, which grows with r as the velocity components $(v_r)_{\eta=\eta_1}$ and $(v_\varphi)_{\eta=\eta_1}$ decrease. To ensure self-similarity of the function P in Eq. (5.18), the quantity P must denote the excess pressure p–p_∞,

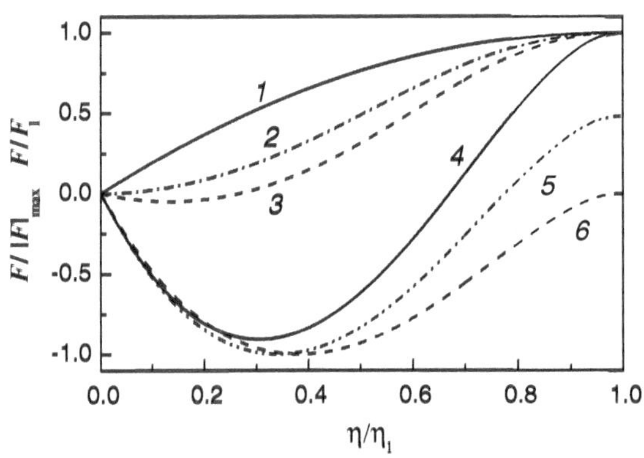

Fig. 5.11 Profiles of the radial velocity F/F_1 (*1–4*) or $F/|F|_{max}$ (*5, 6*) in a gap between a disk and a cone [9]. $\boldsymbol{G_1 = 0}$: *1*—$F_1 = 2$; *2*—$F_1 = 63$; *3*—$F_1 = 90$. $\boldsymbol{G_1 = 97.96}$: *4*—$F_1 = 20$; *5*—$F_1 = 10$, $|F|_{max} = 20.66$; *6*—$F_1 = 0$, $|F|_{max} = 24.28$

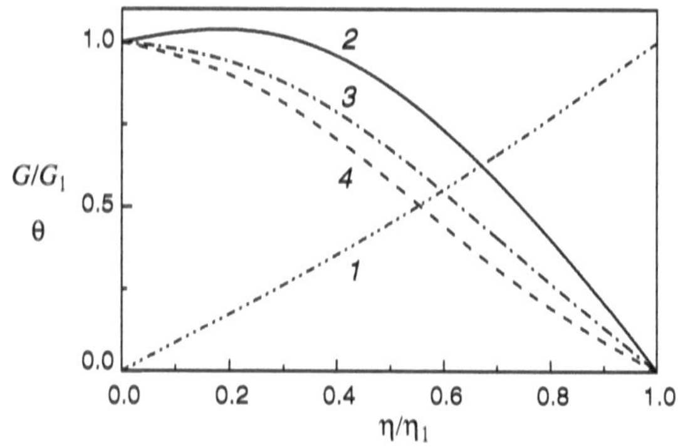

Fig. 5.12 Profiles of the tangential velocity component G/G_1 and temperature θ in the gap between a cone and a disk [9]. *1*—G/G_1 for $F_1 = 30$, $G_1 = 97.96$; 2—θ for $G_1 = 97.96$ and $F_1 = 10$; *3*—θ for $G_1 = 97.96$ and $F_1 = 30$; *4*—θ for $G_1 = 97.96$ and $F_1 = 60$ ($Pr = 0.71$, $n_* = 2$)

where $p = p_\infty = \text{const.}$ for $r \to r_\infty$. Thus, the parameter P shows the pressure recovery level; in non-swirling flow ($G_1 = 0$), P increases with F_1 (Fig. 5.13).

Flow swirl $G_1 = 97.96$ entails noticeable additional augmentation of the pressure recovery parameter P, whereas the contribution of F_1 in the range $F_1 = 0$–20 is rather insignificant. As can be seen from Fig. 5.14, curves $Nu(F_1)$ computed by Eq. (5.32) for $F_1 = 50$–63 demonstrate maxima at $n_* = 2$ and 0 and minima at $n_* = -1$.

For non-swirling flow ($G_1 = 0$) and non-zero inlet radial velocity F_1, the Nusselt numbers increase together with the exponent n_* (curves 1–3, Fig. 5.14).

If the exponent n_* remains within the range $n_* = 0$–2 and the radial velocity F_1 increases, the Nusselt numbers (a) demonstrate a trend of augmentation under the conditions of non-separating centrifugal flow, (b) stay practically constant, if the function F_1 approaches the onset of separation, and (c) show a reduction for centripetal secondary flow over the disk (curves 1 and 2 in Fig. 5.14). These trends become rather insignificant at $n_* = 0$.

Fig. 5.13 Static pressure drop in the gap for $G_1 = 0$ (curves *1*–*4*) and $G_1 = 97.96$ (curves *5*, *6*) [9]. *1*, *6*—$F_1 = 20$; *2*—$F_1 = 45$; *3*—$F_1 = 63$; *4*—$F_1 = 90$; *5*—$F_1 = 10$

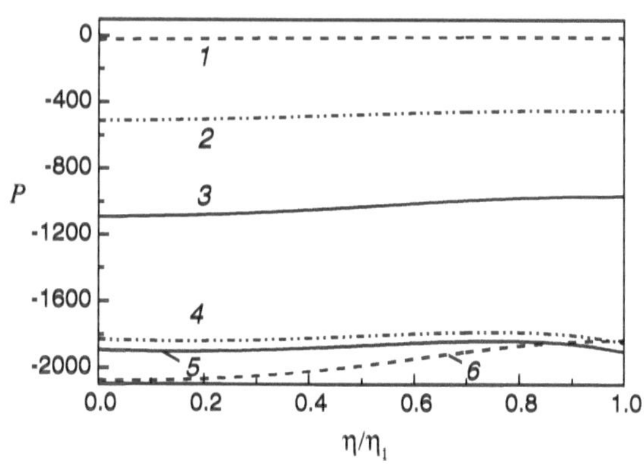

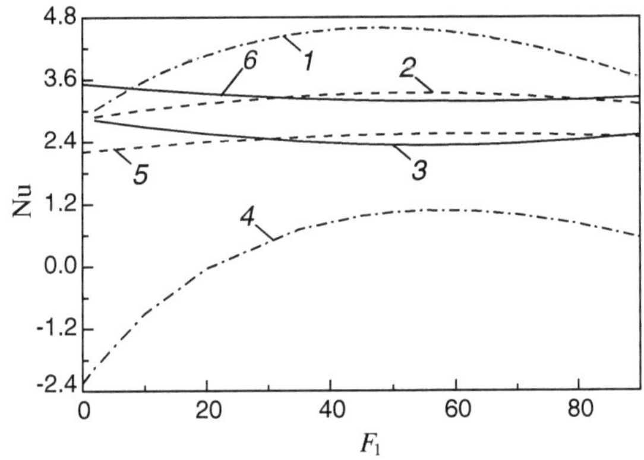

Fig. 5.14 Nusselt numbers in the gap for $G_1 = 0$ (curves *1–3*) and $G_1 = 97.96$ (curves *4–6*) [9]. *1, 4*—$n_* = 2$; *2, 5*—$n_* = 0$; *3, 6*—$n_* = -1$

Flow with initial swirl $G_1 = 97.96$ demonstrates different signs of v_r and dT_w/dr accompanied with reduced Nusselt numbers at $n_* = 2$ and 0 (curves 4 and 5 in Fig. 5.14) as compared to the flow without swirl. Given $n_* = -1$, signs of v_r and dT_w/dr become the same, accompanied with increased Nusselt numbers (curve 6 in Fig. 5.14) as compared to non-swirling fluid. Although, for $G_1 = 97.96$ and increasing F_1, the radially inward flow persists in the vicinity of the disk, the shapes of the curves 4, 5, and 6 for $Nu(F_1)$ are analogous to curves 1, 2, and 3 plotted for non-swirling flow for the same values of n_* (see Fig. 5.14).

Temperature profiles 2 and 3 in Fig. 5.12 for swirling flow ($G_1 = 97.96$) and $n_* = 2$ show a decreasing behavior at $F_1 > 21$. For $F_1 \le 21$, the temperature curve 4 in Fig. 5.11 exhibits a maximum near the wall, if fluid flows centripetally in the direction of the decrease in T_w. This causes the Nusselt number curve 4 in Fig. 5.14 to become negative: the disk is heated by the fluid (whereas positive *Nu* numbers mean fluid heated by a disk).

To conclude, in this chapter, self-similar solutions of the Navier–Stokes and energy equations were derived for fluid flow in a conical gap depicted in Figs. 5.1 and 5.2. Simulations were performed for the cases "rotating cone–stationary disk," "rotating disk–stationary cone," "co-rotating or contra-rotating disk and cone," and "non-rotating conical diffuser." Effects of the boundary conditions and various Prandtl/Schmidt numbers on the pressure, velocity, and temperature pattern, as well as on the Nusselt/Sherwood numbers, were studied.

References

1. Schlichting G (1968) Boundary-layer theory. McGraw-Hill Book Company, New York
2. Shevchuk IV (2004) Laminar heat transfer of a swirled flow in a conical diffuser. Self-similar solution. Fluid Dyn 39(1):42–46
3. Fewell ME, Hellums JD (1977) The secondary flow of Newtonian fluids in cone and plate viscometers with small gap angles. Trans Soc Rheol 21(4):535–565
4. Mooney M, Ewart RH (1934) The conicylindrical viscometer. Physics 5:350–354

5. Sdougos HP, Bussolari SR, Dewey CF (1984) Secondary flow and turbulence in a cone-and-plate device. J Fluid Mech 138:379–404
6. Buschmann MH (2002) A solution for the flow between a cone and a plate at low Reynolds number. J Thermal Sci 11(4):289–295
7. Buschmann MH, Dieterich P, Adams NA, Schnittler H-J (2005) Analysis of flow in a cone-and-plate apparatus with respect to spatial and temporal effects on endothelial cells. Biotechnol Bioeng 89(5):493–502
8. Sucosky P, Padala M, Elhammali A, Balachandran K, Jo H, Yoganathan AP (2008) Design of an ex vivo culture system to investigate the effects of shear stress on cardiovascular tissue. Trans ASME J Biomech Eng 130(3): Paper 035001
9. Shevchuk IV (2009) Convective heat and mass transfer in rotating disk systems. Springer, Berlin
10. Shevchuk IV (2004) A self-similar solution of Navier–Stokes and energy equations for rotating flows between a cone and a disk. High Temp 42(1):95–100
11. Shevchuk IV (2011) Laminar heat and mass transfer in rotating cone-and-plate devices. Trans ASME J Heat Transf 133(2): Paper 024502
12. Cebeci T, Bradshaw P (1984) Physical and computational aspects of convective heat transfer. Springer, Berlin
13. Dorfman LA (1963) Hydrodynamic resistance and the heat loss of rotating solids. Oliver and Boyd, Edinburgh
14. Owen JM, Rogers RH (1989) Flow and heat transfer in rotating-disc systems. In: Rotor-stator systems, vol 1. Research Studies Press Ltd., Taunton
15. Shevchuk IV (1998) Simulation of heat transfer in a rotating disk: the effect of approximation of the tangent of the angle of flow swirling. High Temp 36(3):522–524
16. Shevchuk IV (2001) Effect of the wall temperature on laminar heat transfer in a rotating disk: an approximate analytical solution. High Temp 39(4):637–640
17. Shevchuk IV (2002) Laminar heat transfer in a rotating disk under conditions of forced air impingement cooling: approximate analytical solution. High Temp 40(5):684–692

Chapter 6
Heat and Mass Transfer of a Rotating Disk for Large Prandtl and Schmidt Numbers

6.1 Laminar Flow

Convective heat and mass transfer over a single disk rotating in fluid with high Prandtl or Schmidt numbers can be found in many practical and research applications. For instance, in electrochemistry, where the Schmidt numbers are several orders of magnitude larger than unity, rotating disk electrode is involved in measurements of the convective diffusion coefficient [1–14]. Another example is naphthalene sublimation technique often used to measure mass transfer coefficients α_m [15–29].

The differential Eq. (1.28) of convective diffusion, including the time-averaged fluctuating components, is analogous to the energy Eq. (2.5), provided that the temperature T and the thermal diffusivity a are replaced by the concentration C and the diffusion coefficient D_m, respectively. The Navier–Stokes and continuity equations hold, if constant fluid properties are assumed.

If the Schmidt number Sc replaces the Prandtl number, and the nondimensional function θ is written as

$$\theta = (C - C_\infty)/(C_w - C_\infty), \tag{6.1}$$

then the self-similar Eqs. (2.32)–(2.36) for steady-state axisymmetric laminar flow become valid for convective mass transfer.

Surface concentration on the disk does not vary; thus $C_w = \text{const}$. Therefore, rewritten convective diffusion Eq. (2.36) reduces to

$$\theta'' - ScH\theta' = 0. \tag{6.2}$$

The following equations (analogous to Eq. (3.4)) can be used for estimation of the Sherwood number

I.V. Shevchuk, *Modelling of Convective Heat and Mass Transfer in Rotating Flows*, Mathematical Engineering, DOI 10.1007/978-3-319-20961-6_6

$$Sh = K_1 Re_{\omega}^{n_R}, \quad Sh_{av} = K_2 Re_{\varphi}^{n_R}. \tag{6.3}$$

The constants K_1 and K_2 in Eq. (6.3) are affected by the boundary conditions, flow type (laminar, transitional, or turbulent) and the Schmidt numbers. The exponent n_R is affected by the flow type, whereas $K_1 = K_2$ and $n_R = 1/2$ in a laminar flow regime.

Thus, the aforementioned analogy between convective heat and mass transfer, enables the use of theoretical solutions or empirical experimental equations simply via replacing C, Sc and Sh with of T, Pr and Nu (or vice versa), accordingly.

For laminar flow, Eqs. (2.32)–(2.36) for $Pr > 1$ and $Sc > 1$ at $N = 0$ and $\beta = 0$ were solved numerically using Mathcad [30]. Table 6.1 shows that the calculated coefficient K_1 is increasing with growing Pr or Sc numbers.

For the same Prandtl number, the constant K_1 is an increasing function of the exponent n_*: the value of K_1 at $Pr = 0.71$, 2.0 and 10^6 becomes 3.3, 2.73 and 2.2 times larger, respectively, if the constant n_* changes from −1 to 3. Thus, at increased Prandtl numbers, the influence of the exponent n_* on the constant K_1 gets less pronounced.

The approximate Eq. (3.6) for the coefficient K_1 for the boundary condition (2.30), $Pr \geq 1$ and nonzero values n_* was derived by Dorfman [31]. Values of K_1 by Eq. (3.6) surpass the exact solution. Equation (3.6) deviates from the exact solution at $n_* \leq 0$ by 16–40 % even for $Pr = 1$. For $n_* = 0$ and $Pr = 1$–3, this deviation reaches 10–11 %. For larger exponents n_* and $Pr = 1$–3, the deviation of

Table 6.1 Values of the constant K_1, exact solution of Eqs. (2.32)–(2.36) for $Pr > 1$ [30]

Pr (Sc)	$n_* = -2$	$n_* = -1.5$	$n_* = -1$	$n_* = -0.5$	$n_* = 0$	$n_* = 1$	$n_* = 2$	$n_* = 3$	$n_* = 4$
1.0	0.0	0.1305	0.2352	0.3221	0.3963	0.5180	0.6159	0.6982	0.7693
1.5	0.0	0.1682	0.2979	0.4028	0.4906	0.6324	0.7450	0.8389	0.9199
2.0	0.0	0.1989	0.3482	0.4669	0.5653	0.7226	0.8466	0.9498	1.0386
2.28	0.0	0.2140	0.3728	0.4982	0.6016	0.7663	0.8960	1.0036	1.0963
2.5	0.0	0.2251	0.3907	0.5209	0.6280	0.7982	0.9319	1.0428	1.1383
3.0	0.0	0.2480	0.4279	0.5680	0.6826	0.8640	1.0061	1.1238	1.2251
5.0	0.0	0.3206	0.5445	0.7153	0.8533	1.0697	1.2382	1.3774	1.4971
10.0	0.0	0.4410	0.7368	0.9577	1.1341	1.4083	1.6206	1.7957	1.9460
15.0	0.0	0.5254	0.8710	1.1268	1.3300	1.6446	1.8877	2.0880	2.2599
20.0	0.0	0.5924	0.9776	1.2610	1.4855	1.8323	2.0999	2.3203	2.5095
50	0.0	0.8536	1.3925	1.7835	2.0909	2.5635	2.9269	3.2260	3.4825
100	0.0	1.1108	1.8009	2.2979	2.6871	3.2840	3.7422	4.1190	4.4421
500	0.0	1.9943	3.2033	4.0644	4.7351	5.7596	6.5442	7.1888	7.7413
1000	0.0	2.5467	4.0802	5.1691	6.0162	7.3083	8.2972	9.1096	9.8057
10^4	0.0	5.6363	8.9846	11.348	13.181	15.971	18.104	19.855	21.356
10^5	0.0	12.291	19.548	24.657	28.613	34.632	39.230	43.003	46.236
10^6	0.0	26.626	42.304	53.328	61.860	74.834	84.742	92.873	99.838

Eq. (3.6) from the exact solution is smaller (1–6 %). However, at $Pr \to \infty$, the inaccuracy of Eq. (3.6) abruptly increases [30].

The functional dependence of the constant K_1 on the Schmidt (or Prandtl) number according to Dorfman's Eq. (3.6) and the exact solution for $n_* = 0$ ($T_w =$ const. or $C_w =$ const.) is depicted in Fig. 6.1. The inaccuracies of Eq. (3.6) make it unusable already for $Sc = 1$–3.

Equations (3.7) and (3.8) can be rewritten for mass transfer for $Sc = 0$–∞, respectively

$$K_1 = 0.6109Sc/(0.5301 + 0.3996Sc^{1/2} + Sc)^{2/3}, \quad (6.4)$$

$$K_1 = 0.6Sc/(0.56 + 0.26Sc^{1/2} + Sc)^{2/3}. \quad (6.5)$$

Equations (6.4) and (6.5) result nearly in the same values. Maximal deviation of them from the exact solution is 4 and 5 %, respectively, for $Sc = 5$–20 (Fig. 6.1). For higher Schmidt numbers, inaccuracies of Eqs. (6.4) and (6.5) tend to zero (Table 6.2).

Another expression was derived in the work [13]

$$K_1 = 0.621Sc/(1 + 0.298Sc^{-1/3} + 0.14514Sc^{-2/3}). \quad (6.6)$$

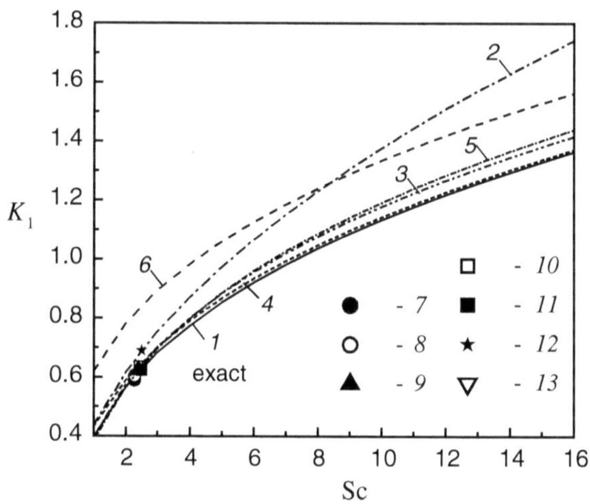

Fig. 6.1 Constant K_1 in Eq. (6.3), laminar flow at $C_w =$ const. [30]. *1*—Exact solution; *2*—Eq. (3.6) for $n_* = 0$; *3*—Eqs. (6.4) and (6.5); *4*—Eq. (6.6); *5*—Eq. (6.7); *6*—Eq. (6.8). Experiments: *7*—$K_1 = 0.59$, $Sc = 2.28$ [15]; *8*—$K_1 = 0.604$, $Sc = 2.28$ [17]; *9*—$K_1 = 0.625$, $Sc = 2.4$ [20, 21, 26]; *10*—$K_1 = 0.636$, $Sc = 2.44$ [19]; *11*—$K_1 = 0.625$, $Sc = 2.5$ [25]; *12*—$K_1 = 0.69$, $Sc = 2.5$ [24]; *13*—$K_1 = 0.628$, $Sc = 2.5$ [22]

Table 6.2 Constant K_1 by Eqs. (6.4)–(6.8), a rotating disk for C_w = const. or T_w = const. [30]

Pr (*Sc*)	Exact	(6.4) [33]	(6.5) [34]	(6.6) [13]	(6.7) [2]	(6.8) [11]
1.0	0.3963	0.3941	0.4025	0.4303	0.3827	0.62
2.0	0.5653	0.5753	0.5864	0.5892	0.5664	0.7812
2.28	0.6016	0.6144	0.6257	0.6238	0.6065	0.816
2.5	0.6280	0.6430	0.6543	0.6491	0.6358	0.8415
5.0	0.8533	0.8839	0.8946	0.8676	0.8855	1.0602
20.0	1.4855	1.5414	1.5414	1.4924	1.5688	1.6829
50	2.0909	2.1552	2.1424	2.0958	2.2019	2.2841
100	2.6871	2.753	2.7278	2.6915	2.8147	2.8778
500	4.7351	4.7885	4.7222	4.7400	4.8890	4.9209
1000	6.0162	6.056	5.9651	6.0218	6.1773	6.2000
10^4	13.181	13.126	12.904	13.192	13.3565	13.358
10^5	28.613	28.332	27.834	28.639	28.7954	28.7779
10^6	61.860	61.074	59.990	61.915	62.0461	62.000

At the expense of a larger deviation from the exact solution for Sc = 1–2 (8 % at Sc = 1 and 4 % already at Sc = 2) Eq. (6.6) ensures only deviations of less than 1–3 % at higher Prandtl or Schmidt numbers (Fig. 6.1; Table 6.2).

For $Sc \to 0$, Eqs. (6.4)–(6.6) reduce to the asymptotic relation $K_1/Sc = 0.885$ [32]. For $Sc \to \infty$, they ensure agreement with another asymptotic $K_1 = 0.62Sc^{1/3}$ [11, 32].

One more relation for K_1 for $Pr = 0{-}\infty$ was designed as a combination of asymptotic solutions for the cases $Pr \to 0$ and $Pr \to \infty$ [2]. Rewritten, using Sc number, this results in

$$K_1 = \left[(0.88447Sc)^{-1.077} + (0.62048Sc^{1/3})^{-1.077}\right]^{-1/1.077}. \tag{6.7}$$

For Sc = 2, Eq. (6.7) merges with the self-similar solution. Deviation of Eq. (6.7) from the exact solution grows up to 3.2 % at Sc = 2.5, exhibits a maximum of 5.6 % at $Sc \approx 20$ and, for larger Schmidt numbers, diminishes to 2.7 % at Sc = 1000 and 0.6 % at $Sc = 10^5$.

Over the range of $Sc < 2$, deviation of Eq. (6.7) from the exact solution changes its sign, and increases in absolute values being 3.4 % at Sc = 1 and 8.2 % at Sc = 0.1 (Table 6.2).

To conclude, preference should be rendered to that of Eqs. (6.4)–(6.7) that ensures the lowest inaccuracy at the Schmidt numbers specific for the problem is to be solved.

Application to electrochemistry problems. Levich [11] derived an asymptotic solution for convective diffusion for very large Schmidt numbers $Sc \gg 1$

$$K_1 = 0.62Sc^{1/3}. \tag{6.8}$$

It coincides with the asymptotic solution for heat transfer for $Pr \gg 1$ given in [32].

Table 6.2 elucidates that Eq. (6.8) correlates well with the exact solution at $Sc > 500$ (deviation for $Sc = 500$ is 3.9 % and reduces to zero for $Sc \to \infty$). Equation (6.8) overruns the exact solution by 7.1 % at $Sc = 100$ and by 56.7 % at $Sc = 1$. Levich's Eq. (6.8) was successfully validated in experimental studies [4, 5, 7, 8, 12, 14] at high Schmidt numbers.

Rotating disk electrodes are intensively employed in experimental electrochemical investigations [1, 11]. Convective diffusion, which displays itself as the diffusion of the electrical current on the electrode, is modeled by Eq. (1.28).

For this case, Eq. (6.3) for laminar flow in view of Eq. (6.8) is usually rewritten as [1, 11]

$$i_L = 0.62nFC_F C_\infty D_m^{2/3} \nu^{-1/6} \omega^{1/2}, \tag{6.9}$$

where i_L is the limiting diffusion current of electrons to the surface of a rotating disk electrode; n is the number of electrons that are involved in the current; F is the disk area; C_F is the Faraday constant (96,485 C/mol); C_∞ is the concentration at infinity, mol/m^3. Based on this, one can ascertain that the mass transfer coefficient can be written as $\alpha_m = i_L/(nFC_F C_0)$, while Eq. (6.9) translates into Eq. (6.8).

In practice, the following tasks are actual: (1) searching a functional dependence of i_L on ω; (2) finding the diffusion coefficient D_m, whereas the value of i_L is measured; and (3) measurements of Volt–Ampere characteristics using a rotating disk electrode.

Naphthalene sublimation technique for experimental determination of the mass and heat transfer coefficients. Convective heat transfer from a surface to air is analogous to convective mass transfer in naphthalene sublimation to air. Naphthalene sublimation has been often employed to measure the average mass transfer of an *entire* disk weighted before and after the measurement to determine the amount of naphthalene lost by the disk as a result of experiments [19–21, 25–27]. Currently, accurate instrumentation is available for local pointwise scanning of the naphthalene layer thickness and subsequent calculation of local mass transfer coefficients for laminar, transitional, and turbulent flow [15–18, 22, 24, 28]. In frames of the analogy between the surface heat and mass transfer, constants K_1 in Eq. (3.4) for the Nusselt number and Eq. (6.3) for the Sherwood number can be expressed as [23]

$$K_1 = C\,Pr^{m_p}, \tag{6.10}$$

$$K_1 = C\,Sc^{m_p}, \tag{6.11}$$

where the coefficient C is identical in both equations. The effects of the Prandtl and Schmidt numbers are described by respective multipliers in Eqs. (6.10) and (6.11).

Equations (6.10) and (6.11) are used for the Prandtl and Schmidt numbers moderately diverging from unity: $Pr = 0.7$–0.74 for air, whereas $Sc = 2.28$–2.5 for naphthalene sublimation in air. Therefore, the constant C is assigned to be equal to the coefficient K_1 at $Sc = 1$ and $Pr = 1$ at $T_w =$ const. or $C_w =$ const. (see Table 6.1), i.e., $C = 0.3963$.

Authors [15–19, 24, 28] used the naphthalene sublimation technique to measure rotating disk mass transfer and set the exponent m_p to be the same for all values of Pr and Sc, which yields a relation between the Nu and Sh numbers

$$Nu/Sh = (Pr/Sc)^{m_p}. \tag{6.12}$$

The scatter of the values of the exponent m_p in the literature amounted up to 45 %: $m_p = 1/3$ [17], $m_p = 0.4$ [15, 17, 18, 20], $m_p = 0.53$ [19], and $m_p = 0.58$ [24].

Erroneous values m_p entail fallacious results of post-processing of the experimental data from the naphthalene sublimation technique aimed at estimation of heat transfer in air. An analysis and recommendation of the proper value m_p were made by the author [23].

Exponent m_p can be detected from the self-similar solution of the problem (Tables 3.1 and 6.1). Table 6.3 lists exponents m_p for the Prandtl/Schmidt numbers moderately deviating from unity [23]. It is evident from here that the function $m_p(Pr)$ exhibits a decreasing trend and varies from $m_p = 0.5723$ to $m_p = 0.5024$, if the Prandtl/Schmidt numbers grow from 0.7 up to 2.5. Consequently, the effective exponent $m_p = 0.53$ suggested in [19] is practically the average m_{p} value weighted over the range $Pr = 0.7$–2.5.

Figure 6.1 depicts different experimental data for the constant K_1 in naphthalene sublimation in air. These data agree well with the self-similar solution (see Table 6.2); only the too large value $K_1 = 0.69$ for $Sc = 2.5$ [24] falls out from the overall picture.

Table 6.3 Value m_p in Eqs. (6.10), (6.11) and (6.12) based on the exact solution of Eqs. (2.32)–(2.36) for laminar flow [23, 30]

Pr (Sc)	0.5	0.6	0.7	0.71	0.72	0.8	0.9	0.95	0.99
m_p	0.5954	0.5827	0.5723	0.5714	0.5705	0.5638	0.5571	0.5551	0.5632
Pr (Sc)	1.05	1.1	1.5	2	2.28	2.4	2.5	3	4
m_p	0.5438	0.5424	0.5264	0.5123	0.5064	0.5041	0.5024	0.4949	0.4841
Pr (Sc)	5	10	20	50					
m_p	0.4765	0.4566	0.4411	0.4251					

Post-processing [23] of the measured results using Eq. (6.12) at $m_p = 0.53$ to reduce them to conditions of heat transfer at $Pr = 0.71$ yields the values $K_1 = 0.325$ [17], $K_1 = 0.328$ [20, 21, 26], $K_1 = 0.331$ [19], $K_1 = 0.321$ [25], $K_1 = 0.322$ [22] that agree well with the exact value $K_1 = 0.326$ for $Pr = 0.71$ and $T_w =$ const. (see Table 3.1) and reliable experimental data (see Chap. 3). Falling out of the overall good conformance are (a) the constant $K_1 = 0.318$ resulting from the low value $K_1 = 0.59$ in naphthalene sublimation at $Sc = 2.28$ measured in [15], and (b) the constant $K_1 = 0.354$ stemming from the high experimental value $K_1 = 0.69$ in naphthalene sublimation obtained in [24] (see Fig. 6.1).

The use of the value $m_p = 0.4$ suggested in [15, 17, 18, 20] and widely used throughout the literature brings for the heat transfer in air at $Pr = 0.71$ [23]: $K_1 = 0.37$ [15]; $K_1 = 0.379$ [17]; $K_1 = 0.384$ [20, 21, 26]; $K_1 = 0.388$ [19]; $K_1 = 0.378$ [25]; $K_1 = 0.380$ [22]; $K_1 = 0.417$ [24]. All these recalculated data are too large as compared to the exact value $K_1 = 0.326$.

Taking the exponent $m_p = 1/3$ [17], one can obtain for $Pr = 0.71$ [23] the constants $K_1 = 0.4$ [15]; $K_1 = 0.409$ [17]; $K_1 = 0.416$ [20, 21, 26]; $K_1 = 0.421$ [19]; $K_1 = 0.411$ [25]; $K_1 = 0.413$ [22]; $K_1 = 0.454$ [24]. They surpass the exact value $K_1 = 0.326$ to an even larger extent.

Involvement of the exponent $m_p = 0.58$ [24] yields for $Pr = 0.71$ the values $K_1 = 0.3$ [15], $K_1 = 0.307$ [17], $K_1 = 0.308$ [20, 21, 26], $K_1 = 0.311$ [19], $K_1 = 0.301$ [25], $K_1 = 0.303$ [22], that are too small [23]. Only the value $K_1 = 0.332$ [24] is acceptable, which is due to the high value $m_p = 0.58$ chosen by the authors [24] to agree with the exact solution $K_1 = 0.326$. However, it is clear that the too large exponent $m_p = 0.58$ results from the too large value $K_1 = 0.69$ in naphthalene sublimation measured in [24], which is discordant with the measurements of the other researchers.

Authors [25] rearranged Eq. (6.12) in the following way

$$Nu/Sh_{Sc=2.5} = f(Pr)Pr^{1/3}. \tag{6.13}$$

The value $K_1 = 0.625$ at $Sc = 2.5$ and function $f(Pr) = 0.576, 0.634, 0.737, 0.842$ and 0.926 at $Pr = 0.1$, 1, 2.5, 10 and 100, respectively, yield jointly the values $K_1 = 0.321$, 0.396, 0.625, 1.134 and 2.686 at the Prandtl numbers mentioned above. This is fully consistent with the self-similar solution at $T_w =$ const. (Table 6.1, $n_* = 0$). The correction function $f(Pr)$ can be recast to incorporate the multiplier $Pr^{1/3}$. Use of Eq. (6.13) ensures higher accuracy than that conveyed by approaches operating with a single value of m_p, though Eq. (6.13) is less practical as Eq. (6.12), because of the involvement of a tabulated function.

To conclude, Eq. (6.12) with the exponent $m_p = 0.53$ [19] can be suggested as the most accurate and practical one for post-processing of the measured laminar mass transfer coefficients of a rotating disk in naphthalene sublimation in air in order to recalculate it to laminar heat transfer in air. As an alternative, Eq. (6.13) (or its modification) can be used.

6.2 Transitional and Turbulent Flow for the Prandtl and Schmidt Numbers Moderately Different from Unity

Values of $Pr \le 5$ and $Sc \le 5$ are considered here as those moderately deviating from unity. The objective is again a validation of the experimental technique dealing with sublimation of naphthalene from a rotating disk in air at $Sc = 2.28–2.5$ [23].

Local Sherwood numbers in naphthalene sublimation experiments in air in transitional and turbulent flow obtained in the recent works [15, 18] together with the data for laminar flow and different empirical approximations are depicted in Fig. 6.2. Recast Eq. (3.13) [15] for transitional flow (corrected range of validity) and empirical equations [15, 18] for turbulent flow look as follows [15, 18]

$$Sh = 2.0 \times 10^{-19} Re_{\omega}^{4} \quad \text{for} \quad Re_{\omega} = (1.9–2.75) \times 10^{5} \text{ (Ref. [15])}, \tag{6.14}$$

$$Sh = 0.0512 Re_{\omega}^{0.8} \quad \text{for} \quad Re_{\omega} \ge 2.75 \times 10^{5} \text{ (Ref. [15])}, \tag{6.15}$$

$$Sh = 0.0518 Re_{\omega}^{0.8} \quad \text{for} \quad Re_{\omega} \ge 2.5 \times 10^{5} \text{ (Ref. [18])}. \tag{6.16}$$

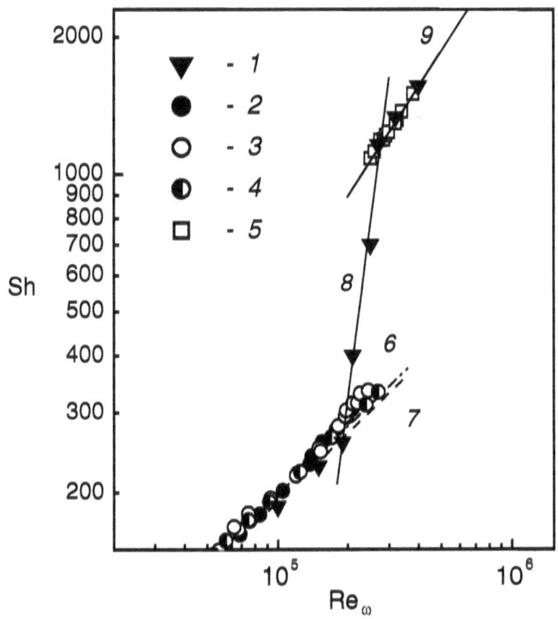

Fig. 6.2 Local Sherwood numbers for naphthalene sublimation in air [23, 30]. Experiments: *1*—$Sc = 2.28$ [15]; *2*—$Sc = 2.4$ [21]; *3*—$Sc = 2.4$ [26]; *4*—$Sc = 2.44$ [19]; *5*—Sc not mentioned [18]. Empirical approximations, Eq. (6.3): *6*—laminar flow, $n_R = 1/2$, $K_1 = 0.625$ [20, 21, 25, 26]; *7*—laminar flow, $n_R = 1/2$, $K_1 = 0.604$ [17]; *8*—transitional flow, $n_R = 4$, $K_1 = 2 \times 10^{-19}$, Eq. (6.14) [15]; *9*—turbulent flow, $n_R = 0.8$, $K_1 = 0.0512$, Eq. (6.15) [15]

In practice, one often needs to estimate average Sherwood numbers $Sh_{\rm av}$ (or average Nusselt numbers $Nu_{\rm av}$) of an *entire* disk, where areas occupied by laminar/transitional flow or laminar/transitional/turbulent flow emerge at the same time. For instance, only surface-averaged mass transfer coefficients of an *entire* disk were measured in [19–21, 25, 26].

Measurements of the average Sherwood number over an *entire* disk covered with areas of laminar, transitional and turbulent flow were performed by [19–21, 25, 26]. Reynolds analogy between mass transfer and fluid flow was involved to derive a quite inconvenient theoretical solution for $Sh_{\rm av}$ for an entire disk [21, 26] incorporating parameters, which were rather difficult to determine by means of the used approach. More promising is the model for $Sh_{\rm av}$ first used in the paper [7] and further generalized by the author of the present work [23, 30], which enables verifications of the recent measurements of the local Sherwood numbers by means of comparisons with the vast database for the average Sherwood numbers for an *entire* disk.

The author [7] assumed that laminar-turbulent transition sets on instantly at a radial coordinate $r_{\rm tr}$ corresponding to the Reynolds number $Re_{\omega,\rm tr}$. Subsequently, the value $Sh_{\rm av}$ for an entire disk can be found using the following integral

$$Sh_{\rm av} = \frac{2}{b}\left[\int\limits_0^{r_{\rm tr}} Sh_{\rm lam}{\rm d}r + \int\limits_{r_{\rm tr}}^{b} Sh_{\rm turb}{\rm d}r\right]. \tag{6.17}$$

Sherwood numbers are presented by Eq. (6.3) accompanied with the constants $K_{1,\rm lam}$ and $n_{\rm R}$ = 1/2 for laminar flow, and $K_{1,\rm turb}$ and $n_{\rm R}$ = 0.8 for turbulent flow.

An integration of Eq. (6.17) yields

$$Sh_{\rm av} = K_{1,\rm lam}Re_{\omega,\rm tr}^{1/2}\left(\frac{Re_{\omega,\rm tr}}{Re_\varphi}\right)^{1/2} + \frac{2}{2n_R+1}K_{1,\rm turb}Re_\varphi^{n_{\rm R}}\left[1-\left(\frac{Re_{\omega,\rm tr}}{Re_\varphi}\right)^{n_{\rm R}+1/2}\right]. \tag{6.18}$$

If $Re_\varphi < Re_{\omega,\rm tr}$, the second summand in Eq. (6.18) must be discarded. Asymptotically at $Re_\varphi \gg Re_{\omega,\rm tr}$, Eq. (6.18) degenerates to Eq. (6.3) for turbulent flow with

$$K_{2,\rm turb} = \frac{2}{2n_{\rm R}+1}K_{1,\rm turb}. \tag{6.19}$$

Given $n_* = 0$, which effectively means T_w = const. and C_w = const., Equation (6.18) translates into Eq. (3.25), while Eq. (6.19) turns to Eq. (3.35) in view of the relation $2n_{\rm R} = 1 + m$ resulting from Eqs. (2.78) and (3.31).

In [12] it is suggested taking into account regions of laminar, transitional and turbulent flow separately. If the transition sets on at the radial location $r_{\rm tr1}$ (or at

$Re_{\omega,\mathrm{tr1}}$) and ends at the radial location r_{tr2} (or $Re_{\omega,\mathrm{tr2}}$), a definite integral for Sh_{av} can be written as

$$Sh_{\mathrm{av}} = \frac{2}{b}\left[\int_0^{r_{\mathrm{tr1}}} Sh_{\mathrm{lam}}\mathrm{d}r + \int_{r_{\mathrm{tr1}}}^{r_{\mathrm{tr2}}} Sh_{\mathrm{tran}}\mathrm{d}r + \int_{r_{\mathrm{tr2}}}^{b} Sh_{\mathrm{turb}}\mathrm{d}r\right]. \tag{6.20}$$

The transitional Sherwood number Sh_{tran} is specified by the first of Eq. (6.3) complemented with experimental values of $K_{1,\mathrm{tran}}$ and $n_{\mathrm{R,tran}}$ for transitional flow. Integration of Eq. (6.20) results in

$$\begin{aligned} Sh_{\mathrm{av}} = {} & K_{1,\mathrm{lam}} Re_{\omega,\mathrm{tr1}}^{1/2}\left(\frac{Re_{\omega,\mathrm{tr1}}}{Re_{\varphi}}\right)^{1/2} + \frac{2}{2n_{\mathrm{R,tran}}+1} K_{1,\mathrm{tran}} Re_{\omega,\mathrm{tr2}}^{n_{\mathrm{R,tran}}}\left(\frac{Re_{\omega,\mathrm{tr2}}}{Re_{\varphi}}\right)^{1/2} \\ & \times\left[1-\left(\frac{Re_{\omega,\mathrm{tr1}}}{Re_{\omega,\mathrm{tr2}}}\right)^{n_{\mathrm{R,tran}}+1/2}\right] + \frac{2}{2n_{\mathrm{R}}+1} K_{1,\mathrm{turb}} Re_{\varphi}^{n_{\mathrm{R}}}\left[1-\left(\frac{Re_{\omega,\mathrm{tr2}}}{Re_{\varphi}}\right)^{n_{\mathrm{R}}+1/2}\right]. \end{aligned} \tag{6.21}$$

Equation (6.21) holds at $Re_{\varphi} \geq Re_{\omega,\mathrm{tr2}}$. If $Re_{\varphi} < Re_{\omega,\mathrm{tr2}}$, the last term in Eq. (6.21) is discarded, whereas the second summand turns to

$$\begin{aligned} Sh_{\mathrm{av}} = {} & K_{1,\mathrm{lam}} Re_{\omega,\mathrm{tr1}}^{1/2}\left(\frac{Re_{\omega,\mathrm{tr1}}}{Re_{\varphi}}\right)^{1/2} \\ & + \frac{2}{2n_{\mathrm{R,tran}}+1} K_{1,\mathrm{tran}} Re_{\varphi}^{n_{\mathrm{R,tran}}}\left[1-\left(\frac{Re_{\omega,\mathrm{tr1}}}{Re_{\varphi}}\right)^{n_{\mathrm{R,tran}}+1/2}\right]. \end{aligned} \tag{6.22}$$

Asymptotically for $Re_{\varphi} \gg Re_{\omega,\mathrm{tr2}}$, Eq. (6.21) transforms to the second of Eq. (6.3), whereas the constant $K_{2,\mathrm{turb}}$ is given by Eq. (6.19). A solution derived in [12] is a particular case of Eq. (6.21), whose empirical constants resulting from experiments [12] at high Sc numbers are fixed numerical values. Hence, the solution [12] as it is can not be used to describe the experimental data for naphthalene sublimation.

Substitution of numerical values of the constants resulting from measurements at naphthalene sublimation in air [15] (see Eqs. (6.14), (6.15) and caption to Fig. 6.1) into the general Eqs. (6.18)–(6.22) yields

(a) applied to Eq. (6.18)

$$Sh_{\mathrm{av}} = 0.59 Re_{\omega,\mathrm{tr}}^{1/2}\left(\frac{Re_{\omega,\mathrm{tr}}}{Re_{\varphi}}\right)^{1/2} + \frac{2}{2.6} 0.512 Re_{\varphi}^{0.8}\left[1-\left(\frac{Re_{\omega,\mathrm{tr}}}{Re_{\varphi}}\right)^{1.3}\right], \tag{6.23}$$

(b) applied to Eq. (6.19)

$$K_{2,\mathrm{turb}} = \frac{2}{2.6} K_{1,\mathrm{turb}} = 0.0394, \tag{6.24}$$

(c) applied to Eq. (6.21)

$$\begin{aligned} Sh_{\mathrm{av}} &= 0.59 \times 1.9 \times 10^5 \times Re_\varphi^{-1/2} + \frac{4}{9} 10^{-19} (2.75 \times 10^5)^{4.5} \\ &\times Re_\varphi^{-1/2} \left[1 - \left(\frac{1.9 \times 10^5}{2.75 \times 10^5} \right)^{4.5} \right] \\ &+ 0.0394 Re_\varphi^{0.8} \left[1 - \left(\frac{2.75 \times 10^5}{Re_\varphi} \right)^{1.3} \right], \quad Re_\varphi \geq 2.75 \times 10^5, \end{aligned} \tag{6.25}$$

(d) applied to Eq. (6.22)

$$\begin{aligned} Sh_{\mathrm{av}} &= 0.59 \times 1.9 \times 10^5 \times Re_\varphi^{-1/2} + \frac{4}{9} 10^{-19} Re_\varphi^4 \left[1 - \left(\frac{1.9 \times 10^5}{Re_\varphi} \right)^{4.5} \right], \\ Re_\varphi &= (1.9\text{–}2.75) \times 10^5. \end{aligned} \tag{6.26}$$

The Reynolds number $Re_{\omega,\mathrm{tr}}$ (instant transition to turbulence) in Eq. (6.23) remains a free parameter to be tuned for a better agreement with particular experiments.

Figure 6.3 shows validations of Eqs. (6.23)–(6.28) by comparison with experimental data. Experimental data 1, 5 and curve 6 for Sh_{av} for purely turbulent flow stem from the works [23, 30] and result from reprocessing of the measured data [15, 18] and Eq. (6.15) using Eq. (6.24). For laminar flow, we have $K_{2,\mathrm{lam}} = K_{1,\mathrm{lam}}$ (curves 7 and 8). Curve 9 combining Eqs. (6.25), (6.26) and incorporating boundaries of transitional flow conforms to experiments [19, 21, 26] for Sh_{av} for an *entire* disk depicted in Fig. 6.3.

In Fig. 6.3, experimental data 1 for Sh_{av} for an *entire* disk were calculated in [23] using Eqs. (6.25), (6.26) and measurements [15] for laminar, transitional and turbulent flow. These data points go beyond curve 9 at respective values of the argument Re_φ.

The replacement of the Reynolds number $Re_{\omega,\mathrm{tr}}$ in Eq. (6.23) (instant transition to turbulence) with its values at the onset and end of transition (i.e., 1.9×10^5 and 2.75×10^5) yields curves 10 and 11 lying above and below curve 9, respectively. Reynolds number of the instant transition to turbulence $Re_{\omega,\mathrm{tr}} = 2.35 \times 10^5$, an arithmetic mean of values $Re_{\omega,\mathrm{tr1}}$ and $Re_{\omega,\mathrm{tr2}}$, substituted into Eq. (6.23) conveyed curve 12, which agrees with curve 9.

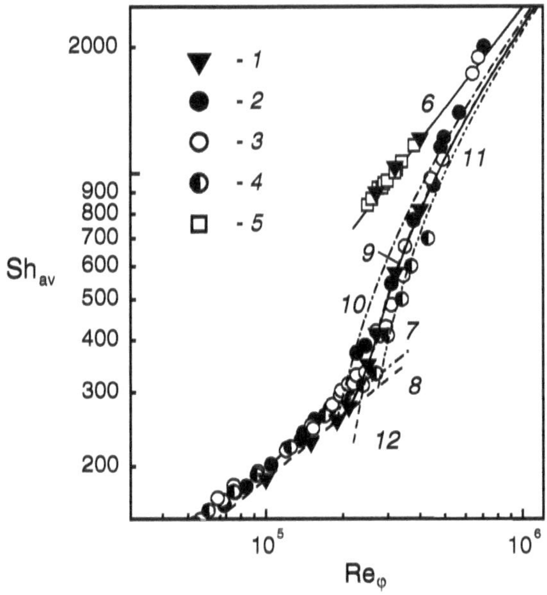

Fig. 6.3 Average Sherwood numbers, naphthalene sublimation in air [23, 30]. Experiments: *1*—$Sc = 2.28$ [15]; *2*—$Sc = 2.4$ [21]; *3*—$Sc = 2.4$ [26]; *4*—$Sc = 2.44$ [19]; *5*—Sc not mentioned [18]. Calculation, Eq. (6.3): *6*—turbulent flow, $n_R = 0.8$, $K_2 = 0.0394$, Eq. (6.24) [15]; *7*—laminar flow, $n_R = 1/2$, $K_1 = 0.625$ [20, 21, 25, 26]; *8*—laminar flow, $n_R = 1/2$, $K_1 = 0.59$ [15]. Calculation of Sh_{av} for the *entire* disk: *9*—Eqs. (6.25) and (6.26); *10*—Eq. (6.23) at $Re_{\omega,tr} = 1.9 \times 10^5$; *11*—Eq. (6.23) at $Re_{\omega,tr} = 2.75 \times 10^5$; *12*—Eq. (6.23) at $Re_{\omega,tr} = 2.35 \times 10^5$

In the asymptotic case of $Re_\varphi \to \infty$, lines 9–12 coincide with curve 6 valid for purely turbulent flow.

Thus, Eqs. (6.21) and (6.22) incorporating terms accounting for the coexistence of laminar, transitional and turbulent flow areas ensure the best agreement with experiments for the Sh_{av} number for an *entire* disk. Equation (6.18) resulting from a simpler model [7] ensures the efficiency similar to that of Eqs. (6.21) and (6.22), if an "effective" Reynolds number $Re_{\omega,tr}$ of the instant transition to turbulent flow is chosen correctly.

Application to the naphthalene sublimation technique. Again, a recalculation of the mass transfer to heat transfer data is performed using Eq. (3.4) and (6.3), with the constants K_1 defined in Eqs. (6.10) and (6.11), accordingly [23]. The factor C is equal to the constant K_1 for $Sc = 1$, $Pr = 1$ under conditions T_w = const. or C_w = const.

Authors [15, 18] used the constant $m_p = 0.4$ for a turbulent flow regime and $Pr = Sc = 0.7$–2.5. Equation (6.12) at $m_p = 0.4$ yields the value $K_1 = 0.0323$ for heat transfer in air at T_w = const. and $Pr = 0.72$, starting from the values $K_1 = 0.0512$–0.0518 (see Eqs. (6.15) and (6.16)) for $Sc = 2.28$ as a base for the recalculation. But, in reality, experiments [35–37] (see Table 3.5) conveyed the value of the constant $K_1 = 0.0188$ at T_w = const. and $Pr = 0.72$. The theoretical model [38, 39] gave the value K_1 = 0.0187 for the same conditions. Thus, also for turbulent flow, an erroneous value m_p leads to fallacious translation of the naphthalene sublimation data to heat transfer in air.

Equations (6.10), (6.11) are to be used for the Prandtl and Schmidt numbers moderately diverging from unity: Pr = 0.7–0.74 for air; Sc = 2.28–2.5 for naphthalene sublimation. Hence, the constant C in Eqs. (6.10), (6.11) must be equal to the coefficient $K_1 = 0.0232$ in turbulent flow at Sc = 1, Pr = 1 and conditions T_w = const., C_w = const. (Table 3.7).

Detecting of the exponent m_p for turbulent flow is performed using experimental data. Only experimental Eqs. (6.15) and (6.16) [15, 18] can serve for this purpose. Based on Eq. (6.15) as well as the values $C = 0.0232$, $K_1 = 0.0188$ (at T_w = const. and $Pr = 0.72$) [35–37], one can transform Eqs. (6.10) and (6.11) as follows

$$K_1 = 0.0232 Pr^{0.64} \quad \text{for } Pr \le 1, \tag{6.27}$$

$$K_1 = 0.0232 Sc^{0.96} \quad \text{for } Sc \ge 1. \tag{6.28}$$

This means that the exponent m_p for turbulent flow is not universal being a function of the Prandtl and Schmidt numbers, which apparently results from different effects of the Pr or Sc larger and smaller than unity. Equations (6.27) and (6.28) yield as a result

$$Nu/Sh = Pr^{0.64}/Sc^{0.96}. \tag{6.29}$$

Using the idea of a correction function, Eq. (6.13), one can transform Eq. (6.29) as

$$Nu/Sh_{Sc=2.28} = f(Pr). \tag{6.30}$$

At $Pr = 0.72$, the correction function $f(Pr)$ takes the value $f(Pr) = 0.367$.

As an alternative, one can use an effective value of the exponent m_p so that

$$Nu/Sh = (Pr/Sc)^{0.87}. \tag{6.31}$$

The exponent $m_p = 0.87$ in Eq. (6.31) is more than twice larger than the value 0.4 mistakenly recommended in [15, 18]. Nevertheless, the value $m_p = 0.87$ must not be used in Eqs. (6.27) and (6.28) to avoid significant errors in predictions of the constant K_1.

The empirical Eq. (3.10) [37] for transitional flow at T_w = const. and $Pr = 0.72$ is the most appropriate to be used jointly with Eq. (6.14) for transitional flow at C_w = const. Thus, Eq. (6.12) should be recast in view of Eqs. (3.10) and (6.14) as follows

$$Nu/Sh = (Pr/Sc)^{0.6}. \tag{6.32}$$

Equation (3.10) is valid for $Re_\omega = 1.95 \times 10^5$–$2.5 \times 10^5$, while Eq. (6.14) holds at $Re_\omega = 1.9 \times 10^5$–$2.75 \times 10^5$. These diffcrences are though rather insignificant.

To conclude, Eqs. (6.29)–(6.31) should be employed to recalculate the data for turbulent mass transfer for naphthalene sublimation in air to the conditions of heat transfer in air. Equation (6.32) should be applied for transitional flow for the same purpose [23].

6.3 Transitional and Turbulent Flow at High Schmidt Numbers

High values of the Schmidt numbers can be encountered in electrochemistry problems: Sc = 34–10,320 [4, 5, 7, 8, 12]. Main objectives of this section are validation and development of recommendations for the further use of the experimental and theoretical data of different authors [40].

Experimental data [7] for average Sherwood numbers for an entire disk at $Re_\varphi = 0.278 \times 10^6$–$1.8 \times 10^6$, Sc = 930–10,320 were described by a relation

$$Sh_{\rm av} = Sc^{1/3} Re_\varphi^{-1/2} [0.62 Re_{\omega,\rm tr} + 1.08 \times 10^{-2} (Re_\varphi^{1.37} - Re_{\omega,\rm tr}^{10.37})]. \tag{6.33}$$

Here, Eq. (6.18) at $K_{1,\rm lam} = 0.62 Sc^{1/3}$, $K_{1,\rm turb} = 0.0148 Sc^{1/3}$, $Re_{\omega,\rm tr} = 2.78 \times 10^5$ and $n_{\rm R} = 0.87$ was taken into account. In the transitional region at $Re_\omega = 2.3 \times 10^5$–$2.9 \times 10^5$, Eq. (6.33) lies below the experimental data [7] (in analogy to curve 12 in Fig. 6.3), which results from simplifications incorporated in model (6.18) and mentioned in Sect. 6.2.

A reduced form of Eq. (6.33) for purely turbulent flow [7] and an equation for the local Sherwood numbers derived in [30, 40] have the following form

$$Sh_{\rm av} = 1.08 \times 10^{-2} Re_\varphi^{0.87} Sc^{1/3}, \tag{6.34}$$

$$Sh = 1.48 \times 10^{-2} Re_\omega^{0.87} Sc^{1/3}. \tag{6.35}$$

Measurements [4] of the average Sherwood numbers for an entire disk performed at $Re_\varphi = 5 \times 10^4$–$1.8 \times 10^6$, Sc = 345–6450 (transition at $Re_\varphi = 2.3 \times 10^5$–$2.9 \times 10^5$) for the region of turbulent flow were described by the relation

$$Sh_{\rm av} = 0.0725 Re_\varphi^{0.9} Sc^{0.33}. \tag{6.36}$$

The authors [5] measured local Sh and average $Sh_{\rm av}$ numbers at laminar, transitional, and turbulent flow for $Re_\omega = 4 \times 10^4$–$2.2 \times 10^6$, Sc = 680–7200 (transition at $Re_\omega = 2.2 \times 10^5$–$3.0 \times 10^5$). Sherwood numbers for the turbulent flow [5] and average values for an entire disk (approximated in [30]) are given by the following relations, respectively

$$Sh = 1.09 \times 10^{-2} Re_{\omega}^{0.91} Sc^{1/3}, \tag{6.37}$$

$$Sh_{\mathrm{av}} = 7.67 \times 10^{-3} Re_{\varphi}^{0.91} Sc^{1/3}, \tag{6.38}$$

$$Sh_{\mathrm{av}} = Sc^{1/3} Re_{\varphi}^{-1/2} [0.62 Re_{\omega,\mathrm{tr}} + 7.67 \times 10^{-3} (Re_{\varphi}^{1.41} - Re_{\omega,\mathrm{tr}}^{1.41})], \tag{6.39}$$

where the Reynolds number of the abrupt transition was $Re_{\omega,\mathrm{tr}} = 2.78 \times 10^5$ [7].

Experiments [8] for Sh_{av} for an entire disk were performed at $Re_{\varphi} = 10^4$–1.18×10^7, $Sc = 34$–1400. For purely turbulent flow at $Re_{\varphi} = 8.9 \times 10^5$–$1.18 \times 10^7$, authors [8] obtained

$$Sh_{\mathrm{av}} = 1.17 \times 10^{-2} Re_{\varphi}^{0.896} Sc^{0.249}. \tag{6.40}$$

Experiments for the local Sherwood numbers in transitional flow at $Re_{\omega} = 2.0 \times 10^5$–$3.0 \times 10^5$ and $Sc = 1192$–2465 were described by the empirical Eq. (3.14) [12].

The authors [12] deduced empirical equations for Sh_{av} for turbulent flow (based on experiments [4]), simultaneous existence of laminar and transitional flow, as well as simultaneous existence of laminar, transitional and turbulent flow, respectively

$$Sh_{\mathrm{av}} = 7.8 \times 10^{-3} Re_{\varphi}^{0.9} Sc^{1/3}, \tag{6.41}$$

$$Sh_{\mathrm{av}} = Sc^{1/3} Re_{\varphi}^{-1/2} [0.89 \times 10^5 + 9.7 \times 10^{-15} Re_{\varphi}^{3.5}], \tag{6.42}$$

$$Sh_{\mathrm{av}} = Sc^{1/3} Re_{\varphi}^{-1/2} [7.8 \times 10^{-3} Re_{\varphi}^{1.4} - 1.3 \times 10^5]. \tag{6.43}$$

Equations (6.42) and (6.43) are particular cases of Eqs. (6.22) and (6.21), accordingly, with Eq. (6.8) used for laminar, Eq. (3.14) for transitional and Eq. (6.41) for turbulent flow.

Theoretical solutions for the local turbulent Sherwood numbers at high Schmidt numbers derived in [9, 41] can be presented as follows, respectively,

$$Sh_{\mathrm{av}} = 7.07 \times 10^{-3} Re_{\varphi}^{0.9} Sc^{1/3}, \tag{6.44}$$

$$Sh_{\mathrm{av}} = 5.93 \times 10^{-3} Re_{\varphi}^{0.91} Sc^{0.34}. \tag{6.45}$$

A theoretical solution obtained in [10] coincides with Eq. (6.41). The solution obtained in [42] has a form of Eq. (6.41) with the coefficient changed to 6.43×10^{-3}.

In [43] a theoretical solution claimed to be valid for $Sc = 0.72$–∞ has been proposed, however, as demonstrated in [30, 40], this relation is inaccurate.

Some of the theoretical and empirical relations for the Sherwood numbers for purely turbulent flow, as well as for average Sh_{av} numbers for an entire disk

Fig. 6.4 Average Sherwood numbers at high Schmidt numbers [30]. Approximation of experiments: *1*—laminar flow, Levich's Eq. (6.8); *2*—Eq. (6.33) for an *entire* disk [7]; *3*—Eq. (6.34) [7]; *4*—Eq. (6.36) [4]; *5*—Eq. (6.41) [10, 12]. Theoretical solution: *6*—Eq. (6.44) [9]

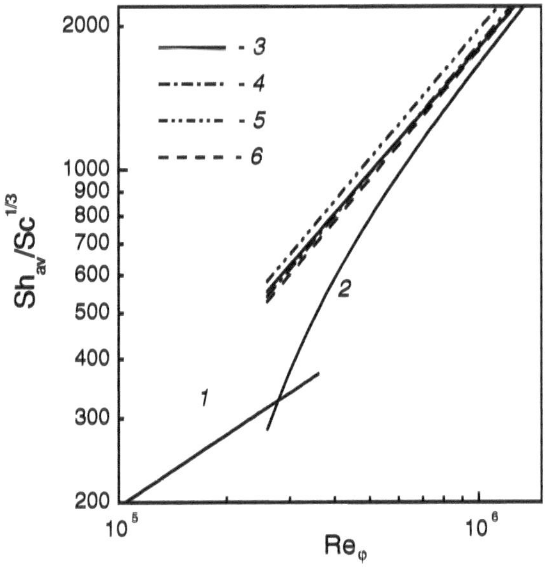

simultaneously occupied by laminar, transitional, and turbulent flow areas agree well with each other. Curves by Eqs. (6.34), (6.36), (6.44), and (6.45) practically merge (see Fig. 6.4). Equation (6.41) significantly surpasses original experiments [4]; corrected coefficient 6.43×10^{-3} [42] shifts predictions by Eq. (6.41) 9 % below those by Eq. (6.44). Empirical Eqs. (6.34) and (6.36) practically coincide, which corroborates the reliability of these experiments.

Equation (6.38) for turbulent flow and Eq. (6.39) for an entire disk significantly surpass Eqs. (6.33) and (6.34), respectively (see Fig. 6.5). Only in Eq. (6.40) [8], exponent 0.249 at the Schmidt number is not equal to 1/3. The large scatter of experiments around the approximation curve [8] is rather an evidence that the exponent 0.249 is erroneous. As demonstrated in [30, 40], differences between the curves by Eq. (6.40) plotted in the relation $Sh_{av}/Sc^{1/3}$ versus Re_φ for different Sc values revealed in experiments [8] is rather significant. Hence, Eq. (6.40) should be discarded as too inaccurate.

The exponent for the Reynolds number Re_φ in Eq. (6.35) diverges from those in Eqs. (6.15) and (6.16). Equation (6.35) can be recast to make the exponent for Re_φ equal to 0.8. This yields for the entire disk [30, 40]

$$Sh_{av}Sc^{-1/3} = 0.62Re_{\omega,tr}^{1/2}\left(\frac{Re_{\omega,tr}}{Re_\varphi}\right)^{1/2} + \frac{2}{20.6}0.0365Re_\varphi^{0.8}\left[1 - \left(\frac{Re_{\omega,tr}}{Re_\varphi}\right)^{1.3}\right]. \tag{6.46}$$

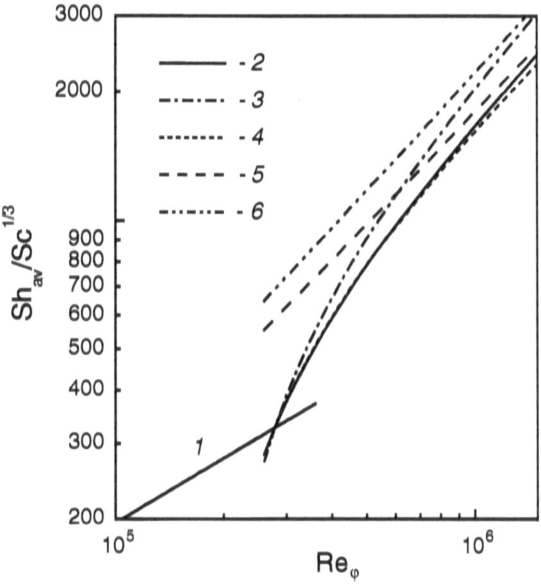

Fig. 6.5 Average Sherwood numbers at high Schmidt numbers, approximation of experiments [30]. *1*—laminar flow, Levich's Eq. (6.8); *2*—Eq. (6.33) for an *entire* disk [7]; *3*—Eq. (6.39), *entire* disk; *4*—Eq. (6.46), *entire* disk; *5*—Eq. (6.34) [7]; *6*—Eq. (6.38) [5]

Figure 6.5 depicts curves 2 and 4 plotted by Eqs. (6.33) and (6.46), respectively. Here again $Re_{\omega,\mathrm{tr}} = 2.78 \times 10^5$, like in Eq. (6.33). Curves 2 and 4 in fact merge for $Re_\varphi \le 9.0 \times 10^5$; deviations start to become visible for $Re_\varphi > 9.0 \times 10^5$.

For purely turbulent flow, Eq. (6.46) reduces asymptotically to the relations

$$Sh = 3.65 \times 10^{-2} Re_\omega^{0.8} Sc^{1/3}, \tag{6.47}$$

$$Sh_{\mathrm{av}} = 3.65 \times 10^{-2} \frac{2}{2.6} Re_\varphi^{0.8} Sc^{1/3} = 2.81 \times 10^{-2} Re_\varphi^{0.8} Sc^{1/3}. \tag{6.48}$$

Figure 6.6 demonstrates that Eq. (6.35) [7] used at $Sc = 2.28$ predicts Sherwood numbers close to the experiments [15, 18] for naphthalene sublimation in air and their approximation Eq. (6.15). Equations (6.35) (curve 10) and (6.15) (curve 9) correlate well at larger Reynolds numbers $Re_\omega = 0.6 \times 10^6$–$2.0 \times 10^6$. Curve 11, Eq. (6.47), lies in the vicinity of curve 10 at smaller Reynolds numbers $Re_\omega \le 0.7 \times 10^6$. Equation (6.47) yields $K_1 = 0.048$ at $Sc = 2.28$, which is only 6.7 % below the value $K_1 = 0.0512$ in Eq. (6.15). Curve 12 by Eq. (6.37) goes noticeably beyond experiments and approximation curve 9 in Fig. 6.6.

Dependence 13 in Fig. 6.6 plotted by experimental Eq. (3.14) [12] for transitional flow at $Sc = 2.28$ conforms well to Eq. (6.14) and experiments [15].

To conclude, the most reliable empirical relations for developed turbulent flow and an *entire* disk relying on the analysis made above are Eqs. (6.33)–(6.36).

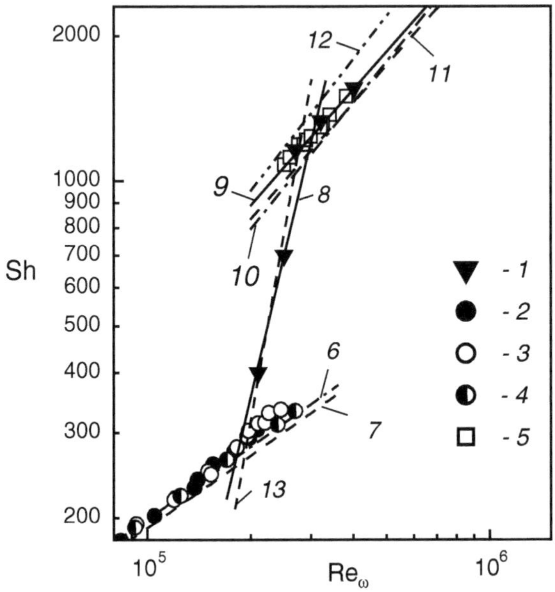

Fig. 6.6 Local Sherwood numbers for naphthalene sublimation [30]. Experiments: *1*—Sc = 2.28 [15]; *2*—Sc = 2.4 [21]; *3*—Sc = 2.4 [26]; *4*—Sc = 2.44 [19]; *5*—Sc not mentioned [18]. Empirical Eq. (6.3): *6*—laminar flow, $n_R = 1/2$, $K_1 = 0.625$ [20, 25, 26]; *7*—laminar flow, $n_R = 1/2$, $K_1 = 0.604$ [17]; *8*—transitional flow, $n_R = 4$, $K_1 = 2 \times 10^{-19}$ [15]; *9*—turbulent flow, $n_R = 0.8$, $K_1 = 0.0512$ [15]. Developed turbulent flow, Sc = 2.28: *10*—Eq. (6.35) [7]; *11*—Eq. (6.47); *12*—Eq. (6.37) [5]. Transitional flow, Sc = 2.28: *13*—Eq. (3.14) [12]

6.4 An Integral Method for *Pr* and *Sc* Numbers Much Larger Than Unity

Model with a constant value $\Delta \ll 1$. The thickness of the thermal (or diffusion) boundary layer at very high Pr or Sc numbers is much smaller than the thickness of the velocity boundary layer (i.e., $\Delta \ll 1$.). Hence, in Eq. (3.40) obtained for $\Delta =$ const. and $T^+ \equiv T^+(y^+)$, all summands in the parentheses in its left-hand part but a_* tend to zero

$$\Delta^{2n+1} a_* = \frac{4+m}{4+m+n_*}(a_* - 2b_* + c_*)Pr^{-n_p}. \tag{6.49}$$

Relying on Eq. (6.49), one can derive analytical solutions for constants Δ and K_1

$$\Delta = \left[\frac{4+m}{4+m+n_*}\left(1 - \frac{2D_3}{C_2}\right)\right]^{\frac{1}{2n+1}} Pr^{-\frac{n_p}{2n+1}}, \tag{6.50}$$

$$K_1 = K_3\left[\frac{4+m}{4+m+n_*}\left(1-\frac{2D_3}{C_2}\right)\right]^{\frac{-n}{2n+1}} Pr^{1+n_p\left(\frac{n}{2n+1}-1\right)}, \tag{6.51}$$

where the coefficients C_2 and D_3 are described in the comments to Eqs. (2.68) and (2.69).

The cumulative exponent at the Prandtl number in Eq. (6.51) for $Pr \gg 1$ must be equal to 1/3 (see Sect. 6.3), which yields the following expression for n_p

$$n_p = \frac{2}{3}\cdot\frac{2n+1}{n+1}. \tag{6.52}$$

As a result, the constants K_1 and K_2 in view of Eq. (3.35) can be written as

$$K_1 = K_3\left[\frac{4+m}{4+m+n_*}\left(1-\frac{2D_3}{C_2}\right)\right]^{\frac{-n}{2n+1}} Pr^{1/3}, \tag{6.53}$$

$$K_2 = K_3\left[\frac{4+m}{4+m+n_*}\left(1-\frac{2D_3}{C_2}\right)\right]^{\frac{-n}{2n+1}} \frac{n_*+2}{2+n_*+m} Pr^{1/3}. \tag{6.54}$$

To enable validations against electrochemical experiments, let us further treat the Sherwood numbers rather than the Nusselt numbers and replace *Pr* with *Sc*.

In Fig. 6.7, Eq. (6.54) for Sh_{av} (at $n_* = 0$) is validated against the empirical Eq. (6.34) [7] and theoretical Eq. (6.44) [9]. Curves 5 and 6 predicted by Eq. (6.54) at $n = 1/7$ and 1/9 lie 20–30 % below the curves 3 and 4 suggested by Eqs. (6.34) and (6.44), accordingly. Such a discrepancy between theory and measurements is

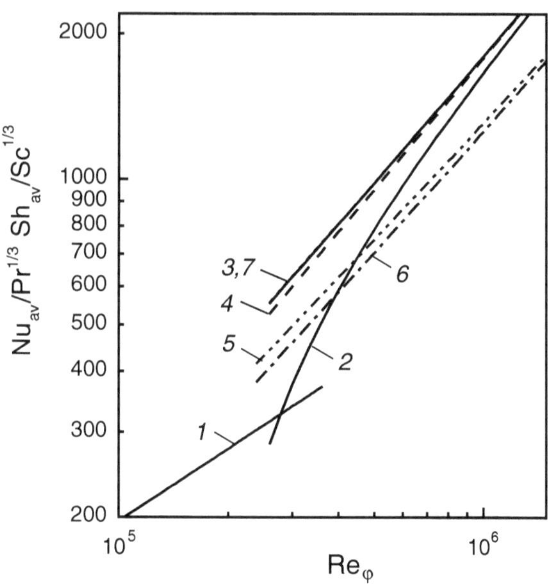

Fig. 6.7 Average Sherwood numbers for high Schmidt numbers [30]. Approximation of experiments: *1*—laminar flow, Levich's Eq. (6.8) [11]; *2*—Eq. (6.33) for an *entire* disk [7]; *3*—Eq. (6.34) [7] Theoretical solutions: *4*—Eq. (6.44) [9]; *5*—Eq. (6.54) for $n = 1/7$; *6*—Eq. (6.54) for $n = 1/9$; *7*—Eq. (6.73)

too high. In addition, the slope of the curves 5 and 6 (exponents at Re_φ being 0.8 and 0.833, constants K_2 being 0.0207 and 0.0126, accordingly) distinctly deviates from the slope of curves 3 and 4 (exponents at Re_φ being 0.87 and 0.9, constants K_2 being 0.0207 and 0.126, accordingly). Therefore, some model approaches incorporated in the present integral method partially fail at high Pr and Sc numbers and need to be improved. In Eq. (3.32) for the coefficient K_1, the total exponent at the Reynolds number can be increased, provided that the relative thickness Δ is assigned to be a decreasing function of the local Reynolds number Re_ω.

Model with a variable value of $\mathbf{\Delta}$. The present integral method incorporates a model, in frames of which a boundary layer consists two parts. In the vicinity of the wall, a viscous and heat conduction sub-layers emerge, where the velocity and temperature profiles are described by Eq. (2.62). In the main part of the boundary layer (outside of the viscous sub-layer), velocity components are described by the power-law functions (see Chaps. 2 and 3). If Prandtl and Schmidt numbers are slightly different from unity, the thermal/diffusion and velocity boundary layers have a thickness of the same order of magnitude [30, 40]. Hence, integration of Eq. (2.23) for the thermal boundary layer has been performed over the *entire* velocity boundary layer. Viscous and heat conduction sub-layers are not taken into account in this integration, because they are negligibly thin in comparison with the overall boundary layer thickness. Velocity profiles in Eqs. (2.17)–(2.19) are integrated in the same way [31, 38, 39, 44, 45].

At very high Prandtl and Schmidt numbers, the boundary layer structure changes drastically. A very thin thermal/diffusion boundary layer is fully incorporated inside

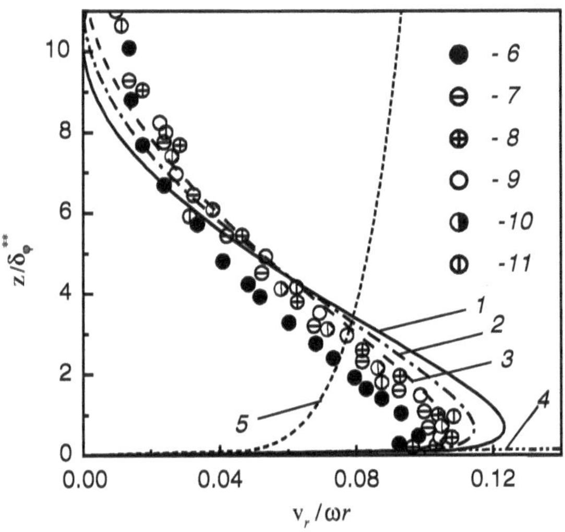

Fig. 6.8 Radial velocity profiles in the turbulent boundary layer over a free rotating disk [30]. *1*—$n = 1/7$; *2*—1/8; *3*—1/9. Equation (2.41), [44]: *4*—Eq. (6.55) at $Re_\omega = 1.0 \times 10^6$; *5*—Eq. (3.27) at $\sigma = 0$, $n = 1/7$ (see Fig. 3.7). Experiments: *6*—$Re_\omega = 0.4 \times 10^6$; *7*—$0.65 \times 10^6$; *8*—$0.94 \times 10^6$; *9*—$1.6 \times 10^6$ [46]; *10*—0.6×10^6; *11*—1.0×10^6 [47]

the viscous sub-layer of the velocity boundary layer; here the radial velocity profile varies linearly depending on the coordinate z (curve 4 in Fig. 6.8). This fact is taken into account in theoretical models [5, 9–11, 32, 42] for large *Pr* and *Sc* numbers.

Next to the wall, the radial velocity v_r varies as a linear function

$$v_r = \frac{\tau_{wr}}{\mu} z = \frac{\tau_w \alpha}{\mu(1+\alpha^2)^{1/2}} z = \frac{\rho V_*^2 \alpha}{\mu(1+\alpha^2)^{1/2}} \frac{c_f}{2} z = \alpha(1+\alpha^2)^{1/2} \omega A_c Re_{\omega}^{n_R} z. \quad (6.55)$$

Here the constant n_R is defined in Eq. (3.31).

The coordinate of the boundary of the viscous sub-layer z_1^+, where the linear model (6.55) holds, can be written as

$$\frac{z_1}{\delta} = \frac{z_1^+}{\gamma(1+\alpha^2)^{1/2} A_c^{1/2} Re_{\omega}^{1/(1+3n)}}, \quad (6.56)$$

where $z_1^+ = 12.54$; 13.44; 14.23 and 15.09 for $n = 1/7$; 1/8; 1/9 and 1/10, respectively (see Chap. 2). According to Eq. (6.56), this corresponds to $z_1/\delta = 0.01$–0.02. Figure 6.8 confirms the validity of the model (6.55) up to $z/\delta_{\varphi}^{**} = 0.2$, $z/\delta = 0.02$, or $\Delta = \delta_T/\delta = 0.02$.

In the power-law model, the Stanton number is given by Eq. (2.64). In Sect. 2.4.3, the model assumption $\left(z_{1T}^+/z_1^+\right)^{n_T-1} Pr^{-n_T} = Pr^{-n_p}$ completes Eq. (2.64), whereas validations of this model against experiments deliver the value of the exponent n_p.

At high Prandtl or Schmidt numbers, the entire thermal/diffusion boundary layer is included inside the viscous sub-layer of the velocity boundary layer. Hence, one can assume that the relation between the coordinates z_1^+ and z_{1T}^+ (viscous and heat conduction sub-layer) can be recast as

$$\left(z_{1T}^+/z_1^+\right)^{n_T-1} Pr^{-n_T} = K_{\alpha} Pr^{-n_p}. \quad (6.57)$$

Validations of the model (6.57) against experiments for the *Nu* or *Sh* numbers enable finding the coefficient K_{α} and exponent n_p. Consequently, given $n = n_T$, Eqs. (2.66) and (2.67) turn to

$$St = (c_f/2)\Delta^{-n} Pr^{-n_p} K_{\alpha} = A_c Re_{\omega}^{-2n/(3n+1)} \Delta^{-n} Pr^{-n_p} K_{\alpha}, \quad (6.58)$$

$$Nu = St Re_{\omega} Pr (1+\alpha^2)^{1/2} = A_c (1+\alpha^2)^{1/2} Re_{\omega}^{n_R} \Delta^{-n} Pr^{1-n_p} K_{\alpha}. \quad (6.59)$$

Substituting Eqs. (2.53), (6.55), (6.58) and (6.59) into Eq. (2.20), one can transform the latter to the following notation:

$$\frac{n}{2(n+2)} \alpha\omega \frac{\mathrm{d}}{\mathrm{d}r}\left[r\delta^2 \Delta^2 Re_{\omega}^{n_R} \Delta T\right] = K_{\alpha} \Delta^{-n} Pr^{-n_p} Re_{\omega}^{n_R} \nu \Delta T, \quad (6.60)$$

where Eq. (2.77) determines the boundary layer thickness δ, while $\Delta T = T_w - T_{\infty}$.

The condition $\Delta = \text{const.}$ is inapplicable to Eq. (6.60), otherwise the exponents at the variable r on the left- and right-hand sides of Eq. (6.60) are not equal to each other.

Let us assume the parameter Δ to be a power-law function

$$\Delta(r) = C_\Delta r^k. \tag{6.61}$$

Substituting Eq. (6.61) into Eq. (6.60) and keeping in mind Eqs. (2.30), (2.77), (2.78) and (3.31), one can finally obtain

$$\Delta = C_{\Delta *} Re_\omega^{k/2}, \tag{6.62}$$

$$C_{\Delta *} = C_{\Delta **} Pr^{-n_\mathrm{p}/(2+n)}, \tag{6.63}$$

$$C_{\Delta **} = \left[\frac{K_\alpha 2(n+2)/n}{\alpha\gamma^2(1 - nk + n_* + 2n_\mathrm{R})}\right]^{1/(n+2)}, \tag{6.64}$$

$$k = -2m/(2+n). \tag{6.65}$$

Equation (6.59) for the Nu number and the expression for Nu_{av} can be written as follows:

$$Nu = K_1 Re_\omega^{n_{\mathrm{R}*}}, \tag{6.66}$$

$$Nu_{\mathrm{av}} = K_2 Re_\varphi^{n_{\mathrm{R}*}}, \tag{6.67}$$

$$n_{\mathrm{R}*} = n_\mathrm{R} + mn/(2+n), \tag{6.68}$$

$$K_1 = K_\alpha K_3 C_{\Delta **}^{-n} Pr^{1/3}, \tag{6.69}$$

$$K_2 = 2K_1/(2n_{\mathrm{R}*} + 1), \tag{6.70}$$

$$n_\mathrm{p} = (2+n)/3. \tag{6.71}$$

Equation (6.71) takes into account the fact that the total exponent at the Pr number in Eq. (6.69) must be equal to 1/3.

Thus, in Eqs. (6.66) and (6.67), the total exponent $n_{\mathrm{R}*}$ at the Reynolds number is larger than that in Eq. (3.31) due to the additional term $mn/(2+n)$ (see Eq. (6.68)). This summand emerges as a result of the model with the *variable parameter* Δ being a subsiding function of the coordinate r or, in other words, local Re_ω (see Eq. (6.62)).

The values of the exponent $n_{\mathrm{R}*}$ are: $n_{\mathrm{R}*} = 0.84$ at $n = 1/7$, and $n_{\mathrm{R}*} = 0.868$ at $n = 1/9$. The latter agrees well with the exponent 0.87 at the Re_φ number in the experiment-based Eq. (6.35) [7]. To bring Eq. (8.73) at $n_{\mathrm{R}*} = 0.868$ into agreement with Eq. (6.35), the constant K_α must be equal to $K_\alpha = 1.254$, which yields at $n_* = 0$

$$Nu = 1.52 \times 10^{-2} Re_{\omega}^{0.868} Pr^{1/3}, \tag{6.72}$$

$$Nu_{\rm av} = 1.11 \times 10^{-2} Re_{\varphi}^{0.868} Pr^{1/3}, \tag{6.73}$$

$$\Delta = 18.31 Re_{\omega}^{-0.3158} Pr^{-1/3}. \tag{6.74}$$

In Fig. 6.7, curve 7 by Eq. (6.73) and curve 3 by Eq. (6.34) merge. Equations (6.72) and (6.35) are also practically identical.

The parameter range in experiments [7] is $Re_{\varphi} = 0.278 \times 10^6$–$1.8 \times 10^6$, $Sc = 930$–$10{,}320$. At minimal values $Sc = 930$ and $Re_{\omega} = 0.278 \times 10^6$ [7], Eq. (6.74) yields $\Delta = 0.036$. Parameter Δ is a decreasing function of the Schmidt and Reynolds numbers. Thus $\Delta = 0.015$ at $Sc = 10{,}320$ and $Re_{\omega} = 0.278 \times 10^6$, whereas $\Delta = 0.02$ at $Sc = 930$ and $Re_{\omega} = 1.8 \times 10^6$. This conforms to the limit $\Delta \le 0.02$ restricting validity of the linear model of the radial velocity profile.

Model with variable Δ and profile T^+ depending on Re_{ω}. In the theoretical works [9–11, 42], the Nusselt number at high *Pr* values is described by a relation

$$Nu = K_{\rm N}(1+\alpha^2)^{1/2}(c_{\rm f}/2)^{1/2} Re_{\omega} Pr^{1/3}, \tag{6.75}$$

where $K_{\rm N}$ is an empirical constant; Eq. (2.82) at $n = 1/7$ was used for $c_f/2$. Local and average Nusselt numbers take a form of Eqs. (6.66) and (6.67), respectively, with

$$K_1 = K_{\rm N}(1+\alpha^2)^{1/2} A_{\rm c}^{1/2} Pr^{1/3}, \tag{6.76}$$

$$n_{\rm R*} = (2n+1)/(3n+1), \tag{6.77}$$

while the constants K_1 and K_2 are related with Eq. (6.70). At $n = 1/7$, Eq. (6.77) brings $n_{\rm R*} = 0.9$.

Matching Eqs. (6.44) and (6.75) in view of Eq. (6.70), one can obtain $K_{\rm N} = 0.05986$.

Substitution of Eq. (6.61) into the thermal boundary layer equation yields again Eq. (6.62) for Δ with

$$C_{\Delta *} = C_{\Delta **} Pr^{-1/3}, \tag{6.78}$$

$$C_{\Delta **} = \left[\frac{K_{\rm N} 2(n+2)/n}{\alpha\gamma^2(1+2m+2k+n_* + 2n_{\rm R})A_{\rm c}^{1/2}}\right]^{1/(n+2)}, \tag{6.79}$$

$$k = (2n-1)/(3n+1). \tag{6.80}$$

Setting the values $n = 1/7$ and $n_* = 0$ into Eqs. (6.62), (6.78)–(6.80) one can obtain

$$\Delta = 12.54 Re_{\omega}^{-1/4} Pr^{-1/3}. \tag{6.81}$$

At the lower experimental limit of $Sc = 930$ and $Re_{\omega} = 0.278 \times 10^6$ [7], the value of Δ in view of Eq. (6.81) reduces to $\Delta = 0.037$. For the conditions $Sc = 10{,}320$ and $Re_{\omega} = 0.278 \times 10^6$: $\Delta = 0.016$. For $Sc = 930$ and $Re_{\omega} = 1.8 \times 10^6$: $\Delta = 0.023$. These values for Δ conform to the data obtained by Eq. (6.74) and the upper limiting boundary $\Delta \leq 0.02$ of the validity of the linear model for the radial velocity.

According to the *models with a constant and variable value of* Δ, the function T^+ in wall coordinates, defined by the power-law Eq. (2.23) at $n = n_{\mathrm{T}}$ does not depend on the Reynolds number, which is consistent with the results presented in [48, 49].

Model incorporating Eq. (6.75) results in the profile of T^+ being a function of Re_{ω}

$$T^+ = (z^+)^n (1 + \alpha^2)^{-n/2} \gamma^{-n} A_{\mathrm{c}}^{-n/2} C_{\Delta *}^{-n} K_{\mathrm{N}}^{-1} Pr^{2/3} Re_{\omega}^{-0.5(2n^2+n)/(3n+1)}. \tag{6.82}$$

To conclude, a novel methodology for simulations of temperature/concentration profiles for the values of Pr and Sc much larger than unity was outlined in this section. *An original integral method enabled evaluating a relative thickness Δ of the thermal/diffusion boundary layers that has not been attained by the other investigators. It was demonstrated that the model with a subsiding function $\Delta(r)$ yields a new summand in the expression for the exponent at the Reynolds number, which determines functional dependence of* Nu *or* Sh *numbers on the local radius* r. Consequently, theoretical relations obtained for Nusselt and Sherwood numbers are in a good consistency with the selected empirical equations.

References

1. Alden J (1994) Computational electrochemistry. D. Phil. thesis. Chapter 7, Oxford University, Oxford, UK
2. Awad MM (2008) Heat transfer from a rotating disk to fluids for a wide range of Prandtl numbers using the asymptotic model. Trans ASME J Heat Transfer 130(1):014505
3. Barcia OE, Mangiavacchi N, Mattos OR, Pontes J, Tribollet B (2008) Rotating disk flow in electrochemical cells: a coupled solution for hydrodynamic and mass equations. J Electrochem Soc 155(5):D424–D427
4. Daguenet M (1968) Etude du transport de matière en solution, a l'aide des électrodes a disque et a anneau tournants. Int J Heat Mass Transfer 11(11):1581–1596
5. Deslouis C, Tribollet B, Viet L (1980) Local and overall mass transfer rates to a rotating disk in turbulent and transition flows. Electrochim Acta 25(8):1027–1032
6. Dong Q, Santhanagopalan S, White RE (2008) A comparison of numerical solutions for the fluid motion generated by a rotating disk electrode. J Electrochem Soc 155(9):B963–B968
7. Dossenbach O (1976) Simultaneous laminar and turbulent mass transfer at a rotating disk electrode. Berichte der Bunsen-Gesellschaft—Physical Chemistry Chemical Physics 80 (4):341–343
8. Ellison BT, Cornet I (1971) Mass transfer to a rotating disk. J Electrochem Soc 118(1):68–72

9. Kawase Y, De A (1982) Turbulent mass transfer from a rotating disk. Electrochim Acta 27 (10):1469–1473
10. Law CG, Jr, Pierini P, Newman J (1981) Mass transfer to rotating disks and rotating rings in laminar, transition and fully-developed turbulent flow. Int J Heat Mass Transfer 24(5):909–918
11. Levich VG (1962) Physicochemical hydrodynamics. Prentice-Hall Inc, Englewood Cliffs
12. Mohr CM, Newman J (1976) Mass transfer to a rotating disk in transitional flow. J Electrochem Soc 123(11):1687–1691
13. Newman JS (1991) Electrochemical systems, 2nd edn. Prentice-Hall Inc, Englewood Cliffs, New York
14. Şara ON, Erkmen J, Yapici S, Çopur M (2008) Electrochemical mass transfer between an impinging jet and a rotating disk in a confined system. Int Commun Heat Mass Transfer 35 (3):289–298
15. Chen Y-M, Lee W-T, Wu S-J (1998) Heat (mass) transfer between an impinging jet and a rotating disk. Heat Mass Transfer 34(2–3):101–108
16. Cho HH, Rhee DH (2001) Local heat/mass transfer measurement on the effusion plate in impingement/effusion cooling systems. Trans ASME J Turbomach 123(3):601–608
17. Cho HH, Won CH, Ryu GY, Rhee DH (2003) Local heat transfer characteristics in a single rotating disk and co-rotating disks. Microsyst Technol 9(6–7):399–408
18. He Y, Ma LX, Huang S (2005) Convection heat and mass transfer from a disk. Heat Mass Transfer 41(8):766–772
19. Janotková E, Pavelek M (1986) A naphthalene sublimation method for predicting heat transfer from a rotating surface. Strojnícky časopis 37(3):381–393 (in Czech)
20. Koong S-S, Blackshear PL Jr (1965) Experimental measurement of mass transfer from a rotating disk in a uniform stream. Trans ASME J Heat Transfer 85:422–423
21. Kreith F, Taylor JH, Chong JP (1969) Heat and mass transfer from a rotating disk. Trans ASME J Heat Transfer 81:95–105
22. Mabuchi I, Kotake Y, Tanaka T (1972) Studies of convective heat transfer from a rotating disk (6th report, experiment on the laminar mass transfer from a stepwise discontinuous naphthalene disk rotating in a uniformed forced stream). Bull JSME 15(84):766–773
23. Shevchuk IV (2008) A new evaluation method for Nusselt numbers in naphthalene sublimation experiments in rotating-disk systems. Heat Mass Transfer 44(11):1409–1415
24. Shimada R, Naito S, Kumagai S, Takeyama T (1987) Enhancement of heat transfer from a rotating disk using a turbulence promoter. JSME Int J, Ser B 30(267):1423–1429
25. Sparrow EM, Chaboki A (1982) Heat transfer coefficients for a cup-like cavity rotating about its own axis. Int J Heat Mass Transfer 9(25):1334–1341
26. Tien CL, Campbell DT (1963) Heat and mass transfer from rotating cones. J Fluid Mech 17:105–112
27. Wang M, Zeng J (2011) Convective heat transfer of the different texture on the circumferential surface of coupling movement (rotating speed coupling with air velocity) disk. Proc 2011 ICETCE. Lushan, China, pp 1128–1132
28. Wu Y, Wu M, Zhang Y, Wang L (2014) Experimental study of heat and mass transfer of a rolling wheel. Heat Mass Transfer 50(2):151–159
29. Zeng JM, He Y, Wang MH, Ma LX (2011) Analogy study on convection heat transfer on the circumferential surface of rotating disc with naphthalene sublimation. J Qingdao Univ Sci Technol (Natural Sci Ed) 32(5):514–517
30. Shevchuk IV (2009) Convective heat and mass transfer in rotating disk systems. Springer, Berlin
31. Dorfman LA (1963) Hydrodynamic resistance and the heat loss of rotating solids. Oliver and Boyd, Edinburgh
32. Sparrow EM, Gregg JL (1959) Heat transfer from a rotating disc to fluids at any Prandtl number. Trans ASME J Heat Transfer 81:249–251
33. Cheng W-T, Lin H-T (1994) Unsteady and steady mass transfer by laminar forced flow against a rotating disk. Wärme und Stoffübertragung 30(2):101–108

34. Lin H-T, Lin L-K (1987) Heat transfer from a rotating cone or disk to fluids at any Prandtl number. Int Commun Heat Mass Transfer 14(3):323–332
35. Bogdan Z (1982) Cooling of a rotating disk by means of an impinging Jet. Proc 7th IHTC "Heat Transfer 1982". Munich, Germany, vol 3, pp 333–336
36. Elkins CJ, Eaton JK (1997) Heat transfer in the rotating disk boundary layer. Stanford University, Department of Mechanical Engineering, Thermosciences Division Report TSD–103. Stanford University (USA)
37. Popiel CO, Boguslawski L (1975) Local heat-transfer coefficients on the rotating disk in still air. Int J Heat Mass Transfer 18(1):167–170
38. Shevchuk IV (2000) Turbulent heat transfer of rotating disk at constant temperature or density of heat flux to the wall. High Temp 38(3):499–501
39. Shevchuk IV (2005) A new type of the boundary condition allowing analytical solution of the thermal boundary layer equation. Int J Thermal Sci 44(4):374–381
40. Shevchuk IV (2009) An integral method for turbulent heat and mass transfer over a rotating disk for the Prandtl and Schmidt numbers much larger than unity. Heat Mass Transfer 45 (10):1313–1321
41. Wasan DT, Tien CL, Wilke CR (1963) Theoretical correlation of velocity and eddy viscosity for flow close to a pipe wall. AIChE J 9(4):567–569
42. Paterson JA, Greif R (1973) Transport to a rotating disk in turbulent flow at high Prandtl or Schmidt number. Trans ASME J Heat Transfer 95(4):566–568
43. Mishra P, Singh PC (1978) Mass transfer from spinning disks. Chem Eng Sci 33 (11):1449–1461
44. von Karman Th (1921) Über laminare und turbulente Reibung. Z Angew Math Mech 1 (4):233–252
45. Owen JM, Rogers RH (1989) Flow and heat transfer in rotating-disc systems. In: Rotor-stator systems, vol 1. Research Studies Press Ltd., Taunton, Somerset, England
46. Littel HS, Eaton JK (1994) Turbulence characteristics of the boundary layer on a rotating disk. J Fluid Mech 266:175–207
47. Itoh M, Hasegawa I (1994) Turbulent boundary layer on a rotating disk in infinite quiescent fluid. JSME Int J Ser B 37(3):449–456
48. Kader BA (1981) Temperature and concentration profiles in fully turbulent boundary layers. Int J Heat Mass Transfer 24(9):1541–1544
49. Suga K (2007) Computation of high Prandtl number turbulent thermal fields by the analytical wall-function. Int J Heat Mass Transfer 50(25–26):4967–4974

Chapter 7
Convective Heat Transfer in a Pipe Rotating Around a Parallel Axis

7.1 Experiments and Simulations of Different Authors

Introduction. Rotors of electrical machines used in railway transport, electric generators, etc., are subject to internal heating due to high electromagnetic losses. This involves installation of internal air cooling systems incorporating rotating straight pipes [1–4]. This chapter is based on the results of the CFD simulations of such a cooling geometry published in [5]. Cross sections of the studied electric motor and respective CFD model are schematically depicted in Fig. 7.1.

Channels of an air cooling configuration are uniformly installed along the circumference inside the rotor (see Fig. 7.1a, b). Rotation of the pipes causes secondary flows, which are absent in non-rotating channels, provided that the inlet and thermal boundary conditions are the same. Consequently, heat transfer and hydraulic resistance cannot be predicted with the help of standard correlations used for non-rotating pipes.

The geometrical configuration employed in [5] is shown in Fig. 7.1c. In channels rotating parallel to the rotation axis, centrifugal buoyancy, and Coriolis forces emerge [1, 3, 4, 6]. The Coriolis force is defined as a vector product of the absolute flow velocity and the angular velocity of a rotating channel [3, 6, 7]. Non-collinear vectors result in nonzero Coriolis forces, whose absolute value is defined by the magnitudes of the flow velocity and angular velocity vectors, as well as angles between them [3, 6, 7]. If flow develops through the entrance hydrodynamic length of a pipe, velocity profiles are non-stabilized with rather high-velocity components perpendicular to the vector of the angular velocity (and channel walls) resulting in rather high Coriolis forces. In case where centrifugal buoyancy is negligible and channel flow attains a fully hydrodynamically developed state, the streamwise velocity component by far surpasses the radial and tangential components. Thus, the axial and angular velocity vectors become collinear (Fig. 7.1c), whereas the Coriolis force defined by their vector product tends to zero. An account for centrifugal buoyancy is needed, if centrifugal acceleration and wall heat flux are

I.V. Shevchuk, *Modelling of Convective Heat and Mass Transfer in Rotating Flows*, Mathematical Engineering, DOI 10.1007/978-3-319-20961-6_7

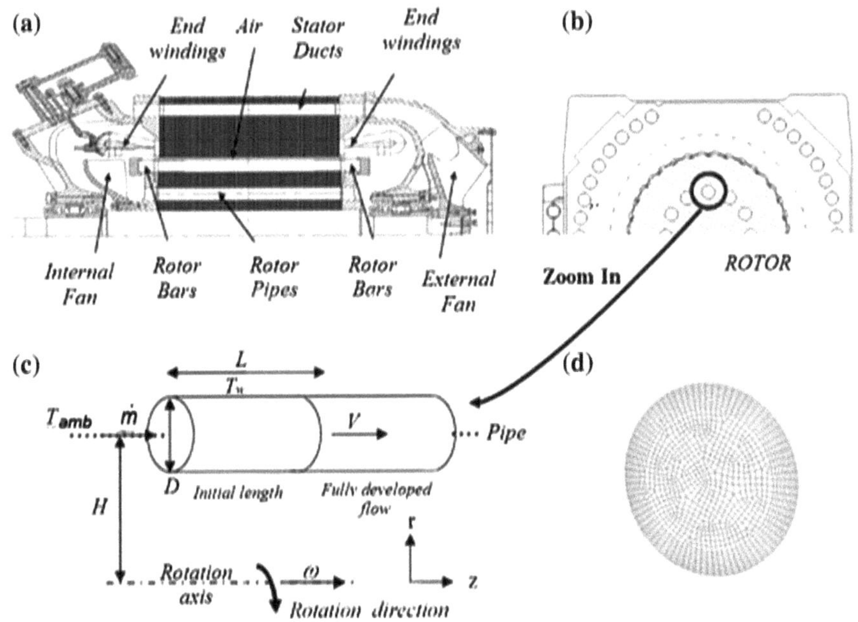

Fig. 7.1 Studied electrical motor: **a** side view; **b** front view; **c** schematic representation of the problem; **d** mesh at the inlet surface [5]

sufficiently large to cause macro-vortex flow occupying the entire channel cross section, with Coriolis forces only distorting the main secondary flow.

Rotation effects in the considered configuration are described by four similarity criteria: (1) Rossby number $Ro = \omega D/\bar{V}$, a ratio of the Coriolis forces (manifested by the angular velocity ω) to inertia forces (represented by the average streamwise flow velocity in the channel $\bar{V}$), while a characteristic length is the pipe diameter D; (2) rotational Reynolds number $Re_{\omega} = \omega D_{\mathrm{h}}^2/\nu$; (3) eccentricity parameter H/D_{h}; and (4) Rayleigh number $Ra_H = \omega^2 H^3 D_{\mathrm{h}} \beta \Delta T/(2\lambda\nu)$ that stands for centrifugal buoyancy. Channel cross section (circular, square, rectangular, or elliptic) also makes a noticeable influence on fluid flow and heat transfer [3, 5–15]. For relatively low Rayleigh numbers, buoyancy effects were vanishingly small. The coolant in experiments considered in this chapter was air.

Circular pipes. In the work [6], heat transfer in turbulent flow over the initial length of the pipe was investigated at ω = 280–2200 rpm and Re = 5000–20,000. Experiments demonstrated that the local Nusselt number increased with increasing $\omega^2 H$.

Experimental data [14] collapsed to an empirical relation (scatter ±14 %)

$$\overline{Nu} = 0.19 Re^{0.56} Re_{\omega}^{0.1}, \tag{7.1}$$

where L/D = 19.3, H/D = 25.19, Re = 5500–33,000, Re_ω = 600–4000, $Ro \leq 0.75$, and ω = 280–2200 rpm. In the non-rotating pipe of the same geometry [14], obtained that $\overline{Nu}_0 = 0.031Re^{0.79}$, which together with Eq. (7.1) yields [3]

$$\overline{Nu}/\overline{Nu}_0 = 6.129Re^{-0.23}Re_\omega^{0.11}. \tag{7.2}$$

Experiments [12] were described by an equation (scatter ±12 %)

$$\overline{Nu} = 0.015Re^{0.78}(Re_\omega/8)^m, \tag{7.3}$$

with m = 0.25. However, authors [3] revealed that an exponent m = 0.2 ensures better agreement with experiments [12]. Here, H/D = 24.02, L/D = 36.65 or H/D = 48.03, L/D = 69.03, Re = 5000–20,000, and Re_ω = 120–1200 (ω = 0–1000 rpm) with $Ro \leq 0.14$.

Experimental data [9] were approximated by an empirical equation

$$\overline{Nu} = 0.018Re^{0.8}(1 - 2D/L)(1 + 0.6RoH/D), \tag{7.4}$$

with H/D = 23.3, Re = (1–4.8) × 10^4, $RoH/D \leq 1.6$, $Ro \leq 0.07$, ωH = 0–30 m/s, and ω = 0–18.33 s^{-1}.

Equations (7.2) and (7.3) for $Ro \leq 0.14$ were generalized in the following form [3]

$$\overline{Nu} = \overline{Nu}_0(1 + 28Ro)^{0.5}. \tag{7.5}$$

Experiments of Baudoin [8] yield the following empirical equations

$$\overline{Nu}/\overline{Nu}_0 = 1 + 0.46(RoH/D)^{1.24}, \tag{7.6}$$

$$\overline{Nu}_0 = 0.023Pr^{1/3}Re^{0.8}\left(1 + (D/L)^{0.7}\right), \tag{7.7}$$

where H/D = 3, L/D = 11.6, Re = 3000–25,000, ω = 0–150 s^{-1}, and $T_{ref} = (T_i + T_{m-out})/2$. With a 14 % increase in the eccentricity, heat transfer increased in experiments by 7 % [8].

Square, rectangular, and elliptic pipes. In [13], experiments have been performed in a square channel, which were described by an empirical equation (scatter ±14 %)

$$\overline{Nu} = 0.011Re^{0.78}Re_\omega^{0.11}, \tag{7.8}$$

where L/D_h = 48, H/D_h = 32–48, Re_ω = 100–1000, and $Re \leq 2500$. Equation (7.8) predicts noticeably lower heat transfer rates than the one in a circular pipe, Eq. (7.3) [12]. The exponent 0.78 for the Reynolds number in the range Re = 100–1000 is an evidence of an earlier transition to turbulence excited by the rotation effects.

Eccentricity effect was studied experimentally by Mori et al. [11] in a rectangular channel. Eccentricity was equal to 6 in a fully heated channel and to 30 or 60 in a channel with only a half cross section heated. Angular velocity was set to vary from 800 to 3000 rpm. At the eccentricity parameter equal to 6, Mori et al. [11] obtained an empirical equation for the mean Nusselt number

$$\overline{Nu} = 2.1 Re_{\omega}^{0.11} \left[Re (D_{\mathrm{h}}/L)^{1.6} \right]^{0.43} Ra_H^{0.021}, \tag{7.9}$$

where Re = 2000–10,000 and $Ra_H = 10^8$–2.0×10^{10}. The effect of the centrifugal buoyancy was very weak. Mori et al. [11] assert that Eq. (7.9) agrees well with experiments [6].

The 3D fluid flow and heat transfer were numerically modeled in the works of [1, 4, 6, 10, 15] in channels with circular, square, rectangular, and elliptic cross sections with the help of different CFD codes and RANS turbulence models. In simulations of Mahadevappa et al. [10], the highest heat transfer enhancement was attained in circular pipes, while elliptic channels are in an intermediate standing between circular and rectangular channels.

To conclude, the empirical equations considered above quite noticeably disagree with each other. The effect of the eccentricity parameter H/D was considered only in the works of [8, 9], where values of H/D were very different, $H/D = 23.3$ in [9] and $H/D = 3$ in [8]. Maximal angular velocity $\omega = 18.33\ \mathrm{s}^{-1}$ in [9] was by a factor of 8 smaller than $\omega = 150\ \mathrm{s}^{-1}$ in [8]. To compare, maximal angular velocity was 50 s^{-1} in [15] and much lower in the rest of the referenced works. Thus, the only experiments [8] were performed under conditions close to those studied in our CFD investigation [5]. Also, perspectives of heat transfer enhancement at the expense of morphing of the pipe cross section to fit into a real electrical motor remained not elucidated.

Therefore, the *objective* of the study presented here was to use the CFD methodology to investigate the effects of the pipe cross-sectional design on the convective heat transfer in airflow in the channels rotating parallel to a rotation axis. Two types of pipes were studied having an elliptic cross section located radially or circumferentially with respect to the rotation radius. A separate study was performed on the effect of the flow angle of attack at the pipe inlet never investigated in the published literature. Experiments of Baudoin [8] were employed to validate the results of the numerical simulations.

7.2 Computational Model

Dimensions of the circular or elliptic pipe depicted in Fig. 7.1c are similar to a real design of a railway transportation motor [5]. Air is fed into the pipe at a given mass flow rate $\dot{m}$ and inlet temperature $T_{\mathrm{i}} < T_w$ (T_{i} is equal to the ambient temperature). The distance between the rotation and the pipe axes stands for the eccentricity, denoted as H.

7.2.1 Simulation Parameters

The pipe geometry was meshed using the preprocessor GAMBIT. The volume mesh incorporated a 10-cell prism layer at the pipe wall, quad surface cells outside of the boundary layer (Fig. 7.1d), and hexahedral cells forming the volume mesh. Having validated four different volume meshes of 100,000–500,000 elements by way of simulating the average surface heat transfer, it was proved that mesh independence was attained with the mesh size of about 350,000 elements, consisting of 250 cells in the axial, 17 cells in the radial, and 80 cells in the tangential (angular) direction.

In case if the interaction between stationary and rotating parts of geometry is weak, it is suggested by FLUENT [16] to use a rotating reference frame approach. In the new frame, the pipe becomes non-rotating, while Coriolis and centrifugal terms emerge explicitly in the Navier–Stokes equations. This approach was used in the present work.

Turbulence properties were constant at the inlet cross section and set to 10 % for turbulence intensity and 10 for the turbulent viscosity ratio (see definitions in [16]). These values were suggested by previous simulations of the entire motor [5]. Lower levels of the turbulence properties cause underestimated values of the Nusselt number.

The air temperature T_{i} at the inlet to the pipe and the wall temperature T_w were constant: $T_{\mathrm{i}} = 323$ K and $T_w = 473$ K. The mass flow rate $\dot{m}$ was chosen based on the measured data for a real engine (see below). Flow velocity at the inlet was mostly normal to the inlet face. A non-normal inlet velocity was assigned while investigating effects of the angle of attack. The range of variation of the angular velocity of rotation was $0 \le \omega \le 145\ \mathrm{s}^{-1}$.

Magnetic flux in the electric motor exhibits a strongly non-uniform 3D distribution that results in a nonuniform surface distribution of the heat flux per unit area q_w in the cooling channels. If the longitudinal distribution of the wall temperature T_w in straight channels used in the air cooling systems is moderately different from the condition T_w = const., the cross-sectional averaged Nusselt number is close to that for T_w = const. (see, e.g., [17]). This phenomenon is widely used both in experimental measurements of the local Nu numbers by means of TLCs at arbitrary variation of the local surface temperature (see [18, 19]) and in the CFD or analytical modeling of heat transfer at the boundary condition T_w = const. (see [18, 20]).

Rayleigh numbers were estimated as $Ra_H = 2 \times 10^8$, which in view of the Reynolds number $Re = 3500$ in the present investigation delivers conditions with vanishingly small centrifugal buoyancy effects [7, 11].

At the pipe outlet, a boundary condition of mass flow conservation was assigned.

7.2.2 Choice and Validation of the Turbulence Model

Five different turbulence models implemented in FLUENT were investigated in the present study:

Table 7.1 Geometrical setup in the benchmark experiments and the validation tests [5]

	Baudoin [8]	Present configuration
D (mm)	24	25
H (mm)	72	82.5
L (mm)	280	285
$\dot{m}$ (kg/s)	0.006	0.0066
ω s^{-1}	0–150	0–145
L/D	11.66	11.4
H/D	3	3.3

- *k-ε* model with an *enhanced wall treatment,*
- *k-ε* RNG model with an *enhanced wall treatment,*
- realizable *k-ε* model with an *enhanced wall treatment,*
- standard *k-ω* model, and
- *k-ω* SST model.

All five models require a mesh with the first mesh point very close to the wall ($y+ \sim 1$). The mesh with 350,000 elements resulted in the wall values $0.5 \leq y+ \leq 5$ for all turbulence models [5].

The geometry used for the turbulence model testing was close to that used by Baudoin [8]. Geometrical parameters, mass flow rate, and angular velocity of rotation for our and Baudoin's [8] configurations are specified in Table 7.1.

In our simulations [5], *Ro* number varied from 0.008 to 0.35 at *Re* = 14,000. Calculations in order to select an appropriate turbulence model were made at *Ro* = 0.088. In Table 7.2, the mean Nusselt numbers obtained from the simulations are compared with the value 56.4 from Baudoin's Eq. (7.6). The best agreement delivered the standard *k–ω* turbulence model with a relative error not exceeding 10 %. Use of the rest of the models entailed much more significant deviations from experiments (up to 60 % in the worst case). Hence, the standard *k–ω* turbulence model was selected for further simulations.

As seen from Fig. 7.2, simulations based on the standard *k–ω* turbulence model correlate well with Baudoin's empirical Eq. (7.6) [8]. This enabled further use of the developed CFD model in simulations of the effects of the pipe geometry and inlet boundary conditions stated above in the objectives of the present research.

Table 7.2 Mean Nusselt numbers in simulations [5] and experiments [8] for the circular pipe

	Nu	Difference (%)
Baudoin [8]	56.4	–
k-ε standard	90.9	60.6
k-ε RNG	86.8	53.9
k-ε realizable	80.2	42.2
k-ω standard	51.8	−8.2
k-ω SST	66.1	17.2

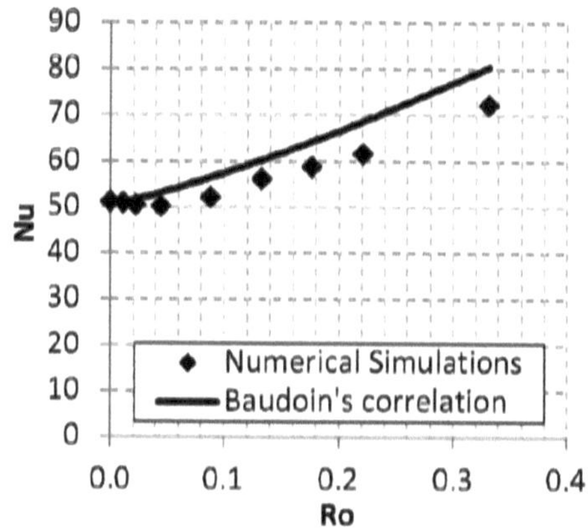

Fig. 7.2 Mean Nusselt numbers versus the Rossby number, simulations using the standard k–ω model (*points*) and Baudoin's empirical Eqs. (7.6) and (7.7) (*line*) [5]

7.3 Circular Pipe: Effect of the Angle of Attack

Parameters of the model. The circular pipe involved in Sect. 7.2.2 in the simulations (see Table 7.1) was employed also here as a reference pipe at $Re = 3500$ and $Ro = 2.24$. This value of the Re number can be recalculated to a total mass flow rate of $\dot{m} = 0.018$ kg/s in 12 pipes with $D = 0.025$ m (i.e., $\dot{m} = 0.0015$ kg/s in each pipe) and an angular velocity of 2200 rpm chosen in conformance with specifications for an industrial motor prototype. Temperature boundary conditions were $T_i = 323$ K and $T_w = 473$ K. Results for the reference case (velocity profile orthogonal to the inlet) were juxtaposed against simulations of the cases with different flow angle of attack or morphed cross-sectional shape.

Different angles of attack. This part of the study elucidates the effects of different inlet velocity profiles (Fig. 7.3) at the same total mass flow rate for all studied cases.

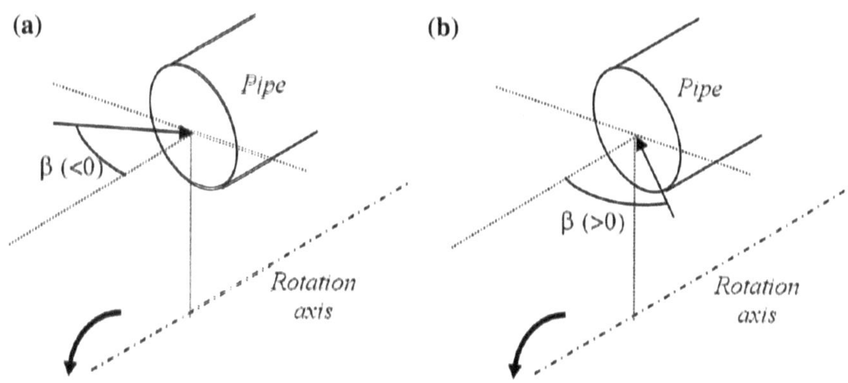

Fig. 7.3 Definition of the angle of attack; **a** negative angles; **b** positive angles [5]

If the velocity of rotation of air is smaller than that of the pipes, the flow angle of attack at the inlet is negative (Fig. 7.3). Axial v_z and tangential v_φ velocity components of the airflow in the simulations were constant at the inlet cross section, with the angle of attack β defined as $v_\varphi = v_z \tan(\beta)$, $v_z = 4\dot{m}/\rho\pi D^2$. Six different angles of attack were used: −80°, −58°, −28°, 28°, 58°, and 80°. An angle of 0° means an orthogonally entering flow. Negative angles denote that the flow at the inlet is contra-rotating with respect to the pipe.

Consequently, the axial velocity component at pipe inlet was calculated as $v_z/(\omega H) = 0.134$, while the resulting velocity $V_{\text{tot}} = \sqrt{v_z^2 + v_\varphi^2} = v_z\sqrt{1 + \tan^2\beta}$ varied over the range $V_{\text{tot}} = 2.54{-}14.63$ m s^{-1}.

If an electrical motor is rotating, airflow fed to the cooling system is non-orthogonal to the inlet of the rotating channels. The flow angle of attack (i.e., the shape of the velocity profile) influences the heat transfer inside the pipe either enhancing or suppressing it.

Table 7.3 represents a comparison of the mean Nusselt numbers predicted for the conditions with different inlet velocity profiles. As seen from Table 7.3, heat transfer is enhanced only if the air at the inlet to the pipe is contra-rotating with respect to the pipe wall (negative angle of attack). An increase in the mean Nusselt numbers reached 31 % for the largest negative angle of attack −80°. When the incoming flow and the pipe were corotating, mean heat transfer reduced in comparison with the reference case of β = 0°.

Figure 7.4a depicts the local cross-sectional averaged Nusselt numbers as a function of the axial coordinate $z^* = z/D_{\text{h}}$ in the rotating pipe (for the highest and lowest angle of attack) and the stationary reference pipe. Variation of turbulence intensity (also cross section averaged) in the streamwise direction is outlined in Fig. 7.4b. Figure 7.4c represents results for the relative Nusselt number $(Nu - Nu_{\text{ref}})/Nu_{\text{ref}}$ for both angles of attack.

Turbulence intensity and local Nusselt numbers demonstrate similar trends: Plotted against the reference case of β = 0°, they are augmented at negative and decreased at positive angles of attack β (see Fig. 7.4). For the positive value β = 80°, the local Nusselt number at z^* > 1 demonstrates a rather weak variation, which differs from the curves for the other values of β. For the negative angle β = −80° at z^* = 0–3.6 (initial hydrodynamic length), the Nusselt number is high and goes beyond the stabilized value by up to 100 %.

Table 7.3 Influence of the angle of attack on mean heat transfer in the circular pipe [5]

Angle β	Nu	Difference (%)
−80°	53.1	31.1
−58°	45.8	13.1
−28°	43.7	7.9
Reference case 0°	40.5	–
28°	39.3	−2.9
58°	38.1	−5.9
80°	29.3	−27.7

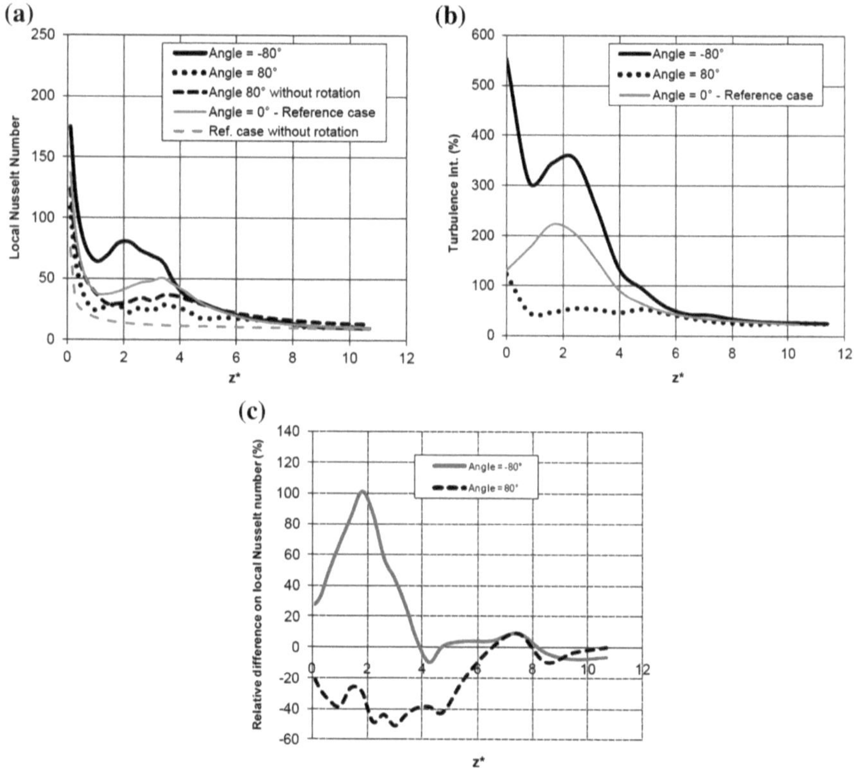

Fig. 7.4 Effect of the angle of attack: **a** local Nusselt number; **b** turbulence intensity; **c** relative difference of the Nusselt number with the reference case [5]

For all angles of attack, the Nusselt number noticeably decreased and reached for $z^* > 6$ nearly constant values of about 10–12. The FLUENT code predicts the Nusselt number $Nu_0 = 9.5$ at the end of a non-rotating pipe (zero rotation reference case in Fig. 7.4a). One can conclude from here that rotation still causes a weak effect on convective heat transfer at the end of the rotating pipe, whose length is insufficient to ensure independency of the local Nusselt number from rotation, which is otherwise expected to collapse to the non-rotating pipe value $Nu_0 = 9.5$. Dittus-Boelter equation for the developed flow in a non-rotating pipe [21]

$$Nu_0 = 0.023 Re^{0.8} Pr^{0.4} \tag{7.10}$$

predicts the value $Nu_0 = 14$, which is about 47 % higher than $Nu_0 = 9.5$ by FLUENT. Authors [20] also noticed that the Nusselt number resulting from the simulations with the help of FLUENT is smaller than those predicted by Eq. (7.10).

As seen from Fig. 7.4, once the flow reaches a certain axial location z^*, it turns to be *quasi-stabilized*, i.e., independent of the inlet velocity profile, with all the curves asymptotically tending to a constant level lying still above the correspondent values for a non-rotating pipe for identical *Re* numbers.

Quasi-stabilized flow with zero angle of attack at the inlet is subject to internal swirl in the end of the pipe (see Fig. 7.5) that is apparently the reason for the Nusselt number to surpass its benchmark value in the reference non-rotating pipe.

Only the initial length of the pipe is subject to tangible effects of rotation. Figure 7.4b reveals high turbulence intensity levels for the negative angle $\beta = -80°$ right after the inlet to the pipe. The reason for this phenomenon is strong shear stresses engendered by collision of the flow and the pipe wall due to their contra-rotation at the pipe inlet.

Having passed the location $z^{*} = 6$, the flow turns to be quasi-stabilized and attains practically the same turbulence intensity levels for all cases depicted in Fig. 7.4b. This testifies that the Rayleigh number (i.e., centrifugal buoyancy) has no effect on convective heat transfer in the rotating pipe considered in the present work.

Figure 7.6 depicts perspective and longitudinal views of the surface heat flux distributions in the pipe for the reference and extreme cases of $\beta = 80°$ and $-80°$. As seen from Fig. 7.6, the heat transfer enhancement next to the inlet of the pipe lies behind the overall heat transfer enhancement in the pipe. This conclusion conforms to [6, 12], who believed that namely the flow pattern immediately downstream of the pipe inlet determines overall peculiarities of convective heat transfer in the pipe. The effect of the angle of attack on heat transfer in comparison with the reference case exhibits itself only next to the pipe inlet, where local heat transfer rates are highest.

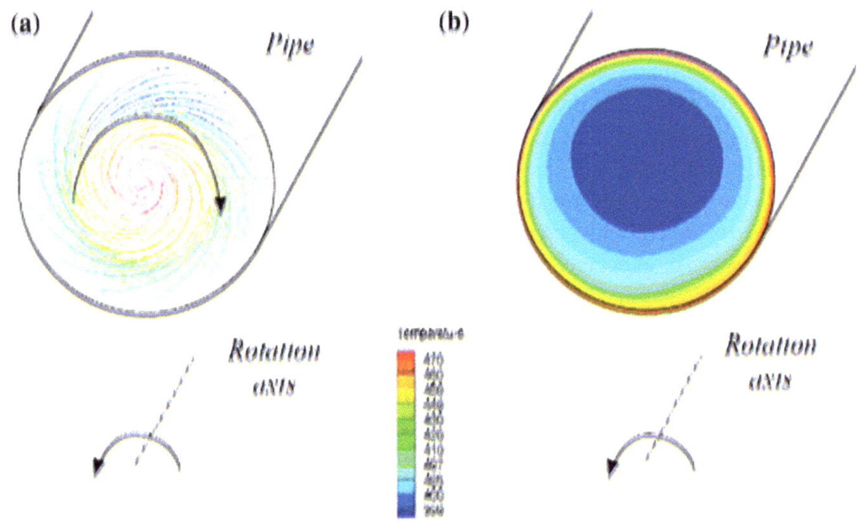

Fig. 7.5 Flow at the outlet of the rotating circular pipe with $\beta = 0°$ (reference case): **a** quasi-stabilized swirling flow, **b** temperature distribution [5]

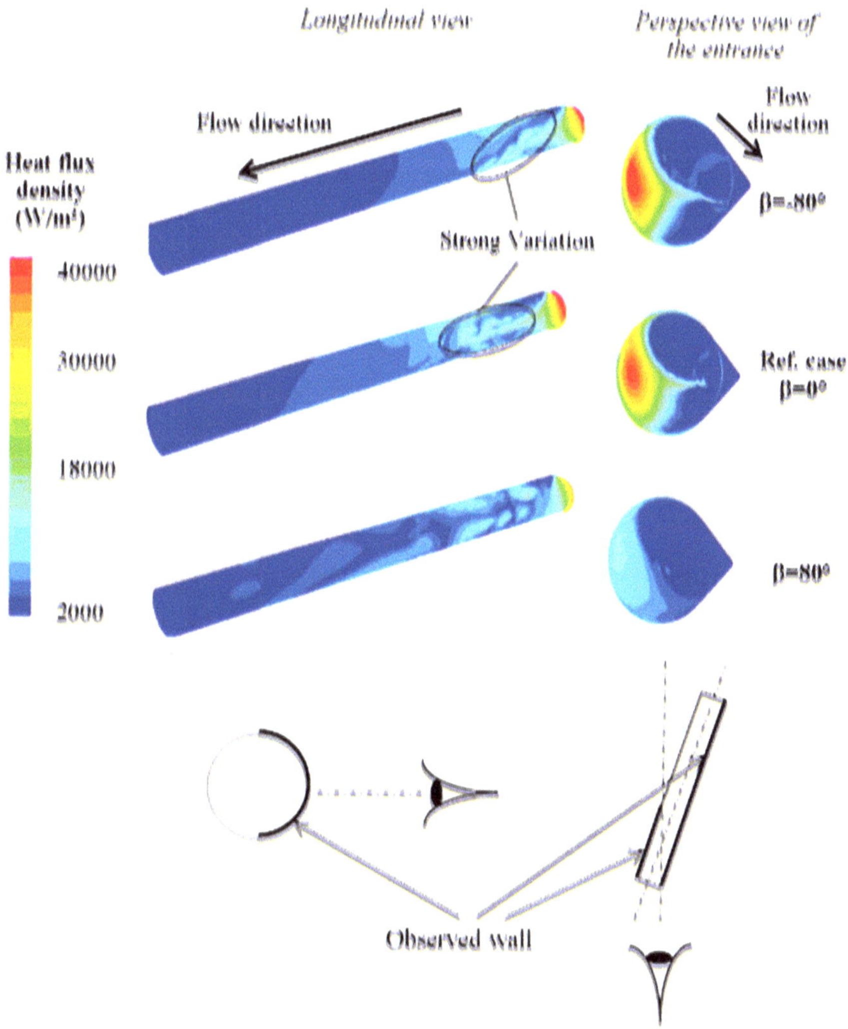

Fig. 7.6 Influence of the angle of attack on the local surface heat transfer [5]

Strong variation of the local surface heat transfer seen in the cases with $\beta = -80°$ and 0° disappears in the case with β = 80° representing the worst cooling case (Fig. 7.6).

Positive angle of attack β (corotating flow and pipe wall) entails noticeable diminution of the local heat transfer next to the pipe inlet, because of the much weaker flow impingement onto the pipe wall. The difference in the circumferential velocities of the air and pipe wall next to the inlet is rather small, which causes small shear stresses and heat transfer on the pipe wall. On the contrary, a negative angle of attack β (contra-rotating flow and pipe wall) causes stronger air impingement onto the pipe wall and higher heat transfer.

To conclude, the efficiency of the pipe cooling is significantly affected by the angle of attack of the flow fed into the pipe. The incoming flow and the pipe itself must be contra-rotating to boost the impingement effects that entail shear stress and heat transfer augmentation next to the pipe inlet. In doing so, heat transfer enhances by 30 % at the largest angle of attack of $\beta = -80°$ of those studied in this work. In practice, contra-rotation of flow and pipe at the inlet can be ensured via installing flow deflectors.

7.4 Elliptic Pipe

Convective heat transfer in a cooling system can be enhanced by replacing the circular cross section of the pipe with an elliptic one. Elliptic cross section was positioned circumferentially or radially, i.e., perpendicular or aligned with the rotation radius (Fig. 7.7), with the hydraulic or equivalent diameter of the pipe retained unchanged.

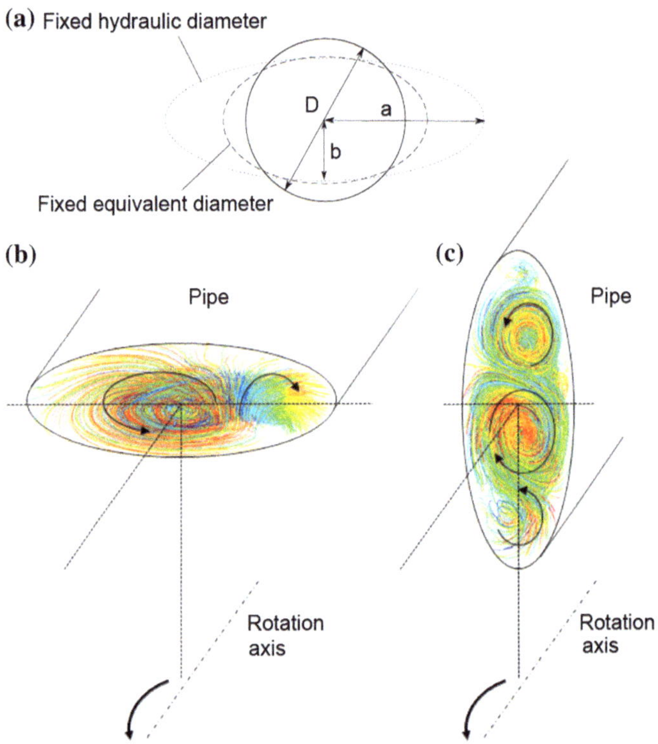

Fig. 7.7 Schematics of the circular and both elliptic cross sections (**a**); path lines at the pipe outlet of elliptic pipes with the fixed hydraulic diameter for the circumferential pipe (**b**) and radial pipe (**c**) [5]

7.4.1 Fixed Hydraulic Diameter

It was assumed that an elliptic pipe and the reference circular pipe (Table 7.1) have the same hydraulic diameter, which in the elliptic pipe was defined as $D_h = 4S/P_e = 25$ mm. Two basic dimensions of the ellipse a and b, shown in Fig. 7.7a, were assigned to be $a = D = D_h = 25$ mm and $b = 9.45$ mm. If the hydraulic diameter D_h remains unchanged, the elliptic area $S = \pi ab$, as compared to circular area $S = \pi a^2/4$, enlarged by a factor of $4b/a = 1.512$. As a result, the elliptic perimeter (representing the contact surface between the flow and the pipe) $P = 4S/D_h = 4S/a = 4\pi b$ also enlarged compared to the circular pipe perimeter $P = \pi D = \pi a$ by a factor of $4b/a = 1.512$. In a single pipe, the total mass flow rate was the same 0.0015 kg/s. The eccentricity parameter also remained unchanged $H = 82.5$ mm (Table 7.1).

Streamwise distributions of the local cross-sectional averaged Nusselt numbers and turbulence intensity in the elliptic pipes are plotted in Fig. 7.8a, b, c against

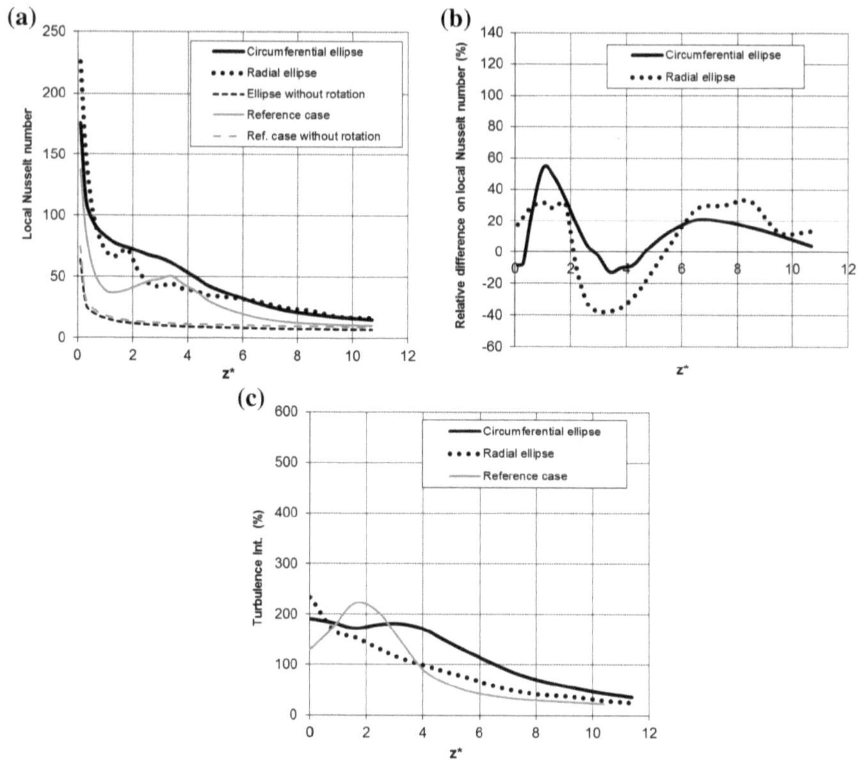

Fig. 7.8 Comparisons between the two ellipses and the reference case of the circular pipe, with the hydraulic diameter fixed: **a** local Nusselt numbers; **b** turbulence intensity; **c** relative difference from the reference case [5]

those in the reference circular pipe. It is evident that in elliptic pipes, the flow at $z^* > 9$ turns to be quasi-stabilized, which occurs at a larger distance from the inlet than in circular pipes. All curves flatten asymptotically to constant levels that are larger than those in circular pipes.

The functions of the local Nusselt numbers in Fig. 7.8a, c in both elliptic pipes look differently from those in the reference circular pipe. Next to the pipe inlet, the turbulence intensity is rather high (Fig. 7.8b). The highest turbulence intensity can be observed at $z^* > 3$ in the pipe with a circumferential elliptic cross section.

As mentioned above in the introduction, over the entrance length of a pipe, the radial and tangential velocities are rather large because of the initial flow swirl, which engenders large Coriolis forces.

Streamlines and temperature fields at the outlets of circumferential and radial elliptic pipes are shown in Figs. 7.7 and 7.9. Approaching the outlet of the elliptic rotating pipes flow turns to be quasi-stabilized and is accompanied with secondary flows caused by Coriolis forces. The nearly constant Nusselt number and turbulence intensity here are larger than those in stationary pipes. Mixing is better in the circumferential elliptic pipe (Fig. 7.7b), and air is in good contact with the pipe. On the contrary, three contra-rotating vortices emerge in the radial elliptic pipe with the middle vortex isolated (Fig. 7.7c). Therefore, air temperature fields are more uniform in Fig. 7.9a than in Fig. 7.9b, where air is involved in a closed circulation loop in the center of the channel. In the radial elliptic pipe, shear stresses are smaller due to the effects of three contra-rotating vortices [5].

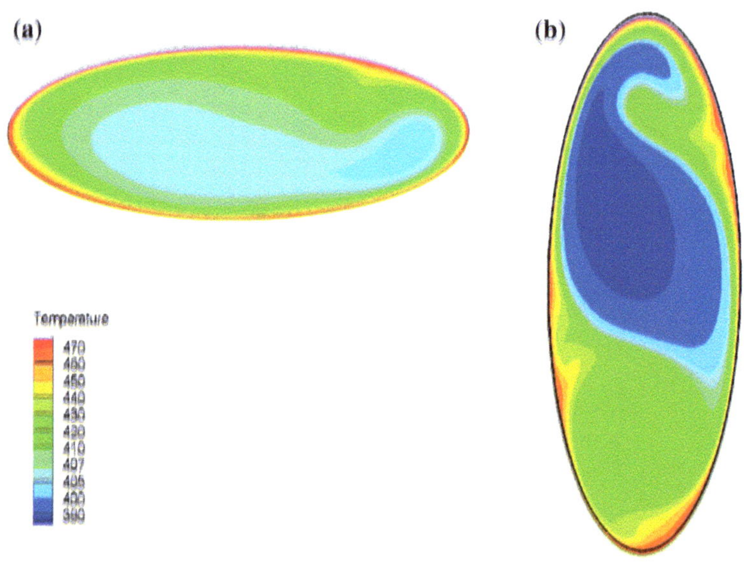

Fig. 7.9 Temperature distributions at the pipe outlet in elliptic pipes with the fixed hydraulic diameter: **a** circumferential pipe; **b** radial pipe [5]

Table 7.4 Cross-sectional geometry effect on heat transfer (hydraulic diameter fixed) [5]

	Mean *Nu*	Difference in *Nu* (%)	*h*	Contact surface *S*	$h \cdot S$	Difference in $h \cdot S$ (%)
Circular pipe	40.5	–	38.7	0.0224	0.87	–
Circumferential ellipse	44.5	+10	42.6	0.0338	1.44	+66.4
Radial ellipse	41.6	+2.8	39.7	0.0338	1.34	+54.8
Total heat transfer, 12 circular pipes			38.7	0.2688	10.40	–
Total heat transfer, 8 circumferential ellipses			42.6	0.2704	11.52	+10.8
Total heat transfer, 8 radial ellipses			39.7	0.2704	10.72	+3.1

As seen from Table 7.4, the mean Nusselt number increased in both circumferential and radial elliptic pipes in comparison with the circular reference pipe. However, the circumferential elliptic pipes increased the heat transfer by 10.0 %, i.e., more efficiently, than the radial elliptic pipes, which provide only 2.8 % heat transfer enhancement.

The overall heat transfer depends on a product of the heat transfer coefficient α and a channel surface S contacting the flow. In the case of an elliptic pipe, the contact surface is enlarged by 51.2 % as compared to the circular pipe. Hence, the product $\alpha \cdot S$ presented in Table 7.4 demonstrates 54.8–66.4 % increase in the radial and circumferential elliptic pipes, respectively, (compared to the circular pipe). This increase entails improvement of the entire cooling system efficiency.

However, the use of the elliptic pipes of the enlarged cross section instead of the circular pipes complicates their packaging in a motor in view of the necessity to ensure acceptable mechanical strength and electromagnetic flux distribution. As a result, only eight elliptic pipes can be installed in a real rotor instead of 12 circular pipes. Table 7.4 presents the data for overall amount of heat, in which cooling systems of 12 circular, 8 radial, and 8 circumferential elliptic pipes are able to remove from the rotor. As compared to the cooling system of 12 circular pipes, overall heat removal increases by only 3.1 % for the arrangement with 8 radial elliptic pipes and by 10.8 % for the configuration of 8 circumferential pipes. To remind, the mass flow rate $\dot{m} = 0.0015$ kg/s in a single elliptic pipe remained identical to that in the circular pipe. Thus, in the cooling configuration of 12 circular pipes, the total mass flow rate was $\dot{m} = 0.018$ kg/s in contrast to $\dot{m} = 0.012$ kg/s in the cooling configuration consisting of 8 elliptic pipes.

To correctly estimate the performance of the cooling configuration with 8 elliptic pipes, the overall mass flow rate through them must be enlarged also to $\dot{m} = 0.018$ kg/s, i.e., to $\dot{m} = 0.00225$ kg/s through a single pipe. Table 7.5 shows

Table 7.5 Heat transfer in the elliptic pipes with the increased mass flow rate $\dot{m} = 0.00225$ kg/s providing the same pressure drop [5]

	Mean $\overline{Nu}$	Difference in $\overline{Nu}$ (%)	α	Contact surface S	$\alpha \cdot S$	Difference in $h \cdot S$ (%)
Circumferential ellipse	53.3	+31.6	51	0.0338	6.05	+99.0
Radial ellipse	48	+18.5	45.9	0.0338	5.45	+79.3
Total mass flow rate, 8 circumferential ellipses			51	0.2704	13.8	+32.6
Total mass flow rate, 8 radial ellipses			45.9	0.2704	12.4	+19.3

that for the mass flow rate $\dot{m} = 0.00225$ kg/s in a single pipe, the heat transfer enhancement was 99 % in the circumferential and 79.3 % in the radial elliptic pipe in contrast to the circular pipe with $\dot{m} = 0.0015$ kg/s. The total cooling efficiency in the configuration with 8 elliptic pipes at $\dot{m} = 0.018$ kg/s (whose cumulative heat transfer surface is practically the same as that in the configuration of 12 circular pipes) is 32.6 % for circumferential and 19.3 % for radial elliptic pipes.

To recall, in an elliptic pipe, the cross-sectional area S is 1.512 times larger than that in a circular pipe. In view of the fact that the mass flow rate $\dot{m} = \rho \bar{V} S = 0.00225$ kg/s in an elliptic pipe is 1.5 times larger than $\dot{m} = 0.0015$ kg/s in a circular pipe, one can conclude that the mean axial velocity $\bar{V}$ is identical in both pipe geometries. Hence, the axial Reynolds number $Re = 3500$ is also identical, since the hydraulic diameter and the mean axial velocity remain the same. Therefore, one can expect that the pressure losses in a circular pipe with $\dot{m} = 0.0015$ kg/s must be equal to those in an elliptic pipe with $\dot{m} = 0.00225$ kg/s.

The Blasius equation for the friction factor for fully developed flow in stationary pipes valid for $Re = 3000$–100,000 [22]

$$f_0 = 0.316 Re^{-0.25} \tag{7.11}$$

delivers the value $f = 0.041$ for $\dot{m} = 0.0015$ kg/s. CFD simulations yield $f = 0.047$ for the stationary circular pipe with $\dot{m} = 0.0015$ kg/s, and $f = 0.049$ for the stationary elliptic pipe with $\dot{m} = 0.00225$ kg/s, which results in an only 4 % different value.

Thus, a calculation of the cooling system with 8 elliptic pipes having 51.2 % enlarged cross section disclosed the following disadvantages as compared to the reference configuration of 12 circular pipes: (a) reduced mechanical strength; (b) packaging problems, as elliptic pipes occupy locally more space; (c) larger nonuniformity of the circumferential temperature variation; and (d) deteriorated environment for magnetic flux circulation.

7.4.2 Fixed Equivalent Diameter

Drawbacks of the cooling configuration with 8 elliptic pipes mentioned above served as a motivation for an investigation of a cooling system with 12 elliptic pipes, whose cross-sectional area is identical to that of the reference circular pipe. This means keeping the equivalent diameter of the pipe $D_e = \sqrt{4S/\pi}$ unchanged and morphing it from a circular to an elliptic shape. Keeping in mind that the circular and the elliptic cross-sectional areas must be equal $S = \pi D^2/4 = \pi 0.025^2/4 = \pi ab$, the characteristic dimensions a and b of the ellipse (Fig. 7.7a) were chosen to be a = 16 mm and b = 9.77 mm.

Elliptic pipes were located circumferentially and radially, and the mass flow rate was set to be $\dot{m} = 0.0015$ kg/s. The elliptic cross section of the pipe being the same as that of a reference circular pipe provides a 6 % larger perimeter (i.e., the contact surface).

Figure 7.10a–c depicts functions of the local cross-sectional averaged Nusselt numbers in both elliptic pipes and the reference circular pipe. One can conclude

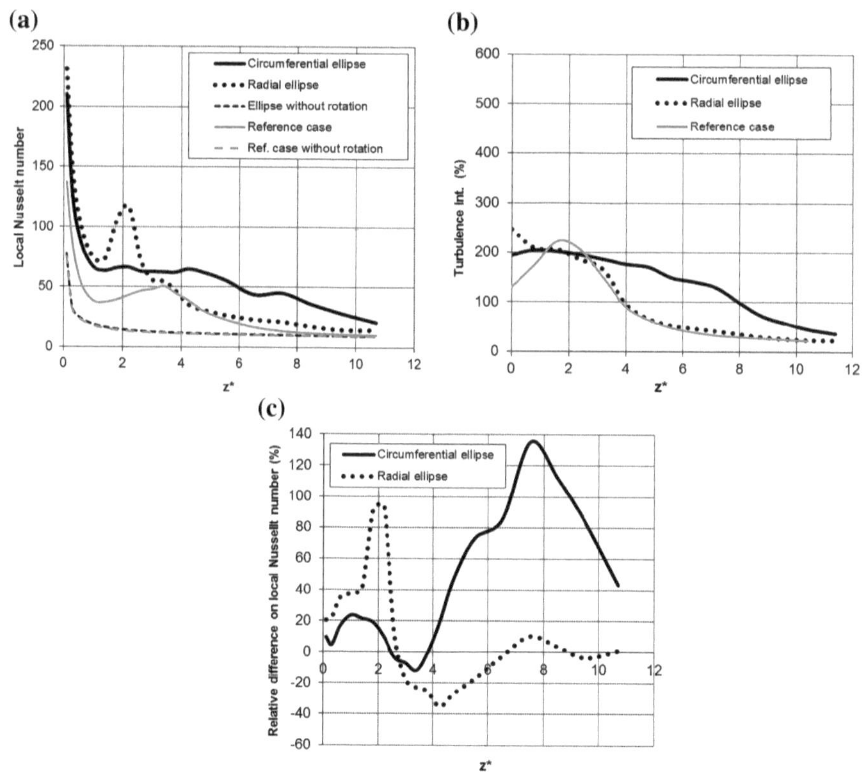

Fig. 7.10 Comparisons between two elliptic and reference circular pipes of the same equivalent diameter; **a** local Nusselt numbers; **b** turbulence intensity; **c** relative difference from the reference case [5]. Nusselt number is based on the equivalent diameter D_e

again that the higher turbulence intensity in the circumferential elliptic pipe ensures better performance than that in the radial elliptic pipe for $z^* > 2.7$. The Nusselt numbers level off to quasi-stabilized values near to the pipe outlet.

Path lines and temperature fields projected to the outlet of circumferential and radial elliptic pipes are shown in Figs. 7.11 and 7.12. Distinct secondary flows persist again over the span of the quasi-stabilized flow that sets on near to the outlet of the elliptic rotating pipes, which cause the Nusselt numbers and turbulence intensity to surpass those in non-rotating pipes though at a constant quasi-stabilized level.

Contrary to flow pattern in Fig. 7.7, only a single vortex in the cross sections of circumferential and radial elliptic pipes is visible in Fig. 7.11, because the cross-sectional shapes in Fig. 7.11 are closer to circular than those in Fig. 7.7.

In the circumferential elliptic pipe (Fig. 7.12a), the temperature field at the outlet cross section is more uniform than that of the vertical elliptic pipe (Fig. 7.12b), which is more symmetrically distributed than that in Fig. 7.9b. This phenomenon is apparently caused by the origin of a single vortex in Fig. 7.11b and three vortices in Fig. 7.7c.

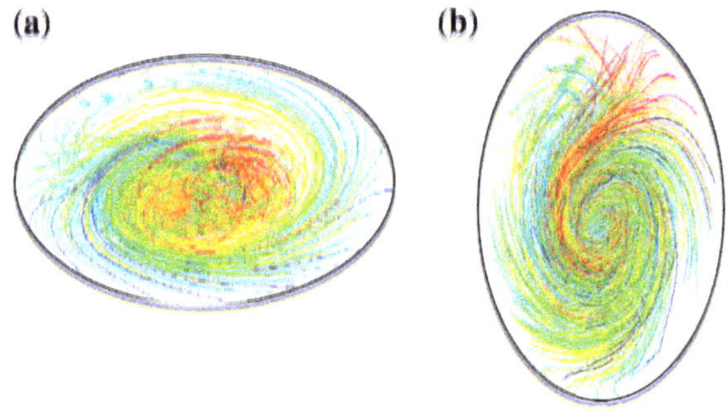

Fig. 7.11 Path lines at the pipe outlet of the elliptic pipes with the fixed equivalent diameter: **a** circumferential pipe; **b** radial pipe [5]

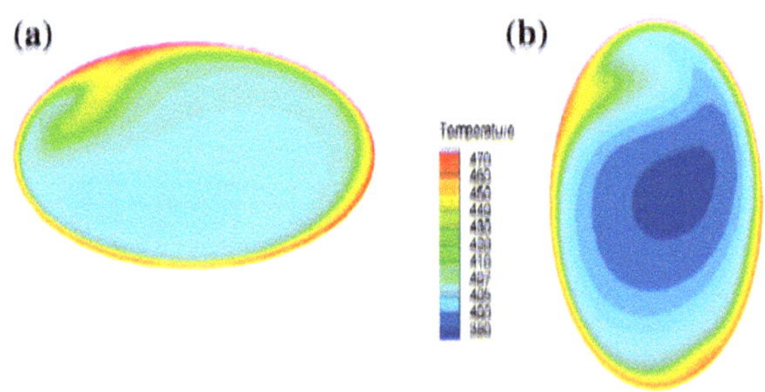

Fig. 7.12 Temperature distributions at the pipe outlet of elliptic pipes with the fixed equivalent diameter: **a** circumferential pipe; **b** radial pipe [5]

Table 7.6 Effect of a cross-sectional shape on heat transfer (equivalent diameter fixed) [5]

	Mean $\overline{Nu}$	Difference in $\overline{Nu}$ (%)	α	Contact surface S	$\alpha \cdot S$	Difference in $\alpha \cdot S$ (%)
Circular pipe	40.5	–	38.7	0.0224	0.87	–
Circumferential ellipse	52.5	+29.6	53.3	0.0237	1.26	+44.8
Radial ellipse	43.7	+7.9	44.3	0.0237	1.05	+20.7

Computations of the mean Nusselt number and the product of the heat transfer coefficient α and the wetted surface S are listed in Table 7.6 for the circular and both elliptic channels. With a fixed equivalent diameter, elliptic geometry brings larger improvement of heat transfer, than in the circular one.

A circumferential elliptic pipe displays again the best performance: 29.6 % enhancement of $\overline{Nu}$ and 44.8 % for the total cooling efficiency $\alpha \cdot S$. The radial elliptic pipe yields only 7.9 % enhancement of $\overline{Nu}$ and 20.7 % of $\alpha \cdot S$. The rate of enhancement of the total heat transfer $\alpha \cdot S$ in a single pipe stays in force also for the entire cooling configuration (as all of them consist of 12 pipes).

The pressure losses in the elliptic and circular pipe are the same, once the equivalent diameter and the mass flow rate do not change [22]. CFD simulations yield the friction factor of f = 0.049 in a non-rotating elliptic pipe for $\dot{m} = 0.0015$ kg/s (Table 7.7). This deviates from the value of f = 0.047 in a non-rotating circular pipe (Table 7.7) by 4 %. Thus, the pressure losses in the considered pipe geometries depend only on the mass flow rate.

From the point of view of the packaging, the cooling system configuration with 12 elliptic pipes, whose cross section has undergone only morphing and is equal to that of the reference circular pipes, is advantageous. These elliptic pipes can be installed in the same locations inside the real rotor, which helps to avoid the problems described above. To conclude, the use of elliptic pipes instead of the

Table 7.7 Calculated friction factors [5]

Configuration	Friction factor f
Stationary circular pipe ($\dot{m} = 0.0015$ kg/s)	0.047
Rotating circular pipe ($\dot{m} = 0.0015$ kg/s)	0.24
Fixed D_h, stationary ellipse ($\dot{m} = 0.0015$ kg/s)	0.060
Fixed D_h, rotating circumferential ellipse ($\dot{m} = 0.0015$ kg/s)	0.788
Fixed D_h, rotating radial ellipse ($\dot{m} = 0.0015$ kg/s)	0.516
Fixed D_h, stationary ellipse ($\dot{m} = 0.00225$ kg/s)	0.049
Fixed D_h, rotating circumferential ellipse ($\dot{m} = 0.00225$ kg/s)	0.408
Fixed D_h, rotating radial ellipse ($\dot{m} = 0.00225$ kg/s)	0.289
Fixed D_e, stationary ellipse ($\dot{m} = 0.0015$ kg/s)	0.049
Fixed D_e, rotating circumferential ellipse ($\dot{m} = 0.0015$ kg/s)	0.362
Fixed D_e, rotating radial ellipse ($\dot{m} = 0.0015$ kg/s)	0.276

circular pipes is beneficial in terms of the cooling system efficiency. From the technological point of view, the most efficient configuration is a *circumferential* elliptic pipe, whose *equivalent* diameter equals to that of a reference circular pipe.

7.4.3 Friction Factor in Rotating Pipes

Computed friction factors in circular and elliptic pipes of both types, rotating and non-rotating, are listed in Table 7.7. Based on these results, one can draw the following important conclusions.

It was proved above that the friction factor remained practically unchanged, once the cross section was subject to morphing (circular toward elliptic) and the axial Reynolds number (based on the hydraulic/equivalent diameter and average axial flow velocity) was kept identical Re = 3500: f = 0.047 for $\dot{m} = 0.0015$ kg/s in a non-rotating circular pipe, f = 0.049 for $\dot{m} = 0.00225$ kg/s in a non-rotating elliptic pipe with the same D_h and f = 0.049 for $\dot{m} = 0.0015$ kg/s in a non-rotating elliptic pipe with the same D_e.

In a non-rotating elliptic pipe with the same D_h and $\dot{m} = 0.0015$ kg/s, the friction factor exhibits an increased value f = 0.06, which can be explained in the following way. According to the Blasius Eq. (7.11), the friction factor f_0 (not the dimensional pressure losses) behaves as an increasing function for the decreased Reynolds number Re = 2333, which corresponds to the elliptic pipes with the same D_h and $\dot{m} = 0.0015$ kg/s. The friction factor is defined as $f = \Delta P D_h/(0.5\rho\bar{V}^2 L)$; for the compared cases, D_h and L are identical, while the ratio of the mean axial velocities $\bar{V}$ in the elliptic and circular pipe is 2333/3500 = 0.667 (Reynolds numbers are Re = 2333 and 3500, respectively). Hence, the ratio of the absolute pressure losses is $\Delta P_{Re=2333}/\Delta P_{Re=3500} = (0.06/0.047) \cdot 0.667^2 = 0.567$.

Table 7.7 shows that the friction factor drastically grows up in a rotating pipe. In a circular rotating pipe and $\dot{m} = 0.0015$ kg/s, the friction factor f is 5.1 larger as compared to a non-rotating pipe.

In elliptic pipes with the same D_h and unchanged Reynolds number Re = 3500 (in other words, for a larger cross section and $\dot{m} = 0.00225$ kg/s), the friction factor is 6.1 time larger in radial elliptic pipes and 8.7 times larger in circumferential elliptic pipes as compared to non-rotating circular pipes.

An analogous trend demonstrates elliptic pipes with the same D_e. In comparison with non-rotating circular pipes, the friction factor is 5.9 times larger for radial elliptic pipes and 7.7 times larger for the circumferential elliptic pipes.

To conclude, in a cooling system with 12 elliptic pipes obtained via morphing of the reference circular pipe and keeping the cross-sectional area unchanged, for the given Reynolds number Re = 3500 *more preferable circumferential elliptic pipes* ensured 45 % of the total heat transfer augmentation, which is the highest among all studied geometries. The friction factor in radial rotating elliptic pipes increased up to maximum 20 % larger than that in circular rotating pipes. Respective growth of the friction factor in the circumferential rotating elliptic pipes was from 51 % (for the same D_e) to 70 % (for the same D_h) larger than that in circular rotating pipes. This confirms the tendency of the highest heat transfer augmentation exhibited by the circumferential elliptic rotating pipes. The heat transfer augmentation in rotating pipes is accompanied with the growth of the pressure losses. At the same axial Reynolds number and rotation rate, the circumferential elliptic rotating pipe for the same D_e exhibited a friction factor growth by 11.3 % smaller than that in the circumferential elliptic rotating pipe for the same D_h (in radial rotating elliptic pipes for the same D_e and the same D_h, this difference is equal to 4.4 %).

References

1. Chiu HC, Jang JH, Yan WM (2007) Combined mixed convection and radiation heat transfer in rectangular ducts rotating about a parallel axis. Int J Heat Mass Transfer 50(21–22): 4229–4242
2. Fasquelle A, Le Besnerais J, Harmand S, Hecquet M, Brisset S, Brochet P, Randria A (2010) Coupled electromagnetics acoustic and thermal-flow modeling of an induction motor of railway traction. Appl Therm Eng 30(17–18):2788–2795
3. Shevchuk IV, Khalatov AA (1996) Heat transfer and hydrodynamics in straight channels rotating about a parallel or inclined axis. High Temp 34(3):455–467
4. Sleiti AK, Kapat JS (2006) Heat transfer in channels in parallel-mode rotation at high rotation numbers. AIAA J Thermophys Heat Transfer 20(4):748–753
5. Fasquelle A, Pellé J, Harmand S, Shevchuk IV (2014) Numerical study of convective heat transfer enhancement in a pipe rotating around a parallel axis. Trans ASME J Heat Transfer 136(5):051901
6. Humphreys JF, Morris WD, Barrow H (1967) Convection heat transfer in the entry region of a tube which revolves about an axis parallel to itself. Int J Heat Mass Transfer 10(3):333–340
7. Morris WD (1981) Heat transfer and fluid flow in rotating coolant channels. Research Studies Press, J. Wiley and Sons
8. Baudoin B (1987) Contribution l'étude des conditions d'écoulement dans le circuit de refroidissement d'un moteur électrique de type ouvert. PhD thesis, Université de Poitiers, France
9. Borisenko AI, Dan'ko VG, Yakovlev AI (1974) Aerodinamika i Teploperedacha v Elekticheskikh Mashinakh (Aerodynamics and heat transfer in electrical machines). Energiya Publ, Moscow
10. Mahadevappa M, Rammohan Rao V, Sastri VMK (1996) Numerical study of steady laminar fully developed fluid flow and heat transfer in rectangular and elliptical ducts rotating about a parallel axis. Int J Heat Mass Transfer 39(4):867–875
11. Mori H, Shiobara R, Hattori K (2000) Heat transfer characteristic of a rectangular channel rotating on a parallel axis (1st report, study on flow and heat transfer characteristics of a large rotating electrical machine). Trans JSME Ser B 66(650):2650–2654
12. Morris WD, Woods JL (1978) Heat transfer in the entrance region of tubes that rotate about a parallel axis. J Mech Eng Sci 20(6):319–325
13. Morris WD, Dias FM (1980) Turbulent heat transfer in a revolving square-sectioned tube. J Mech Eng Sci 22(2):95–101
14. Stephenson PL (1984) An experimental study of flow and heat transfer in a duct, rotating about a parallel axis. Heat and mass transfer in rotating machinery. Hemisphere Publishing Corporation, Washington, DC, pp 39–49

15. Torii S, Yang WJ (1998) Thermal-fluid transport phenomena of a strongly-heated gas flow in parallel tube rotation. Int J Rotat Machinery 4(4):271–282
16. [FLUENT] ANSYS FLUENT User's Guide (2009) Version 12, ANSYS Inc
17. Kays WM, Crawford ME, Weigand B (2005) Convective heat and mass transfer, 4th edn. Mc-Graw-Hill, New York. ISBN: 0-07-246876-9
18. Jenkins SC, Shevchuk IV, von Wolfersdorf J, Weigand B (2012) Transient thermal field measurements in a high aspect ratio channel related to transient thermochromic liquid crystal experiments. Trans ASME J Turbomach 134 (3):031002
19. Jenkins SC, Zehnder F, Shevchuk IV, von Wolfersdorf J, Weigand B, Schnieder M (2013) The effect of ribs and tip wall distance on heat transfer for a varying aspect ratio two-pass ribbed internal cooling channel. Trans ASME J Turbomach 135(2):021001
20. Shevchuk IV, Jenkins SC, Weigand B, von Wolfersdorf J, Neumann SO, Schnieder M (2011) Validation and analysis of numerical results for a varying aspect ratio two-pass internal cooling channel. Trans ASME J Heat Transfer 133(5):051701
21. Çengel YA (1998) Heat transfer: a practical approach. WBC McGraw-Hill, New York
22. Munson BR, Young DF, Okiishi TH, Huebsch WW (2009) Fundamentals of fluid mechanics. Wiley, New York

Chapter 8
Varying Aspect Ratio Two-Pass Internal Ribbed Cooling Channels with 180° Bends

8.1 Experiments and Simulations of Different Authors

Curvilinear flows in turns and bends represent a particular class of rotating flows, where the emergence of centrifugal and Coriolis forces, resulting from streamline curvature in a stationary configuration, increases heat transfer and pressure losses. Ribs installed in such channels cause an additional increase in the heat transfer and pressure losses. A schematic of a typical air cooling system of the gas turbine blades including ribbed two-pass channels is depicted in Fig. 8.1. This chapter elucidates the ways of optimization of such channels and describes the materials published in [1–6] from a unified position.

The increase in the thermodynamic efficiency and specific power is a permanent trend in modern gas turbine development. This requires establishing very high pressures and temperatures at the inlet to a gas turbine. As a result, the melting point of the turbine blade material is far below the gas temperatures in the hot pass. Because of this, internal cooling systems involving ribbed multi-pass channels and 180° bends are used to protect the material of the blades. Ribs are installed in the cooling passages under 30°–90° with respect to the main flow. Packaging restrictions inside the blade cause the height-to-width aspect ratio of a cooling channel to vary from 1:4 to 4:1, often being different in the neighboring passes of the channel. A schematic of a typical ribbed two-pass cooling channel including its geometrical parameters is depicted in Fig. 8.2.

All deviations of the cooling geometry from a smooth straight channel with a constant quadratic aspect ratio cause a significant increase in the pressure losses.

Flows in *multi-pass smooth channels* demonstrate emergence of impingement, recirculation, and flow separation zones induced by 180° bends, which lead to high thermal gradients and increased pressure losses. Authors [7–21] revealed significant heat/mass transfer augmentation in the bend and entrance region of the outlet pass due to secondary flows induced by centrifugal force.

I.V. Shevchuk, *Modelling of Convective Heat and Mass Transfer in Rotating Flows*, Mathematical Engineering, DOI 10.1007/978-3-319-20961-6_8

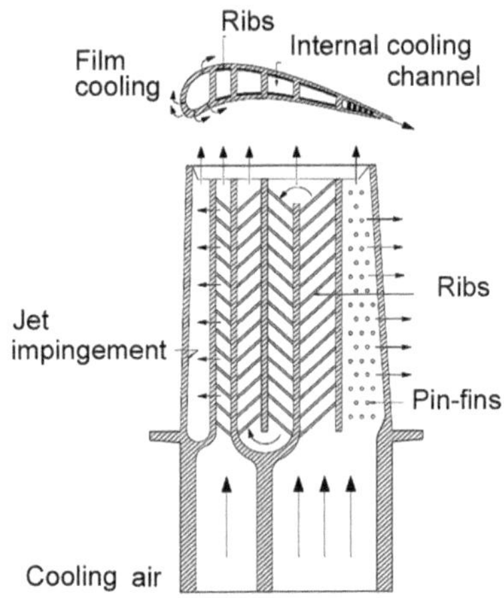

Fig. 8.1 Air cooling configuration of a gas turbine blade including ribbed two-pass channels (modified Fig. 1.25 from [60])

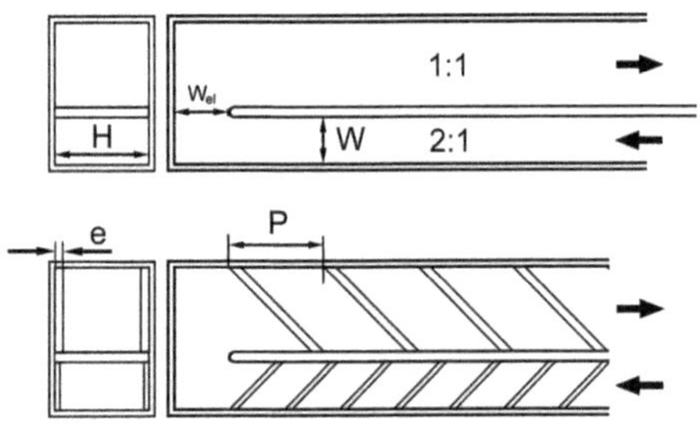

Fig. 8.2 Geometrical parameters of a ribbed two-pass channel [2]

In the ribbed multi-pass channels, rib location affects flow pattern, heat transfer rate, and pressure losses. The flow undergoes separation and reattachment after a rib. Strong secondary flows look as macro-vortex pairs and emerge behind the ribs inclined 30°–70° toward the main flow. Secondary flows noticeably augment the heat transfer; however, pressure losses grow even more significantly.

In sufficiently long ribbed channels, three-dimensional periodic fluid flow, and heat transfer patterns emerge, which repeat themselves over subsequent conventional one-rib periodic segments [1, 2, 22–27], etc.

Experiments [28–30] revealed that 45° and 60° ribs ensure higher thermohydraulic performance than 30° and 90° ribs, whereas the optimum rib spacing is $p/e = 10$ (see Fig. 8.2). There exists an optimal rib height providing rather high heat transfer augmentation together with acceptable rise in the pressure loss [31–33]. Having analyzed experiments of different authors and original simulations, Shevchuk et al. [3] recommended the rib height $e/D_h = 0.1$ as optimal for the aspect ratios $H/W = 4{:}1$ to 1:1, which conforms with conclusions of other authors.

A channel with the aspect ratio $H/W = 4{:}1$ demonstrates the best performance, which decreases with the aspect ratio [3, 34]. In [3, 7, 34–39] it has been demonstrated that thermal efficiency is a decreasing function of the Reynolds number, whereas the developed relations for the Nusselt number and friction factor depend on the parameters p/e, e/D_h, and Re.

In the bend region, centrifugal effects cause a radial pressure gradient, whereas the static pressure is higher close to the tip wall. Centrifugal forces do not compensate the radial pressure gradient in the boundary layers over the top and bottom walls, where the fluid flows toward the divider wall separating the inlet and outlet passes. To close the vortex circulation trajectory, the fluid near the center plane flows back to the tip wall (see, e.g., the PIV measurements [10, 14, 18]). These vortices were named *Dean vortices*. Both in smooth and ribbed channels, Dean vortices entail regions of very high heat/mass transfer in the bend and the entrance to the outlet pass, see, for example, the experiments [40–42]. Local flow impingement causes high heat/mass transfer also on the tip wall (90° turn) and on the external sidewall next to the bend (180° turn).

Experiments in a two-pass ribbed channel (in-line ribs inclined to the outer wall in the inlet pass and to the inner wall in the outlet pass) [40] revealed that the spots of high heat/mass transfer on the tip and outer sidewall shrink, since the Dean vortices conflict with the rib-induced macro-vortices emerging in the inlet pass. On the contrary, ribs at the entrance to the outlet pass reinforce Dean vortices. Nevertheless, downstream of the entrance rib effects destroy the Dean vortices rather fast. Hence, the curvature influence in the bend must be subtracted from the total heat transfer augmentation to clarify the net effects of the ribs.

Channels with *different cross sections of the inlet and outlet passes* have been less intensively studied in the past. Iacovides et al. [41] performed measurements, while Lucci et al. [43] carried out CFD simulations in a smooth channel with a 180° bend, an inlet pass with H/W = 1:1 (30 mm × 30 mm) and an outlet pass with H/W = 1:2 (30 mm × 60 mm). Though all physical effects here are similar to those in a channel with a constant cross section, doubled cross section in the outlet pass causes strong flow deceleration and more distinct separation and an extended recirculation zone over the divider wall at the entrance to the outlet pass.

The distance W_{el} from the tip to the divider wall severely affects the fluid flow and heat transfer pattern in the bend and in the entrance to the outlet pass. In experimental and numerical investigations in a smooth two-pass channel with H/W = 4:1, the distance W_{el} varied over a rather wide range. Increase in the distance W_{el} up to a point W/W_{el} = 1.25 results in reduced overall pressure losses owing to more unrestrained thoroughfare conditions for the flow in the bend. A further increase in W_{el} constitutes favorable conditions for emergence of a recirculation region near the tip wall, which results in a new trend of the increase in the pressure loss accompanied subsequent heat transfer augmentation. Joint effects of the varying channel's cross section and the tip wall distance were first investigated in [2, 4–6] are described below.

Many authors performed *CFD simulations* of convective heat transfer in two-pass channels with 180° bends. Authors [43] simulated the experimental conditions [41] for Re = 100,000. Low Reynolds number k-ε, nonlinear k-ω, and

RSM turbulence models used in [41] performed similarly and yielded results deviating from experiments by 10–100 % in different regions of the channel. In [44], the authors successfully employed an unsteady RANS approach for $Re = 150{,}000$. The steady RANS approach used in [3, 16] jointly with the realizable k-ε turbulence model and an enhanced wall treatment demonstrated good performance in smooth and 45° ribbed channels at $Re = 100{,}000$. High- and low-Reynolds number k-ε models, and V2F k-ε model provided good results in a rectangular 90° bend [45]. DES and LES approaches employed in [46, 47] are complicated and require enormous computing capacities as compared to RANS models and currently enable simulations for $Re < 30{,}000$, whereas in the modern cooling systems for heavy duty gas turbines the Reynolds numbers reach up to $Re = 200{,}000$ and higher.

The choice of the channel aspect is defined by packaging conditions, i.e., the channel location in the blade: a channel with a high aspect ratio H/W is placed in the leading edge or mid-chord of the blade, while a low aspect ratio channel is placed in the trailing edge (see Fig. 8.1). It is therefore important to study the effect of different aspect ratios of the multi-pass cooling channels on heat transfer and pressure drop, as well as the effect of the reduction in the hydraulic diameter of the inlet pass on heat transfer in the outlet pass.

The first objective of the investigation presented in this chapter was to simulate the flow field and heat transfer in a ribbed periodic segment of different aspect ratios $H/W = 4{:}1$, 2:1 and 1:1 [3]. *The second objective* was a simulation of a two-pass channel with a 180° bend, the inlet pass having $H/W = 2{:}1$ [4] or $H/W = 3{:}1$ [5, 6], the outlet pass having $H/W = 1{:}1$, with 45° ribs with $e/D_h = 0.1$ and $p/e = 10$ in both passes, and to validate numerical results against experimental data [1, 2]. The study also sought to (a) analyze effects of the Dean vortices and flow acceleration in the bend in a smooth channel separately from the effects of the ribs and (b) study the effect of changing the tip wall distance on the pressure loss and surface heat transfer to find an optimal value of W_{el}.

8.2 Single Periodic Ribbed Segment with $H/W = 4{:}1$, 2:1 and 1:1

The objectives of the study [3] described below, were to simulate:

(a) Convective heat transfer in a single periodic segment of the channel to validate in detail a CFD model via comparisons with experiments of different authors.
(b) Aspect ratios of $H/W = 4{:}1$, 2:1 and 1:1 at $Re = 100{,}000$, $e/H = 0.025$ to 0.1, $p/e = 10$ and 5 with square 45° ribs installed in-line on the top and bottom walls of a channel.
(c) Effects of the rib height and pitch.

8.2.1 Geometry and Flow Parameters

Two geometries of a periodic segment were investigated in the simulations with an inlet cross section (a) perpendicular to the sidewalls (Fig. 8.3a and b) aligned with 45° ribs (Fig. 8.3b). Structured grids were generated for the geometry in Fig. 8.3b, and hybrid grids for both geometries in Fig. 8.3. Detailed variant simulations proved that analogous grids in both geometries delivered results of similar quality.

In-line setup of the ribs on the top and bottom walls enabled modeling just a half of the computational volume and setting the upper surface to be a symmetry plane (Fig. 8.3).

In case of the $H/W \equiv 4:1$ aspect ratio, dimensions of the periodic segment were $W = 42$ mm, $H = 168$ mm, $H^* = H/2 = 84$ mm, $D_h = (8/5)W = 67.2$ mm, relative rib heights $e/H = 0.025$, 0.03, 0.04, 0.05, and 0.1 or $e/D_h = 0.0625$, 0.075, 0.1, 0.125, and 0.25. Such a channel was investigated experimentally in [48, 49].

In case of the $H/W = 2:1$ aspect ratio, the periodic segment had dimensions $W = 75$ mm, $H = 150$ mm, $H^* = H/2 = 75$ mm, $D_h = 100$ mm and the relative rib heights $e/H = 0.027$, 0.045, 0.067, 0.08, and 0.1 or $e/D_h = 0.04$, 0.067, 0.1, 0.12, and 0.15.

For the $H/W = 1:1$ aspect ratio, the channel dimensions were $W = 150$ mm, $H = 150$ mm, $H^* = H/2 = 75$ mm, $D_h = 150$ mm, together with relative rib heights $e/H = e/D_h = 0.027$, 0.045, 0.067, 0.1, and 0.15. The relative pitch was $p/e = 10$, with some computations for $H/W = 4:1$ performed at $p/e = 5$. The Reynolds number was $Re = 100{,}000$.

To match the thermal boundary conditions in the transient experiments using thermochromic liquid crystals (TLC) [49], the boundary condition $T_w = 350$ K = const. was set on all walls. The inlet temperature was prescribed to be $T_i = 310$ K = const.

Referring to the transient experiments, the assumption of isothermal walls is valid because the test section has a constant temperature when the experiments start, whereas the TLC indication is read for the same temperature value for each

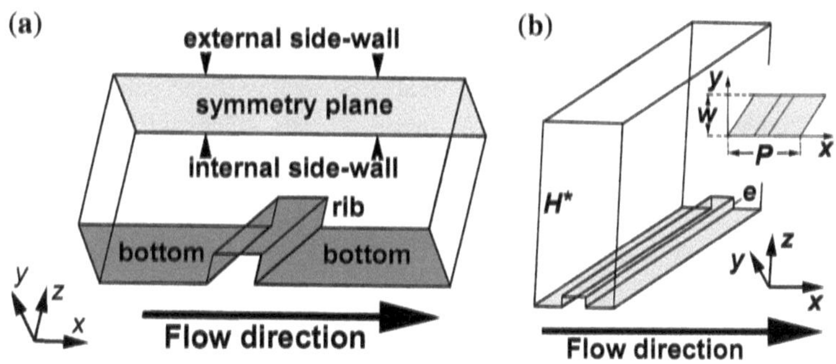

Fig. 8.3 Configurations of the periodic channel's segment with the inlet cross section **a** orthogonal to the sidewalls and **b** aligned with the rib [3]

coordinate. Spatial and temporal bulk-temperature variation can be well modeled assuming timewise isothermal walls [49, 50]. In addition, the effects of a slightly changing wall temperature for an air flow for high *Re* numbers are negligible (see [51]).

8.2.2 Numerical Methodology

All geometries and structured grids were created using the software GAMBIT. Further, CENTAUR® grid generator (see the work [52]) imported GAMBIT-created geometry to generate a hybrid grid. The grid sizes ensuring the necessary near-wall grid resolution amounted from 500,000 to 700,000 cells. Numerous grid-independency studies disclosed that both types of grids delivered practically the same results.

Simulations were performed using two different turbulence models. The first was a realizable *k*-ε model with enhanced wall treatment and a two-layer approach for wall modeling (known to accurately predict separating flows behind inclined ribs). The second was the SST *k*-ω model. Both models were used as they are implemented in the FLUENT code [53]. The realizable *k*-ε model always provided convergent solutions, while simulations using the *k*-ω SST model have not converged in a number of cases. All used grids fulfilled a requirement of the *k*-ε model with an enhanced wall treatment for the "y^+-coordinate" ("wall-law coordinate" perpendicular to the wall) to be $y_1^+ \sim 1$ in the first grid point. In the reality, the condition $y_1^+ \leq 1$ was practically always fulfilled for all grids, while the number of prismatic layers was ≥20. The *k*-ω SST turbulence model has similar requirements. In doing so, the usage of both turbulence models is justified.

Inlet and outlet of the periodic segment were connected through a periodic boundary condition. A mass flow rate corresponding to $Re = 100{,}000$ was assigned as the inlet boundary condition of the periodic segment. A solution was believed converged, once the scaled residuals were less than 10^{-6} or 10^{-9} for the continuity or energy equation, respectively.

All simulations were performed for air, $Pr = 0.72$. Computed average Nusselt number $\overline{Nu}$ and friction factor f were evaluated against their values in a smooth channel (subscript "0"). The friction factor f_0 was calculated from McAdams equation valid for $Re = 3 \times 10^4$–2×10^6 [54–56]

$$f_0 = 0.184\,Re^{-1/5}. \tag{8.1}$$

The Nusselt number Nu_0 was estimated using the Dittus–Boelter Eq. (7.10). Multiple simulations proved that the cross-sectional averaged temperature is practically invariable (with an inaccuracy of 0.4 %) in the streamwise *x*-direction of the ribbed segment. Therefore, as a reference temperature to calculate the Nusselt number, a volume-averaged temperature for each computed case was employed.

8.2.3 Comparative Flow Pattern

The velocity vector field obtained in the simulations of the periodic ribbed segment displays essential flow deceleration right before the rib, acceleration immediately above, and the onset of a separation bubble behind the rib (Fig. 8.4). Having overcome the separation region, flow reattaches to the channel's bottom at a location shifted to approximately two rib heights downstream of the rib, which agrees with the PIV measurements [18].

Figure 8.5 depicts the vector fields describing the secondary flow in the y–z cross section perpendicular to the channel sidewalls and crossing the rib in the middle of its length. The secondary flow has a vortex form: fluid flows lengthwise over the rib from the inner to outer sidewalls (right to left in Fig. 8.5), impinges onto the outer sidewall, turns upwards to the symmetry plane, diverts again backwards (left to right in Fig. 8.5) in the neighborhood of the symmetry plane, and finally flows

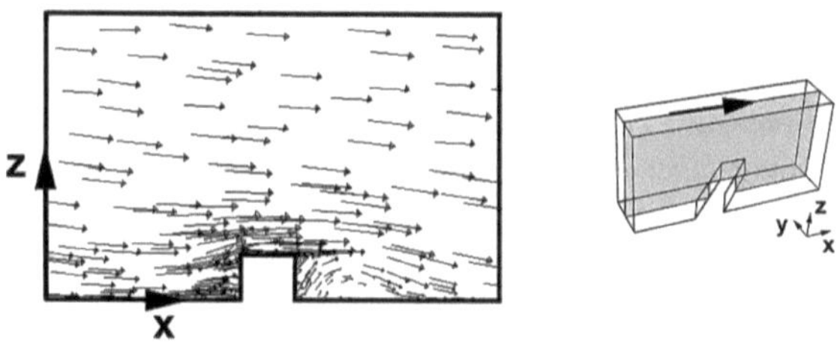

Fig. 8.4 Velocity vectors in the channel vertical x–z mid-plane (parallel to the sidewalls) for the aspect ratio H/W = 2:1 and p/e = 10 [3]

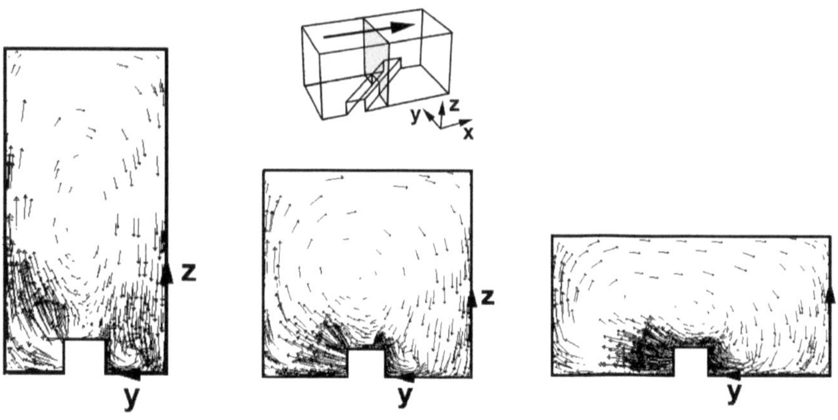

Fig. 8.5 Velocity vectors in the channel y–z cross section orthogonal to the sidewalls for different aspect ratios H/W = 4:1, 2:1 and 1:1, p/e = 10 [3]

downwards to the channel's bottom along the inner sidewall to close the circulation loop (Fig. 8.5). One can also discern the separation bubble behind the rib in Fig. 8.5. Separated and reattached secondary flow immediately after the rib spans over a narrower area for $H/W = 4{:}1$ as compared to $H/W = 1{:}1$, since for $H/W = 4{:}1$ ribs take up a smaller fragment of the channel perimeter (Fig. 8.5). This observation conforms to the findings of [34].

8.2.4 *Heat Transfer and Pressure Drop:* H/W = *4:1*

It is inherent for numerical simulations to deliver information on the local Nusselt numbers over all surfaces of the channel including the rib. Only a few experimental studies, e.g., [31, 33], performed measurements of the Nusselt numbers over four channel walls, whereas the majority took care of heat/mass transfer only at the unribbed bottom.

Figure 8.6a shows the static pressure distribution on the walls of the periodic segment. Flow impinges on the rib and partially diverts to the outer sidewall; therefore a high-pressure region arises in front of the rib. A low-pressure region emerges on the upper facet of the rib and on the bottom right after the rib due to the flow acceleration and subsequent recirculation. Pressure increase is observed on the bottom, where flow reattaches.

Figure 8.6b depicts surface heat transfer distributions in the channel. High heat transfer spots arise over the upper face of the rib due to the flow acceleration, as well as on the bottom and the outer sidewall owing to flow impingement. Heat transfer is low in front of the rib due to flow separation and behind the rib within the recirculation area.

Experimental and numerical maps of the local heat transfer coefficient over the unribbed bottom (Fig. 8.7) demonstrate somewhat better performance of the k-ω

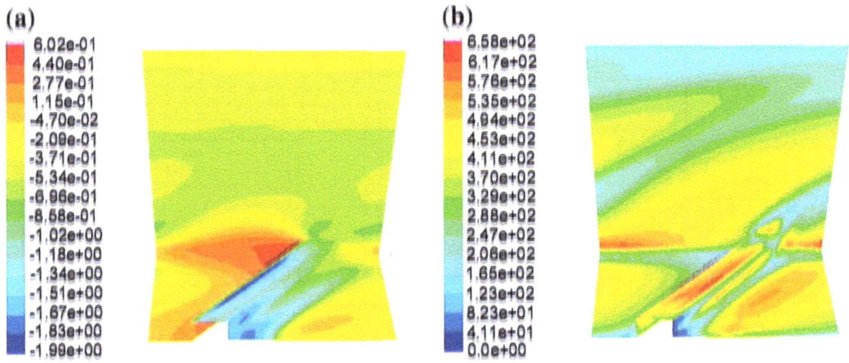

Fig. 8.6 Pressure coefficient C_p **a** and local Nusselt number *Nu* **b** over the external sidewall, rib, and unribbed bottom for $H/W = 4{:}1$, $p/e = 10$. Here $C_p = (P - P_{\rm ref})/(0.5\rho u_b^2)$ [3]

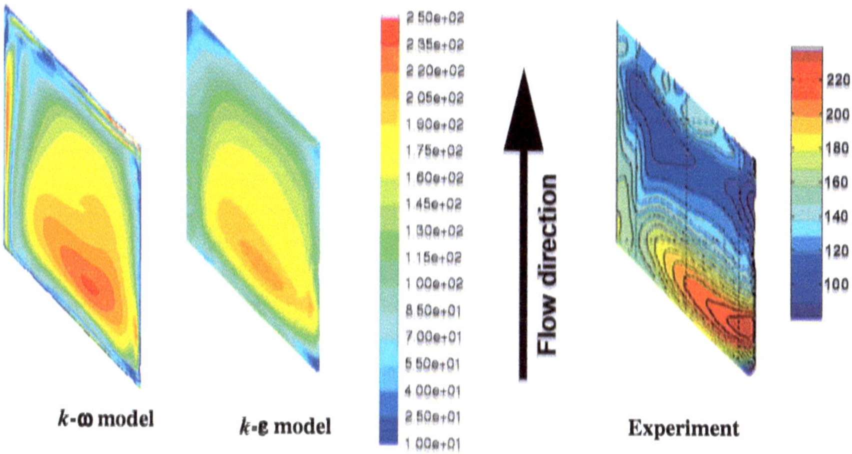

Fig. 8.7 Experimental and numerical data for local distributions of the heat transfer coefficient $\alpha < t$ at the bottom wall of the channel between the ribs at H/W = 4:1, p/e = 10, e/D_h = 0.125. Color range: α = 10–250 W/(m^2 K) [3]

model in capturing the place and magnitude of the heat transfer maximum after the rib. In turn, the k-ε model somewhat better captures the overall local heat transfer distribution over the unribbed bottom until the next rib.

Figure 8.8 displays that the predicted average values of the Nusselt number are in good agreement with experimental data [48, 49] (obtained only for the unribbed bottom wall) despite the evident mismatch in the local maps. Figure 8.8 elucidates the effect of the rib's relative height e/D_h (or e/H) on the average heat transfer at the channel bottom. As can be seen from Fig. 8.8, the average Nusselt number $\overline{Nu}/Nu_0$ normalized by the Dittus–Boelter Eq. (7.10) varies nonlinearly as a function of the relative rib height e/D_h within the range $\overline{Nu}/Nu_0 = 1.6-1.8$. This is accompanied

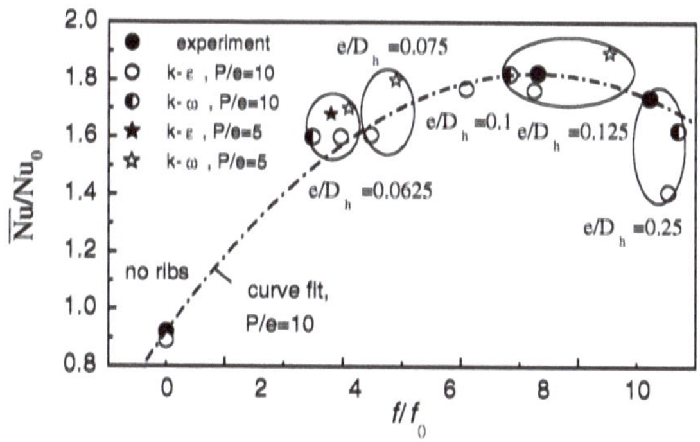

Fig. 8.8 Variation of $\overline{Nu}/Nu_0$ with f/f_0 for different rib heights, H/W = 4:1 [3]. Experiments [48, 49]. Both turbulence models yield similar results. Here Re = 100,000

with an increase in the relative pressure drop f/f_0 from $f/f_0 = 4$ to 10. Over the span of $e/D_h = 0.0625{-}0.25$, the function $\overline{Nu}/Nu_0$ exhibits a maximum at $e/D_h = 0.125$ (or $e/H = 0.05$).

For $e/D_h = 0.125$, the function $\overline{Nu}/Nu_0$ increased by 80 %. Taslim and Liu [32] revealed a maximum of heat transfer at $e/D_h = 0.15$ for $H/W = 1{:}1$ and deterioration of the heat transfer rate for a further enlarged blockage ratio. An interpretation of this phenomenon suggested in [32] is that the flow fails to reattach to the bottom between the large ribs as efficiently, as it does in case of the smaller ribs. Span of the recirculation area between the larger ribs also enlarges. For larger aspect ratios, e.g., $H/W = 4{:}1$, the heat transfer maximum is offset to a smaller value of $e/D_h = 0.125$. Measurements [34] corroborate this statement demonstrating that the heat transfer for $H/W = 4{:}1$ is less sensitive to the rib height effect than at smaller aspect ratios H/W. For $H/W = 4{:}1$ and the same e/D_h, the rib blockage affects the flow in a smaller extent. Below $e/D_h = 0.125$, a smaller relative rib height entails a rather marginal decrease in the heat transfer rate and fast reduction of the pressure loss.

Another influencing parameter is the relative pitch, i.e., the ratio of the spacing between adjacent ribs and the rib height p/e. Figure 8.8 demonstrates that a smaller relative rib pitch $p/e = 5$ entails a 10 % heat transfer augmentation in comparison with the case $p/e = 10$ considered above. This is accompanied with larger pressure losses, which behave as an increasing function of the rib height and reach 19 % at maximum.

8.2.5 *Heat Transfer:* H/W *= 2:1 and 1:1*

Fluid flow and heat transfer in periodic segments with the aspect ratios $H/W = 2{:}1$ and 1:1, including the Nusselt number and static pressure distributions, are qualitatively analogous to those in the case of $H/W = 4{:}1$. Figure 8.5 depicts vector fields in y–z cross section, which turn out to be similar for the three aspect ratios $H/W = 4{:}1$, 2:1, and 1:1.

Figures 8.9 and 8.10 outline functions of the normalized Nusselt numbers $\overline{Nu}/Nu_0$, which increase depending on the rib height over the range $e/D_h = 0.027$–0.15. Both k-ε and k-ω turbulence models yield results for the function $\overline{Nu}/Nu_0$, which agree well with each other and experiments [31] for $Re = 50{,}000$ plotted in Fig. 8.10. To additionally verify the CFD model against experiments [31], a simulation was run at $Re = 50{,}000$ using the k-ε model and resulted in the same magnitude of $\overline{Nu}/Nu_0 = 2.04$ for the unribbed bottom at a smaller normalized friction factor $f/f_0 = 11.05$.

Figures 8.9b and 8.10b exhibit that heat transfer augmentation on the rib surface itself exceeds by 15–20 % (for $H/W = 2{:}1$) and 30–40 % (for $H/W = 1{:}1$) that on the unribbed bottom wall. The overall average heat transfer augmentation taken into account exceeds by 10 % that for the unribbed bottom. Ribs also give rise to heat

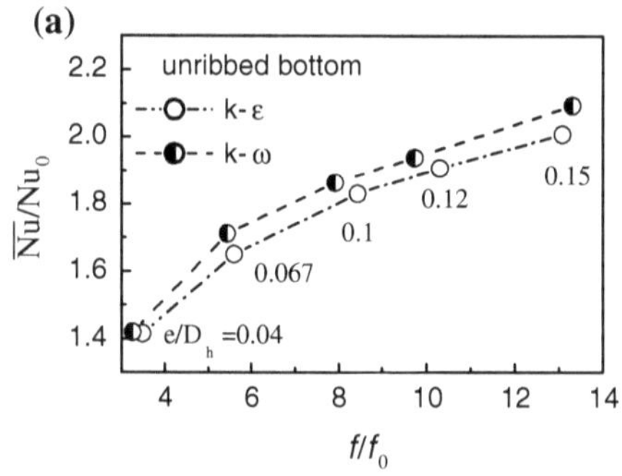

Fig. 8.9 Variation of $\overline{Nu}/Nu_0$ with f/f_0 for different rib heights, H/W = 2:1, p/e = 10: **a** unribbed bottom; **b** unribbed bottom, rib, and total "rib + unribbed bottom" [3]

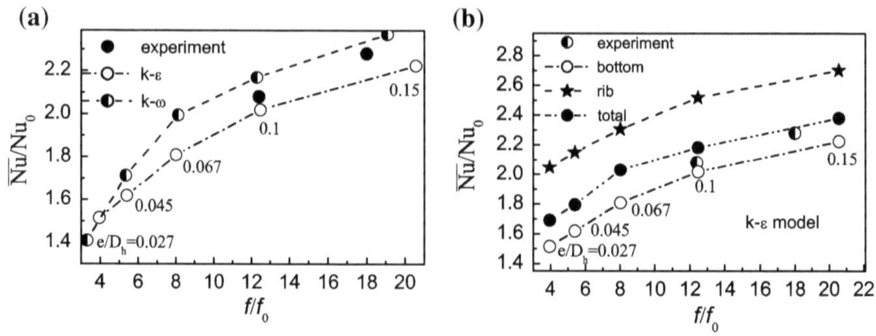

Fig. 8.10 Variation of $\overline{Nu}/Nu_0$ with f/f_0 for different rib heights, H/W = 1:1, p/e = 10: **a** unribbed bottom; **b** unribbed bottom, rib, and total "rib + unribbed bottom" [3]

transfer over the outer sidewall subject to rib-induced flow impingement, which in simulations amounted to $\overline{Nu}/Nu_0 = 1.99$ and 1.96 for Re = 100,000 and 50,000, accordingly. This agrees well the experimental value $\overline{Nu}/Nu_0 = 2.25$ at Re = 50,000 and e/D_h = 0.1 [31].

Thus, simulations [3] in a periodic ribbed segment of a channel presented above resulted in the following conclusions:

1. Realizable k-ε model of turbulence (enhanced wall treatment) and SST k-ω model jointly with hybrid grids with 500,000–700,000 cells enable reliable simulations of convective heat transfer in a periodic segment, whose results are conformant with experiments
2. Local distributions of the computed Nusselt numbers evidently mismatch with experiments for H/W = 4:1. However, the averaged values agreed well with experimental data.

3. For $H/W = 4{:}1$, $Re = 100{,}000$, and an angle of attack of 45°, most beneficial are the ribs with $e/D_h = 0.075$–0.125. The best efficiency at $H/W = 2{:}1$ and 1:1 was attained at $e/D_h = 0.1$ to 0.15. Rib heights exceeding the optimal values entail a faster increase in the pressure loss accompanied with very minor increase in heat transfer rate. The tendency holds independently of the Reynolds number.

8.3 Rectangular Ribbed Channel with *H/W* = 2:1 Inlet, *H/W* = 1:1 Outlet

8.3.1 Geometry and Flow Parameters

Section 8.3 is based on the CFD simulations published in my own paper [4]. The model geometry shown in Fig. 8.11 partially reproduces that used in the experimental study [2]. The inlet and outlet passes were of identical height of $H = 0.15$ m. Channel width and hydraulic diameter, as well as the rib height were $W_{in} = 0.075$ m, $D_h = 0.1$ m, and $e = 0.01$ m in the inlet pass; $W_{out} = 0.15$ m, $D_h = 0.15$ m, and $e = 0.015$ m in the outlet pass; rib pitch $p/e = 10$. To make sure that the condition $e/D_h = 0.1 = \text{const.}$ holds, ribs in the inlet and outlet passes were manufactured with, respectively, different heights. The inlet and outlet faces were designed perpendicular to the bottom/symmetry and sidewalls (Fig. 8.11).

Because the ribs were installed in-line on the bottom and top faces, we simulated only a half-height of the geometry, whereas the upper face of the domain was set to be a symmetry plane (Fig. 8.11). The wall dividing the inlet and outlet passes was $W_{web} = 0.02$ m thick, whose edge in the bend was rounded with a radius $R_w = 0.01$ m.

The other geometrical parameters in the simulations were $W_{el} = 0.075$ m−0.15 m or $W_{el}/W_{in} = 1.0$−2.0 in the smooth channel and $W_{el} = 0.075$ m−0.174 m or $W_{el}/W_{in} = 1.0$−2.32 in the ribbed channel.

Velocity, temperature, and turbulence distributions from the outlet of a periodic segment (see Sect. 8.2 above) were translated to the inlet of the geometry in

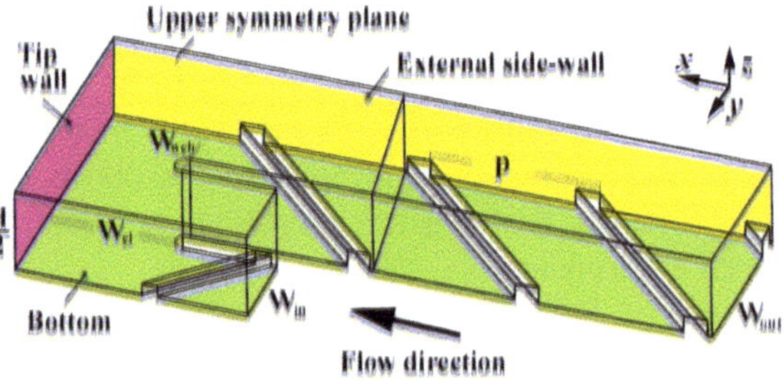

Fig. 8.11 Schematic of the channel geometry in simulations [4]: half-height of the channel of [2], symmetry boundary condition on the upper surface

Fig. 8.11. This modeled experimental conditions [2], where measurements were performed after 13–15 ribbed segments in the inlet pass counting from the inlet to ensure periodical flow pattern right before the entrance to the end.

To evaluate the magnitude of curvature effects in the bend, a smooth channel configuration having the same dimensions as in Fig. 8.11 was additionally simulated.

Like in Sect. 8.2, the boundary condition $T_w = 350$ K = const. was assigned on all walls of the model geometry. TLC measurements [2] were employed in the work [4] for validations of the CFD model. All simulations were done with air at $Pr = 0.72$. The Reynolds number $Re = 100{,}000$ incorporating the hydraulic diameter and bulk velocity at the inlet was used in all simulations.

8.3.2 *Numerical Methodology*

The geometry in Fig. 8.11 and its modifications were built with the help of GAMBIT software. CENTAUR® software [52] (see also Sect. 8.2) was used for the hybrid grid generation. A grid-independency investigation carried out using a coarse mesh of ~1.1 million cells, a medium mesh of 1.9 million cells, a finer mesh of 3.2 million cells, and a very fine mesh of 3.65 million to 3.91 million cells revealed that namely the very fine mesh ensured the sufficient near-wall grid resolution and grid independency for the ribbed channel, while for the smooth channel a grid with 1.15–1.3 million cells was needed. This conforms to the grid sizes of 500,000 to 700,000 cells for the periodic segments employed in the studies [3] (see Sect. 8.2) confirmed also by [16, 57], etc. The second-order CFD solver with double precision was used for all differential equations.

Following Sect. 8.2, a realizable k-ε model with enhanced wall treatment (two-layer approach) at all default values of the constants was involved in the modeling. All grids fulfilled the enhanced wall treatment condition $y_1^+ \sim 1$ (see Sect. 8.2) in the first volume cell over the surface. In fact, y^+-values varied mostly over the range $y_1^+ = 1$–1.8, with the number of prismatic layers ≥20. Convergence criteria set in the simulations were scaled residuals' levels smaller than 10^{-6} for the continuity and 10^{-9} for the energy equation.

8.3.3 *Smooth Channel*

Baseline values: straight channel. In the engineering practice, Eqs. (7.10) and (8.1) are employed to represent the so-called standard conditions, i.e., a straight smooth channel. Strictly saying, Eqs. (7.10) and (8.1) are valid for a circular pipe; hence, it makes sense to model first a smooth channel with identical cross section as that of the ribbed channel and use further the average Nusselt numbers and friction factors from this smooth channel as more realistic standard conditions for the validations purposes. Consequently, a smooth straight segment with $H/W = 2{:}1$ and periodic conditions at the inlet and outlet faces were modeled before the main simulations started. As a result, the averaged Nusselt number *for the entire smooth segment* (total average) obtained in simulations was $\overline{Nu}_{st} = 173.5$ (Table 8.1) with the friction factor for the same segment rated as $f/f_0 = 0.9$. The value $Nu_0 = 201.7$ for $Re = 100{,}000$ suggested by the Dittus–Boelter Eq. (7.10) evidently surpasses the $\overline{Nu}_{st}$.

In the TLC experiments [2], heat transfer in a long smooth inlet pass averaged over a bottom and one sidewall yielded a value $\overline{Nu}_{st}/Nu_0 = 0.82$. Identically averaging the results of the CFD, we obtained the value $\overline{Nu}_{st}/Nu_0 = 0.83$ correlating well with the experiments. Analogous averaging for a smooth periodic segment with $H/W = 1{:}1$ resulted in the values $\overline{Nu}_{st}/Nu_0 = 0.86$ and $f/f_0 = 0.89$, which practically repeat the data for the aspect ratio $H/W = 2{:}1$. An essentially better trade-off of the $\overline{Nu}_{st}$ values from simulations yields a relation for turbulent flow in an isothermal tube given in [51]

$$Nu_1 = 0.021 Re^{0.8} Pr^{0.5}, \tag{8.2}$$

Equation (8.2) suggests the value $Nu_1 = 178.2$ for $Re = 100{,}000$; hence, $\overline{Nu}_{st}/Nu_1 = 0.94$, which agrees with experiments [57] in a $H/W = 2{:}1$ channel. In the experiments [16] in a $H/W = 4{:}1$ channel, the average Nusselt number was $\overline{Nu}_{st}/Nu_1 = 0.92$.

Values $\overline{Nu}_{st}$ instead of Eq. (7.10) were referred as Nu_0 also by Xie et al. [58].

Two-pass channel with constant $H/W = 2{:}1$ (Fig. 8.12). Figures 8.12 and 8.13 show the conventional splitting and numbering of the channel into different parts used in the work [4] which enables treating them separately.

Heat transfer coefficients were evaluated with the help of single values of the reference bulk temperature T_B computed separately for the local sub-volumes of the simulated geometry: inlet pass (faces *1a*, *2*, and *7*), bend (faces *1b*, *3*, *4*, and *6b*) and outlet pass (faces *5*, *6a*, and *8*), which enables more accurate comparisons with the experiments.

In the inlet pass in Fig. 8.12 (bottom *2*, sidewalls *1a* and *7*), the value $\overline{Nu} = \overline{Nu}_{st} = 174.2$ is practically equal to that in a periodic smooth segment, though the backward effect of the bend moderately distorts the local Nusselt number distributions. In the bend region (bottom *3*, tip wall *4*, sidewalls *1b* and *6b*), the normalized Nusselt number is as large as $\overline{Nu}/\overline{Nu}_{st} = 1.94$. The length of the outlet pass is $2.95D_h$ (bottom *5*, sidewalls *6a* and *8*); here the normalized Nusselt number magnitude is $\overline{Nu}/\overline{Nu}_{st} = 2.61$.

Table 8.1 Average Nusselt number in a smooth straight channel with $H/W = 2{:}1$ [4]

Channel part	Bottom	Sidewalls	Total average
$\overline{Nu}_{st}/Nu_0$	0.74	0.93	0.86

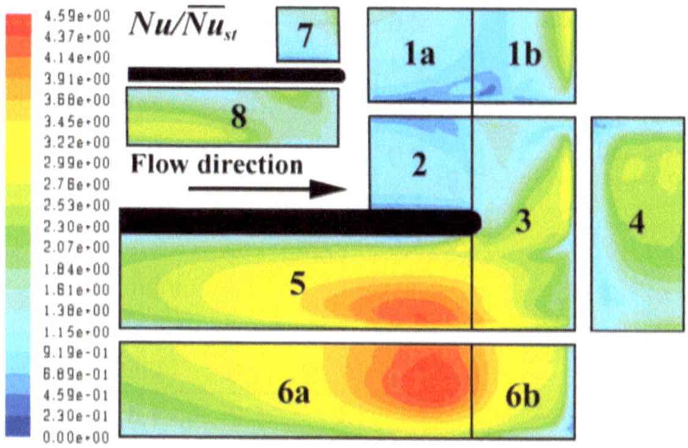

Fig. 8.12 Heat transfer distribution over the bottom, tip, external, and internal (smaller sketch) sidewalls of the smooth channel with constant $H/W = 2{:}1$ (simulations [4])

Two-pass channel with varying H/W = 2:1 to H/W = 1:1 (Fig. 8.13). Figure 8.13 elucidates that the heat transfer in the inlet pass and the bend area of the varying aspect ratio channel is consistent with that in the channel with $H/W = 2{:}1$, and essentially reduced in the outlet pass owing to channel expansion and subsequent flow deceleration and recirculation over the inner wall (see also work [41]). Figure 8.13 displays also spots of augmented heat transfer due to the Dean vortices (bend bottom *3*), and impingement onto three faces: tip wall (*4*), outlet channel bottom (*5*), and outer sidewall (*6*).

Figure 8.14 describes the tip wall distance effect on the average Nusselt numbers for different channel parts. As expected, $\overline{Nu}/\overline{Nu}_{\mathrm{st}} = 1$ in the inlet pass. In the bend, $\overline{Nu}/\overline{Nu}_{\mathrm{st}}$ behaves similarly to that in Fig. 8.12, though somewhat reduces at $W_{\mathrm{el}}/W_{\mathrm{in}} = 2.0$. In the outlet pass, $\overline{Nu}/\overline{Nu}_{\mathrm{st}}$ is linearly decreasing.

In the bend, simulation results exceed experimental data by 22 %. In the outlet pass, both simulations and experiments demonstrate the decreased heat transfer as

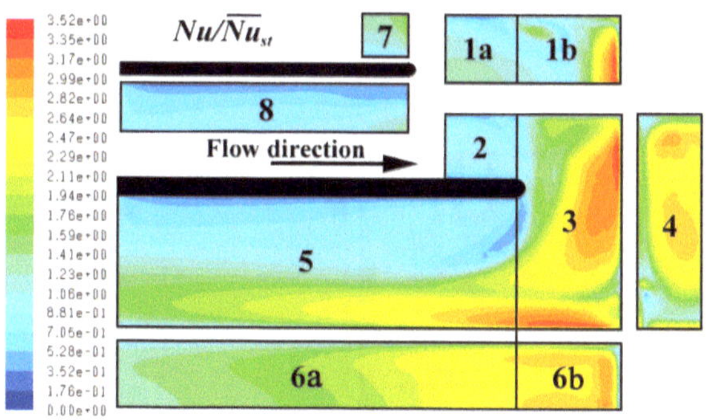

Fig. 8.13 Local heat transfer over the bottom, tip, external, and internal (smaller sketch) sidewalls of the smooth channel with varying aspect ratio of $H/W_{\mathrm{in}} = 2{:}1$ (inlet) and $H/W_{\mathrm{out}} = 1{:}1$ (outlet), $W_{\mathrm{el}} = 112.5$ mm (simulations [4])

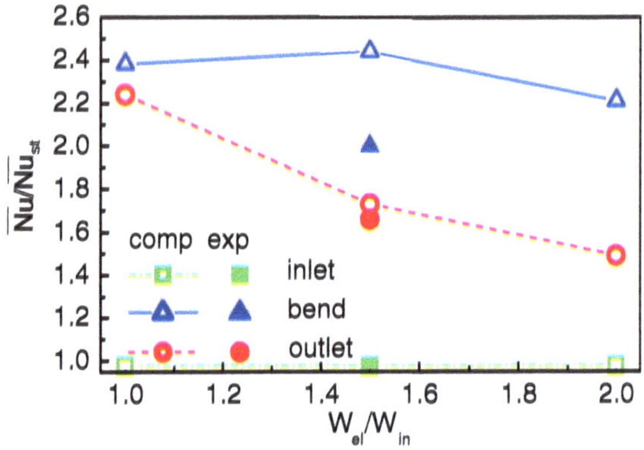

Fig. 8.14 Effect of the W_{el} on average heat transfer in the inlet pass (faces *1a* and *2*), bend (faces *3* and *4*), and outlet pass (faces *6a* and *5*), varying H/W_{in} = 2:1 to H/W_{out} = 1:1 [4]

compared to that in the bend; here the discrepancy between the experiments and simulations is only 4 %.

To conclude, the CFD model described in Sect. 8.3.2 can simulate fluid flow and heat transfer in a two-pass smooth channel with acceptable accuracy.

8.3.4 *Ribbed Channel: Fluid Flow*

Figure 8.15 demonstrates that flow in the ribbed two-pass channel is subject to effects of both bend curvature and the ribs. Impingement spots arise on the tip wall and on the outer sidewall in the outlet pass. On the opposite inner wall of the outlet pass, flow separates and entrains into an intensive recirculation vortex. The through flow at the entrance to the outlet pass is squeezed in the direction of the outer sidewall. Velocity vector fields predicted in the simulations agree with PIV measurements. The sharp turn of the flow near the divider wall (shown by a red arrow in

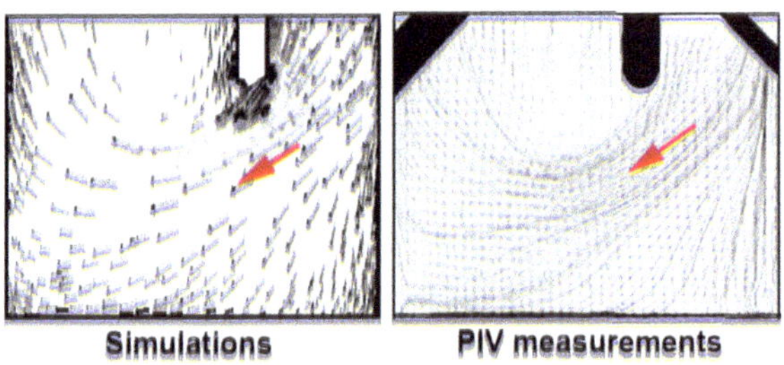

Fig. 8.15 Comparisons of the simulated and measured velocity vector fields in the symmetry plane at W_{el} = 150 mm [4]

Fig. 8.15), the span of the recirculation vortex, and the location of the second impingement area over the outer sidewall are well captured by simulations.

Dean vortices arose regardless of the tip wall distance [4], which however affected the shape and the number of the vortices. For the narrowest tip wall distance $W_{el} = 75$ mm, the main Dean vortex is pushed down to the bottom of the bend and is accompanied with the second smaller counter-rotating Dean vortex (Fig. 8.16).

A much larger tip wall distance $W_{el} = 150$ mm ensures emergence of just one Dean vortex occupying the entire cross section (Fig. 8.16). An outlook and parameters of the Dean vortex predicted in the simulations are in a good consistency with the PIV measurements, which is again indicative that the CFD model described in Sect. 8.3.2 performs well and reliably simulates the complex 3D flow pattern. Hence, this CFD model may be further used for simulations of local and average fluid flow and heat transfer parameters.

With the enlarged tip wall distance W_{el}, the impingement spot arising on the tip wall migrates from the area opposite to the inlet pass to the region opposite to the outlet pass (Fig. 8.17). It is accompanied with an increase of the span of the recirculation vortex on the inner sidewall at the entrance to the outlet pass.

The relative pressure drop ΔP^* curve plotted in Fig. 8.18 using the total ("inlet-to-outlet") difference of the averaged *static pressures* in the model two-pass channel is compared with experimental data for ΔP^* measured at the bottom centerline in the sub-volume of the real test channel corresponding to the geometry in Fig. 8.11.

Figure 8.18 shows that the curves of the relative pressure drop ΔP^* have minima at $W_{el}/W_{in} = 1.75$ in simulations and $W_{el}/W_{in} = 2.0$ in experiments. Hence, having an

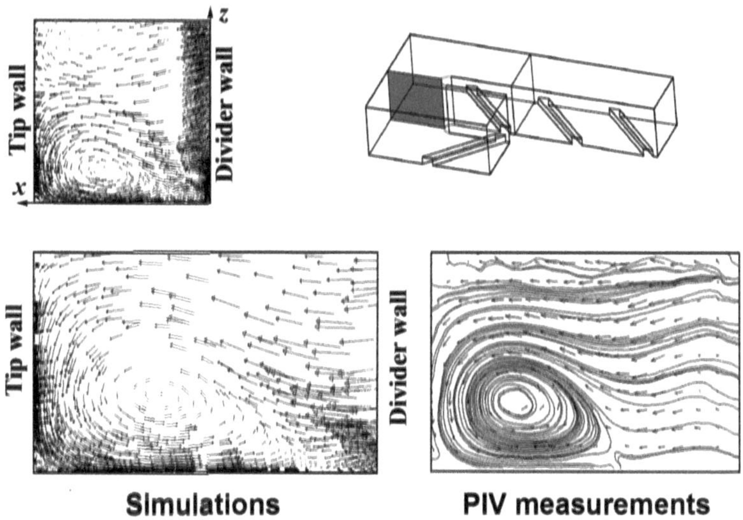

Fig. 8.16 Dean vortices in the bend region for different tip wall distances $W_{el} = 75$ mm (*upper*) and $W_{el} = 150$ mm (*lower*), both simulations and PIV [4]

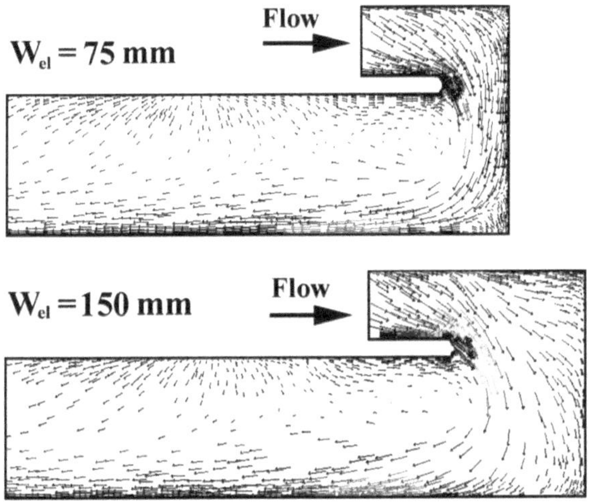

Fig. 8.17 Effect of the W_{el} on the velocity vector field in the symmetry plane (simulations [4])

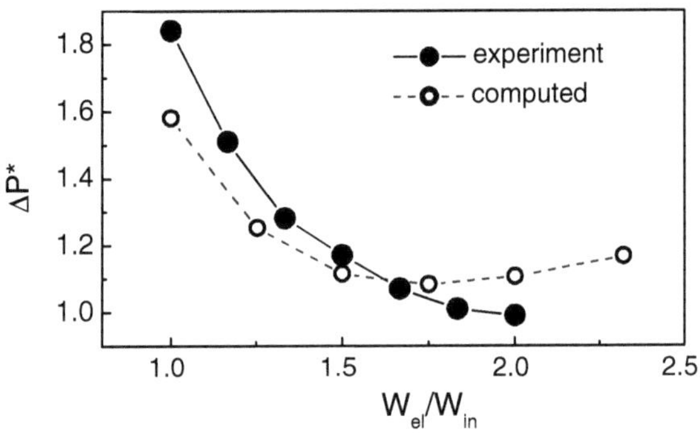

Fig. 8.18 Effect of the W_{el} on the relative pressure drop ΔP^* [4]

objective to minimize the pressure losses, it is expedient to increase the tip wall distance only up to point of minimum of the function ΔP^*. Maximal discrepancy of the simulations and experiments lies within 15 %.

8.3.5 Ribbed Channel: Heat Transfer

Figures 8.19 and 8.20 depict normalized local Nusselt numbers Nu/Nu_0 in the ribbed two-pass channel obtained in simulations and experiments, respectively.

The following reference values were used for normalization of the results:

(a) average Nusselt numbers $\overline{Nu}_{W_{el}=75}$ at the minimum tip wall distance W_{el} = 75 mm to physically analyze the effect of the increasing W_{el};
(b) the value of Nu_0 = 201.7 by the Dittus–Boelter Eq. (7.10).

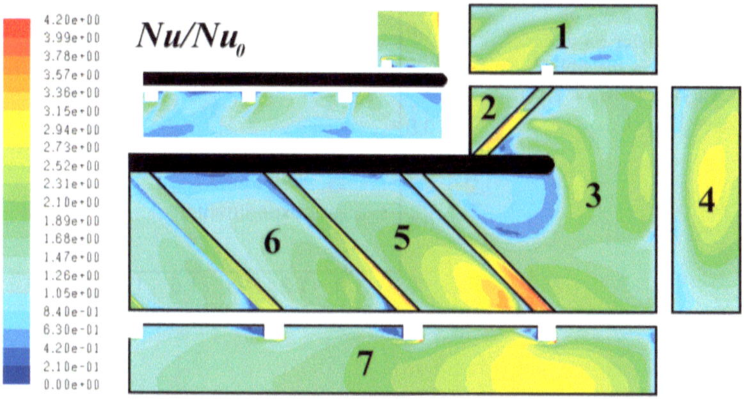

Fig. 8.19 Local heat transfer over the bottom, tip, external, and internal (smaller sketch) sidewalls in the ribbed channel with varying aspect ratio of H/W_{in} = 2:1 in the inlet and H/W_{out} = 1:1 in the outlet, W_{el} = 112.5 mm (simulations) [4]

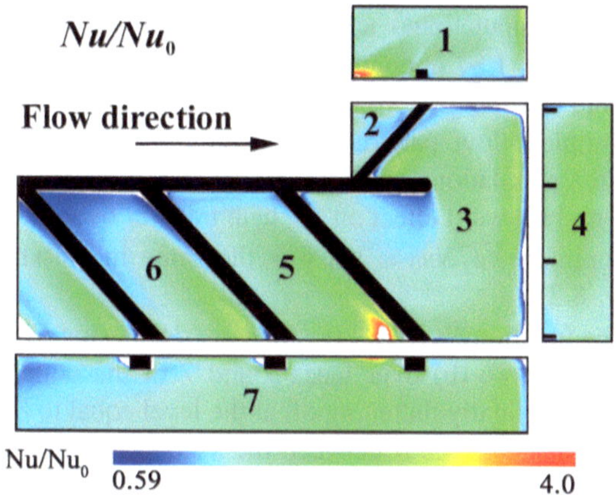

Fig. 8.20 Local heat transfer over the bottom, tip, and external sidewalls of the ribbed channel with varying aspect ratio of H/W_{in} = 2:1 in the inlet and H/W_{out} = 1:1 in the outlet, W_{el} = 112.5 mm (TLC experiments for Nu/Nu_0) [2, 4]

In doing so in the ribbed channel, the effects of the bend and ribs are considered as a joint influence.

Spots of high impingement-induced heat transfer can be seen behind the ribs. The Dean vortices cannot cause this heat transfer enhancement, since Dean vortices arise only at the entrance to the bend (face *3* in Figs. 8.12 and 8.13).

The joint influence of the bend and ribs results in the overall heat transfer augmentation in the bend sub-volume, e.g., over the tip wall and the outer sidewall (face *7*, Fig. 8.19), which agrees with the TLC measurements in Fig. 8.20.

Fig. 8.21 Effect of W_{el} on average heat transfer on the tip wall (face *4*) in the ribbed channel, varying H/W_{in} = 2:1 to H/W_{out} = 1:1 [4]

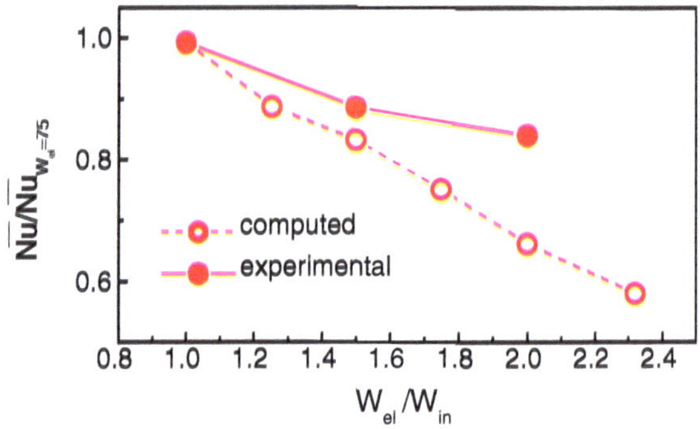

Figures 8.21 and 8.22 elucidate the influence of the tip wall distance. Normalized Nusselt number $\overline{Nu}/\overline{Nu}_{W_{el}=75}$ on the tip wall (Fig. 8.21) and the bend bottom (Fig. 8.22) decreases linearly with the tip wall distance W_{el}, which is confirmed also by experiments. The normalization was performed using the values of $\overline{Nu}_{W_{el}=75}$ estimated individually for the tip wall ($\overline{Nu}/Nu_0 = 2.25$ in experiments, $\overline{Nu}/Nu_0 = 2.28$ in computations) and for the bottom bend ($\overline{Nu}/Nu_0 = 2.11$ in experiments, $\overline{Nu}/Nu_0 = 1.87$ in computations). The higher distinction for the bend bottom arouse from a worse performance of the realizable *k*-ε turbulence model in capturing the complex flow pattern in this region. Nevertheless, the tendency demonstrated in the simulations conforms to the experimental trend.

The bend has no backward effect on the heat transfer over the face *2* of the inlet pass, whose magnitude $\overline{Nu}/Nu_0 = 1.8$ was the same in the simulations and the experiments (Fig. 8.23).

In Fig. 8.24, the function $\overline{Nu}/Nu_0$ (a) is linearly decreasing (by about 30 %) over the inlet sidewall (face *1*), (b) over faces *5* and *7* of the outlet pass it is at first linearly decreasing and finally flattening off at the level equal to 85 % of that for the smallest tip wall distance. Face *6* between the second and third ribs in the outlet pass is practically not influenced by the effect of W_{el}/W_{in}, whereas $\overline{Nu}/Nu_0$ becomes 7 % smaller and levels off at W_{el}/W_{in} = 1.5.

Fig. 8.22 Effect of W_{el} on average heat transfer on the bend bottom (face *3*) in the ribbed channel, varying H/W_{in} = 2:1 to H/W_{out} = 1:1 [4]

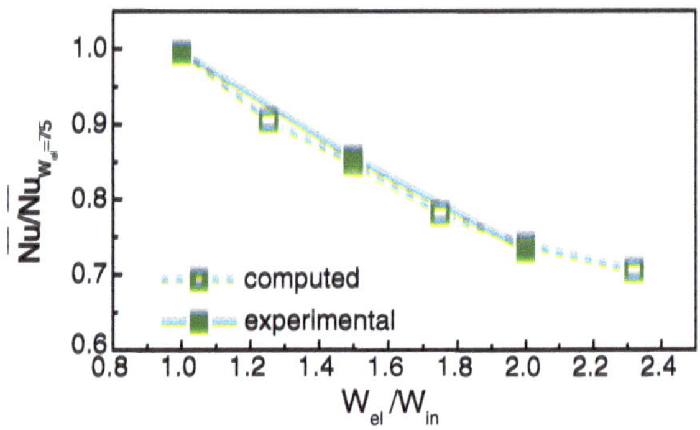

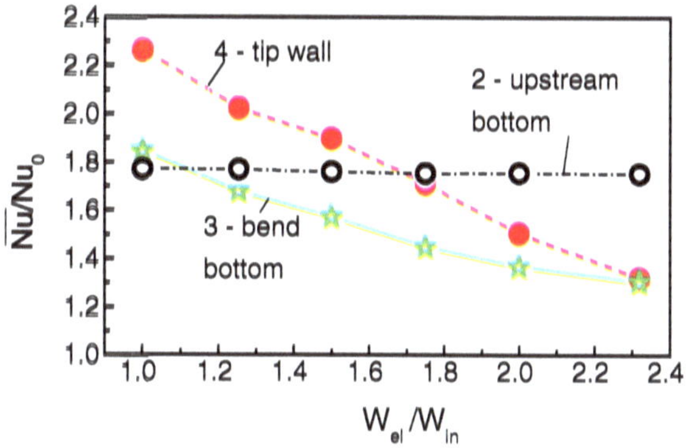

Fig. 8.23 Effect of W_{el} on average heat transfer on the upstream bottom (face *2*), bend bottom (face *3*), and tip wall (face *4*) in the ribbed channel, varying $H/W_{in} = 2{:}1$ to $H/W_{out} = 1{:}1$ (simulations [4])

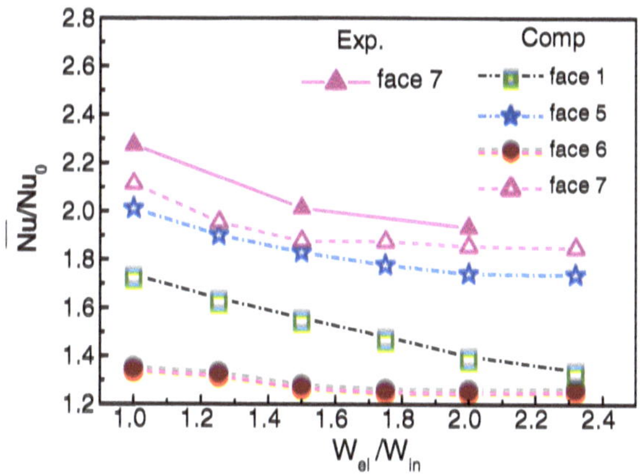

Fig. 8.24 Effect of W_{el} on average heat transfer on the inlet pass sidewall (face *1*), bottom between the first and second ribs in the outlet pass (face *5*), bottom between the second and third ribs in the outlet pass (face *6*) and outlet pass sidewall (face *7*) in the ribbed channel, varying $H/W_{in} = 2{:}1$ to $H/W_{out} = 1{:}1$ [4]

Thus fluid flow and heat transfer in the outlet pass are not influenced by variation of the tip wall distance, once W_{el}/W_{in} goes beyond 1.75. Simulations follow the trends revealed in TLC measurements and lie below the experimental data by only 7 % at the entrance to the downstream pass (face *7*).

It makes also sense to evaluate heat transfer over the rib surface and its share in the overall heat transfer predicted in simulations (and almost never measured in experiments). Over the first rib surface in the outlet pass, the heat transfer augmentation goes by 12–16 % beyond that of the unribbed bottom, face *5* (Fig. 8.25).

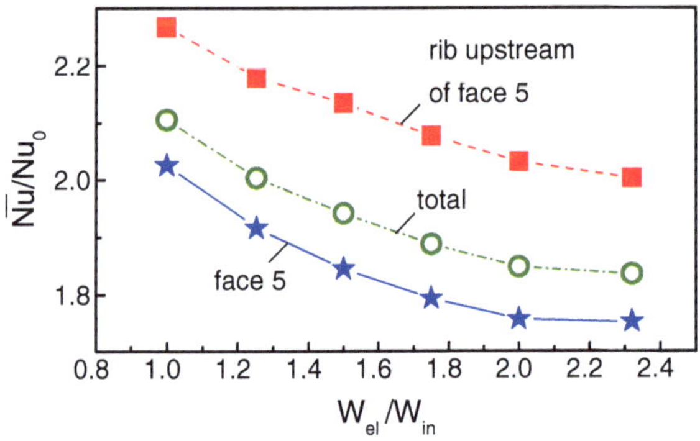

Fig. 8.25 Effect of the W_{el} on average heat transfer in the outlet pass over the bottom between ribs *1* and *2*, (face *5*), rib *1*, and total of face *5* and rib *1* in the ribbed channel, varying $H/W_{in} = 2{:}1$ to $H/W_{out} = 1{:}1$ (simulations [4])

Total heat transfer augmentation due to the rib (face *5* + rib) is 4–5 % higher than that over face *5*.

Aerothermal efficiency is a function that enables a combined estimation of the pressure losses, which accompany attaining a needed rate of the heat transfer augmentation

$$\eta = \frac{Nu/Nu_0}{(f/f_0)^{1/3}}. \tag{8.3}$$

Figure 8.26 depicts the aerothermal efficiency over the bottom bend (face *3*) and tip wall (face *4*). The characteristic length *L* involved in estimating *f* was the total length of the inlet and outlet pass with an addition of a span of the centreline in the bend (Fig. 8.11).

In spite of the increased pressure loss in the channel with the smallest tip wall distance $W_{in} = 75$ mm (Fig. 8.18), it exhibits the highest efficiency due to the high heat transfer in the bend area (see Figs. 8.21, 8.22, 8.23, 8.24, 8.25) although the trend is quite flat until about $W_{el}/W_{in} = 1.5$, where the drop-off increases.

Fig. 8.26 Effect of the W_{el} on aerothermal efficiency η (simulations [4])

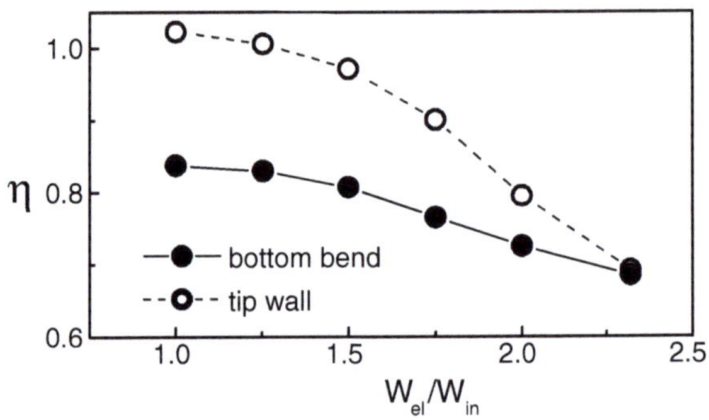

Thus, results of the CFD calculations described in Sect. 8.3.2 and their comparisons with experiments [2, 4] for the average heat transfer and pressure loss through a ribbed two-pass channel with a 180° bend are overall satisfactory. The RANS approach enables simulating the cases with high *Re* numbers [2, 4] in a sufficiently fast and effective way to investigate numerically a much wider span of parameters than it is possible in experiments.

Simulations enabled making the following conclusions:

1. The bend geometry increases heat transfer because of the flow acceleration, impingement on the walls and Dean vortices. The ratio $\overline{Nu}/\overline{Nu}_{st}$ in the smooth channel with $H/W = 2{:}1$ increases almost 2 times in the bend and 2.6 times in the outlet pass. The magnitude of the function $\overline{Nu}/\overline{Nu}_{st}$ in the varying aspect ratio channel reaches the levels of 2.3–2.4 in the bend and behaves as a linearly subsiding function: from 2.3 to 1.5 with the parameter W_{el}/W_{in} increasing from 1.0 to 2.0.
2. In the ribbed channel, at smaller value W_{el}/W_{in} = 1.0, two pairs of the Dean vortices arose in the bend, and only a single pair for W_{el}/W_{in} = 2.0. Relative pressure drop ΔP^* in the entire geometry (Fig. 8.11) behaves as a nonlinear function with a distinct minimum at W_{el}/W_{in} = 1.75, which overall conforms to experiments (deviations within 15 %).
3. In the ribbed channel, the normalized Nusselt number linearly diminishes on the tip wall and bend bottom. Once the parameter W_{el}/W_{in} exceeds 1.75, heat transfer in the outlet pass becomes practically insensitive to the tip wall distance influence. Simulations overall correlate well to experiments.
4. The rib itself in the outlet pass has a heat transfer value of nearly 12–16 % larger than the unribbed bottom, which adds 4–5 % to the total heat transfer augmentation.
5. Aerothermal efficiency is maximal at W_{el}/W_{in} = 1.0–1.5.

8.4 Rectangular Smooth Channel with *H*/*W* = 3:1 Inlet, *H*/*W* = 1:1 Outlet

8.4.1 Geometry and Flow Parameters

Section 8.4 is based on the CFD simulations published in the paper [5]. The objective of this study was to simulate heat transfer and fluid flow in a two-pass smooth channel depicted in Fig. 8.27, with an aspect ratio of the inlet pass H/W_{in} = 3:1 and the rest dimensions being the same as those of the channel with H/W_{in} = 2:1 (Sect. 8.3.1). Therefore, the width of the inlet pass with H/W_{in} = 3:1 was reduced to W_{in} = 0.05 m, with the hydraulic diameter equal to D_{hi} = 0.075 m. The tip wall distance W_{el} was varied from 0.05 to 0.15 m. As the geometry was

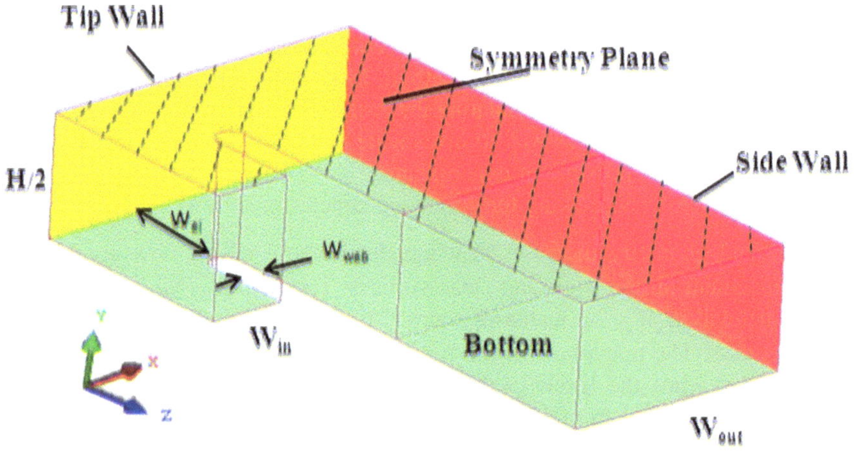

Fig. 8.27 Schematic view of the smooth two-pass channel with H/W_{in} = 3:1 in the inlet pass and H/W_{out} = 1:1 in the outlet pass [5]

symmetrical, only half-height of the model was simulated. This channel aspect ratio has been never studied before.

Boundary conditions. Thermal and regime boundary conditions, as well as physical properties of air were identical to those in Sect. 8.3.1.

8.4.2 *Numerical Methodology*

Overview. ANSYS ICEMCFD was used to design the geometry and to generate the structured mesh, and ANSYS FLUENT was employed as solver. Like in Sect. 8.3.2, the realizable *k*-ε turbulence model with enhanced wall treatment was used in the frame of a RANS approach. The first *y*+ value in the near-wall cell remained in the range $y_1^+ = 1{-}3$ for all simulated cases. Convergence criteria were the same as in Sect. 8.3.2 [5].

Nusselt numbers used for normalization. Nusselt numbers were normalized by Eqs. (7.10) and (8.2). In addition, the correlation of Petukhov et al. [59] was used

$$Nu_{Pe} = \frac{(f/8)Re\,Pr}{1.07 + 12.7(f/8)^{0.5}(Pr^{2/3} - 1)}, \tag{8.4}$$

$$f = (0.79 \ln Re - 1.64)^{-2}. \tag{8.5}$$

Definition of the reference temperature. The channel in Fig. 8.27 was conventionally split into three regions, i.e., inlet pass, bend, and outlet pass. In each of

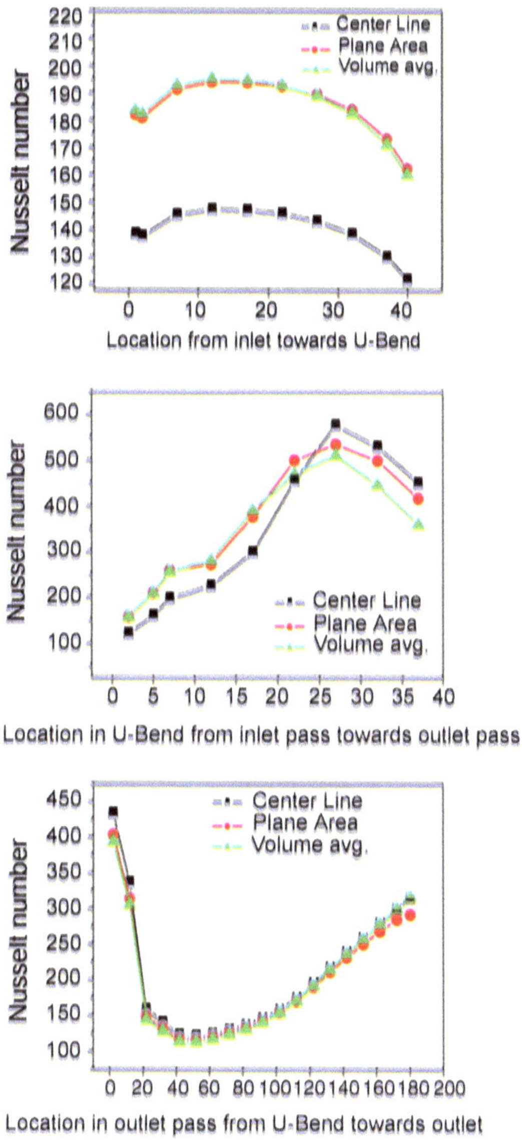

Fig. 8.28 Comparison of three different methods to calculate Nusselt numbers in three different regions of the channel with Re = 100,000, H/W_{in} = 2:1, H/W_{out} = 1:1, W_{el}/W_{in} = 1.5 [5]

them, planes were placed perpendicular to the main flow at equal distances from each other. Cross section averaged Nusselt numbers plotted in Fig. 8.28 were calculated based on the (a) volume-averaged temperature of each region denoted "Volume avg." (as in Sect. 8.3 [4]), area average temperature of each plane denoted "Plane Area" (like in experiments [2]), and centerline temperatures of the selected planes denoted "Center Line".

As seen in Fig. 8.28, at the inlet and outlet passes approach "Volume avg." yields results practically coinciding with the approach "Plane Area" with a maximum discrepancy of about 5 %, although, in the bend region this discrepancy reached 13 %. The "Center Line" approach may be justified only in the outlet pass. In the inlet pass, the centerline temperature is obviously incapable of representing the local bulk temperature. However, calculating the center plane average temperature at each streamwise location requires relatively much effort. Therefore, based on Fig. 8.28, it was decided to follow the approach [4] (used in Sect. 8.3) to further use single values of the volume-averaged temperature at the inlet pass, bend, and outlet pass as a reference temperature.

Grid independence. This study was performed for the case with $W_{el}/W_{in} = 1.5$. Three structured grids were compared with 950,000 (coarse), 1,700,000 (fine), and 2,600,000 (finest) cells, respectively. The near-wall $y+$ values were close to unity. All three grids provided similar results for the Nusselt numbers with the maximum scatter of 0.88 %. Therefore, the most economical coarse grid was selected for the rest of the simulations.

Validation for the case $H/W_{in} = 2{:}1$ and $H/W_{out} = 1{:}1$. The CFD model [5] described above was validated against experiments [2] and simulations [4] for $Re = 100{,}000$, $H/W_{in} = 2{:}1$ and $H/W_{out} = 1{:}1$, $W_{el}/W_{in} = 1.5$, and different tip wall distances. Local Nusselt numbers obtained in the simulations [5] agree well with those presented in Fig. 8.13 [4] and experiments [2].

Figure 8.29 representing average Nusselt numbers for different regions of the two-pass channel replicates Fig. 8.14 from Sect. 8.3 with computations [5] plotted for comparisons. Simulations [5] again agree well with the results [4]. Therefore, the CFD model used in Sect. 8.4.2 was accepted as a baseline model for further investigations in a two-pass smooth channel with $H/W_{in} = 3{:}1$ and $H/W_{out} = 1{:}1$.

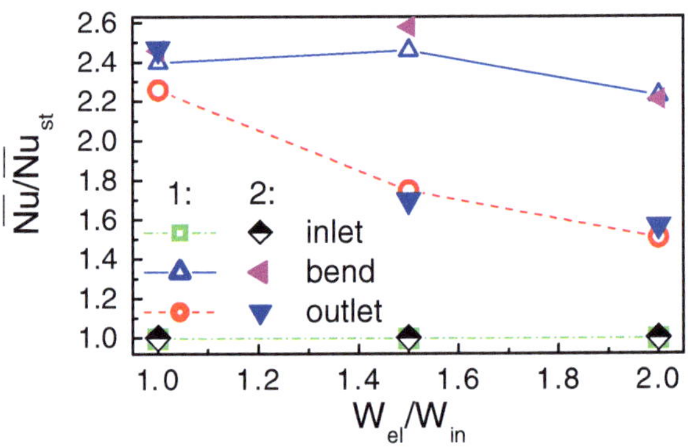

Fig. 8.29 Effect of W_{el} on the average heat transfer in the inlet pass (faces *1a* and *2*), bend (faces *3* and *4*), and outlet pass (faces *6a* and *5*) for the channel with $H/W_{in} = 2{:}1$, $H/W_{out} = 1{:}1$ as predicted in simulations [4, 5]. Symbols *1* [4]; *2* [5]

Table 8.2 Average Nusselt numbers in a smooth straight channel with $H/W_{in} = 3{:}1$ in simulations [5] normalized with Eqs. (7.10), (8.2), and (8.4)

	Bottom	Sidewalls	Total average	Bottom + wall
$\overline{Nu}_{st}$	125	192	175	165
$\overline{Nu}_{st}/Nu_0$	0.62	0.95	0.87	0.82
$\overline{Nu}_{st}/Nu_1$	0.70	1.07	0.98	0.92
$\overline{Nu}_{st}/Nu_{pe}$	0.73	1.13	1.03	0.97

8.4.3 Smooth Periodic Segment

A smooth periodic segment with $H/W_{in} = 3{:}1$ was simulated to find out normalizing values $\overline{Nu}_{st}$ given in Table 8.2. Outlet velocity, temperature, and turbulence profiles from the periodic segment were used as inlet profiles in the main computational domain.

The value $\overline{Nu}_{st}/Nu_0 = 0.82$ for faces "bottom + wall" is in a good agreement with the values 0.83 in the simulations [4] and 0.82 in the experiments [2] for $H/W_{in} = 2{:}1$ (Sect. 8.3). Value $\overline{Nu}_{st}/Nu_1 = 0.92$ [5] agrees with experiments for $H/W_{in} = 4{:}1$ [16].

For the bottom alone, the value $\overline{Nu}_{st}/Nu_0 = 0.62$ for $H/W_{in} = 3{:}1$ is much smaller than the value 0.74 for $H/W_{in} = 2{:}1$. However, larger share of the sidewalls (whose average $\overline{Nu}_{st}/Nu_0 = 0.95$ is almost the same for $H/W_{in} = 3{:}1$ and 2:1) yields $\overline{Nu}_{st}/Nu_0 = 0.87$ as a total average for $H/W_{in} = 3{:}1$ that correlates with the value 0.86 for $H/W_{in} = 2{:}1$ (Table 8.1).

This indicates that the Nusselt number averaged over one sidewall and bottom, as well as the total average, is very weakly affected by the aspect ratio of the channel.

8.4.4 Two-Pass Smooth Channel: Fluid Flow and Heat Transfer

Velocity and temperature fields. Fluid flow and heat transfer patterns in the two-pass smooth channel [5] are analogous to those discussed in Sect. 8.3 for $H/W_{in} = 2{:}1$. In the short inlet pass, bend effects manifest themselves via strong acceleration near the divider wall and deceleration near the outer wall [5]; right upstream of the outlet pass, fluid accelerates near the outer sidewall, and strongly decelerates near the divider wall [5] (analogous to Figs. 8.15 and 8.17). In the center of the bend, one can see a distinct Dean vortex [5] (similar to that in Fig. 8.16).

Pressure drop and heat transfer. As seen in Fig. 8.30, in the channels with $H/W_{in} = 3{:}1$ and $H/W_{in} = 2{:}1$ the *static* pressure drop ΔP^* is a nonlinear increasing function of W_{el}/W_{in}. With the decrease in the aspect ratio from $H/W_{in} = 3{:}1$–$2{:}1$ the

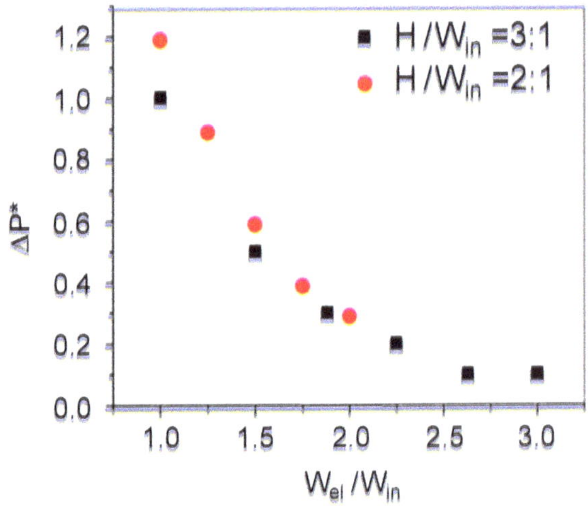

Fig. 8.30 Effect of W_{el}/W_{in} on the relative pressure drop ΔP^* for two-pass smooth channels with H/W_{in} = 3:1 or H/W_{in} = 2:1 and H/W_{out} = 1:1 [5]

pressure drop increases. As the tip wall distance W_{el}/W_{in} grows, the effect of the channel aspect ratio H/W_{in} on the pressure drop ΔP^* practically vanishes. For H/W_{in} = 3:1, the curve for ΔP^* apparently reaches its minimum (extremum) point at W_{el}/W_{in} = 3.0; however, results of simulations are insufficient to detect the minimum point for the aspect ratio H/W_{in} = 2:1.

For the heat transfer analysis, the channel was divided into three regions in accordance with Fig. 8.13: the inlet pass (faces *1a* and *2*), the bend (faces *1b*, *3*, *6b*, and *4*), and the outlet pass (faces *5* and *6a*). Faces *7* and *8* were not taken into consideration. The Nusselt number was defined using a volume (bulk) average temperature for each region separately. The effects of the tip wall distance on the averaged Nusselt number for the channels with H/W_{in} = 3:1 [5] and H/W_{in} = 2:1 [4] are shown in Fig. 8.31. It is evident that with an increase in the inlet aspect ratio from 2:1 to 3:1 the heat transfer rate in the bend and in the outlet pass decreases, because of the more strongly expressed flow deceleration.

In the inlet pass, the heat transfer rate remains at the level of $Nu/\overline{Nu}_{st} = 1$.

In the bend region, in the channel with H/W_{in} = 3:1 the Nusselt number first decreases with the increasing tip wall distance, but afterwards increases and levels off to a constant value at W_{el}/W_{in} = 3.0. The reason for such a behavior in Fig. 8.31, "Bend" is the noticeable enlargement of the enhanced heat transfer area over the tip wall (*4*) and the bend bottom (*3*) in the vicinity of the tip wall following the increase in W_{el}/W_{in} [5]. For the smallest tip wall distances the flow in the bend region makes a 90° turn and causes the strongest impingement over the outer sidewall in the bend region and beginning of the outlet pass followed with the accordingly strongest heat transfer enhancement. For the medium and especially largest tip wall distances, flow in the bend region has enough space to produce the most significant impingement and heat transfer enhancement on the tip wall and the bend bottom

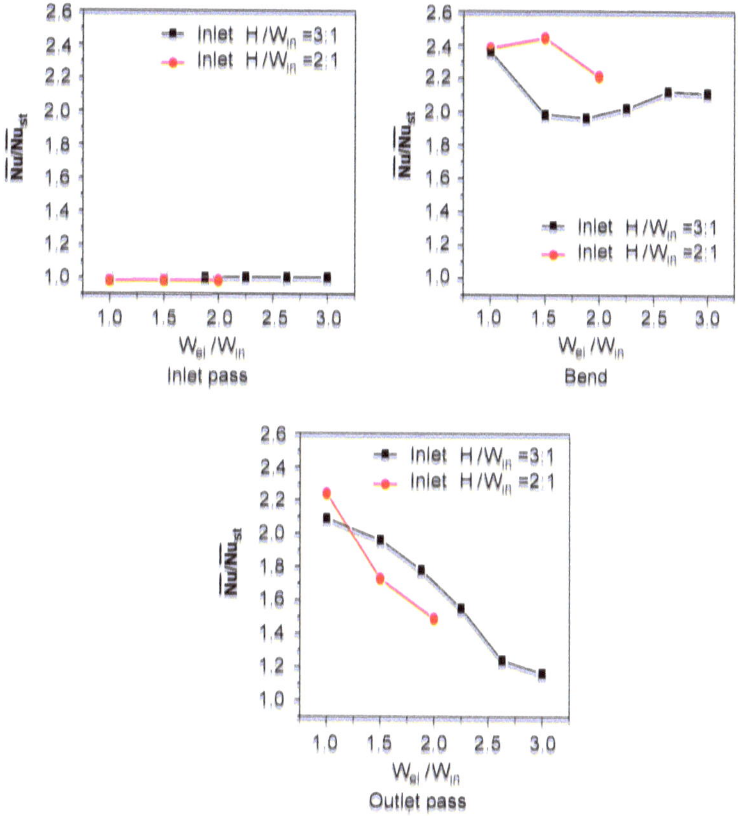

Fig. 8.31 Effects of the W_{el}/W_{in} (or W_{el}/H) on the average heat transfer in the inlet pass (faces *1a* and *2*), bend (faces *3* and *4*), and the outlet pass (faces *5* and *6a*) of the two-pass channels with $H/W_{out} = 1{:}1$, and $H/W_{in} = 3{:}1$ or $H/W_{in} = 2{:}1$ [4, 5]

next to it before it makes the 90° turn, which entails much weaker impingement onto the outer sidewall. For the channel with $H/W_{in} = 2{:}1$, the red line in Fig. 8.31, "Bend" demonstrates a maximum at $W_{el}/W_{in} = 1.5$ and decreased Nusselt numbers for the further increased W_{el}/W_{in}, because the effect of enlargement of the bend region space downstream of the inlet pass is expressed weaker than that in the channel with $H/W_{in} = 3{:}1$. This trend can be expected to appear in the $H/W_{in} = 3{:}1$ channel, if $W_{el}/W_{in} > 3.0$.

In the outlet pass region, in both channels with $H/W_{in} = 3{:}1$ and $H/W_{in} = 2{:}1$, the heat transfer constantly decreases with the increased tip wall distance.

Thus, the CFD simulations described in Sect. 8.4 enable making the following conclusions:

1. With an increase in the W_{el}/W_{in} ratio for $H/W_{in} = 3{:}1$, the relative pressure drop ΔP^* decreased nonlinearly up to a point of minimum. The *static* pressure drop ΔP^* is larger for $H/W_{in} = 2{:}1$; differences between two aspect ratios practically

vanish for $W_{el}/W_{in} > 1.75$. Nondimensionalization of the tip wall distance in order to consider its effect on the pressure drop and heat transfer must be done using the parameter W_{el}/W_{in} rather than W_{el}/H.

2. An analysis of the average heat transfer revealed that:
   (a) *in the inlet pass*, the heat transfer rate remains at the level of $Nu/\overline{Nu}_{st} = 1$.
   (b) *in the bend region*, in the channel with H/W_{in} = 3:1 heat transfer first decreases with the increasing W_{el}/W_{in}, but afterwards increases and levels off to a constant value.
   (c) *in the outlet pass*, in both channels with H/W_{in} = 3:1 and H/W_{in} = 2:1, heat transfer constantly decreases with the increased tip wall distance.

8.5 Rectangular Ribbed Channels with *H*/*W* = 3:1 Inlet, *H*/*W* = 1:1 Outlet

8.5.1 Geometry and Flow Parameters

Section 8.5 is based on the results published in the paper [6]. The objective of this investigation was to undertake a CFD simulation of fluid flow and heat transfer in a two-pass channel of the same dimensions as those depicted in Fig. 8.27. Channel walls are roughened with 45° ribs at $e/D_h = 0.1$ and $p/e = 10$ throughout the channel (Fig. 8.32).

Tip wall distance W_{el} varied from 0.075 to 0.15 m. The ribs were installed in a parallel mode (i.e., in-line) on the top and bottom; therefore, a half-height of the domain was modeled with its upper surface being a symmetry plane. To prevent the back flow, the outlet pass was extruded to a distance equal to its hydraulic diameter.

Boundary conditions. Thermal and regime boundary conditions, as well as physical properties of air were identical to those in Sect. 8.3.1

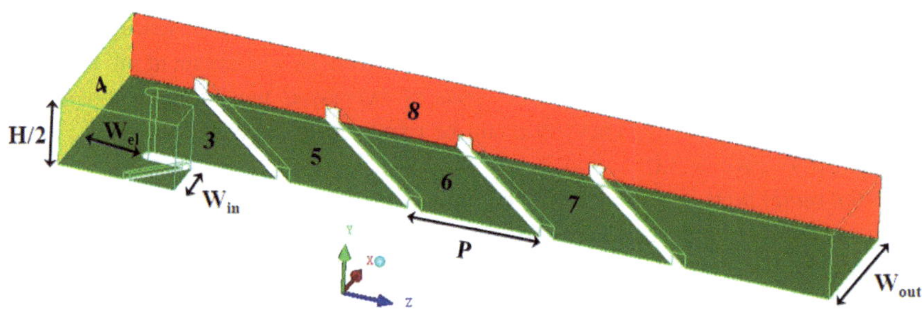

Fig. 8.32 Schematic view of the ribbed two-pass channel with H/W_{in} = 3:1 in the inlet and H/W_{out} = 1:1 in outlet pass and a symmetry boundary condition on the upper surface [6]

8.5.2 Numerical Methodology

Overview. Geometries and structured meshes were created using ANSYS ICEMCFD; ANSYS FLUENT was used as solver. Like in Sects. 8.3.2 and 8.4.2, turbulence modeling was performed using a RANS approach employing the realizable k-ε turbulence model with enhanced wall treatment. The $y+$ value in first near-wall cell was equal to $y_1^+ = 1-3$ for all simulated cases. Convergence criteria were the same as in Sects. 8.3.2 and 8.4.2 [5].

Nusselt numbers used for normalization. Nusselt numbers were normalized by the Dittus–Boelter Eq. (7.10). Faces of the two-pass channel were numbered as shown in Fig. 8.32. As a characteristic length in the Nusselt number, the inlet hydraulic diameter D_{hi} was used at the bend bottom (face *3*) and the tip wall (face *4*), while the outlet hydraulic diameter D_{ho} was utilized for the outlet bottom (face *5* + *6* + *7*) and the sidewall (face *8*).

Grid independence. This study was performed for the geometry with $W_{el} = 131.25$ mm (i.e., $W_{el}/H = 0.88$ or $W_{el}/W_{in} = 2.64$). Three structured grids were investigated with 3.5 (coarse), 4.6 (fine), and 5.7 (very fine) million cells. For all faces except for face *3*, results for the averaged normalized Nusselt numbers $\overline{Nu}/Nu_0$ with the fine grid were coincident with those for the very fine grid. Over face *3*, differences were about 4 %. Therefore, the fine grid was selected for the further analysis and also for all other tip wall distances W_{el}/W_{in}.

Validation for the case of $H/W_{in} = 2{:}1$ and $H/W_{out} = 1{:}1$. Figure 8.33 shows local distributions of the Nusselt number Nu/Nu_0. An agreement between the simulations [6] and measurements [2] is overall good.

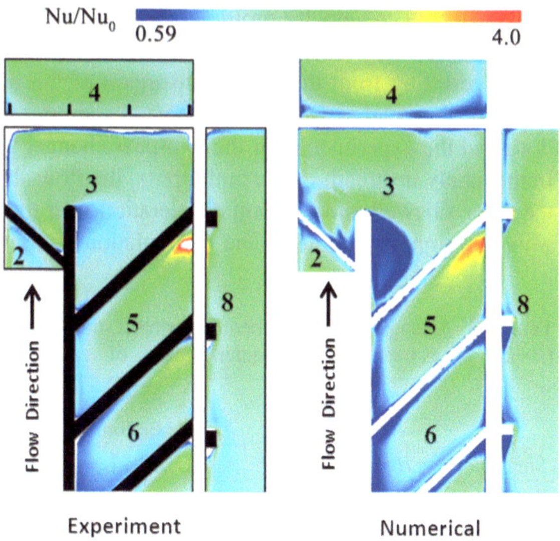

Fig. 8.33 Local Nusselt number distribution by experimental (*left*) [2] and numerical (*right*) results [6] for $H/W_{in} = 2{:}1$, $H/W_{out} = 1{:}1$, $W_{el} = 121.5$ mm, and $Re = 100{,}000$

It was found that the average Nusselt numbers obtained in simulations differed by maximum 12 % from the experiments over almost all surfaces. However, at the bend bottom (*3*) and the tip wall (*4*), the discrepancy reached 27 %, obviously, owing to spots with low heat transfer rates observed locally in these regions. Thus these validations are within the same limits of discrepancy as reported in the literature and hence are plausible.

8.5.3 Ribbed Periodic Segment

A ribbed periodic segment of the aspect ratio $H/W_{in} = 3{:}1$ was simulated in order to compute velocity, turbulence, and temperature profiles to be mapped at the inlet of a two-pass ribbed channel. Local velocity, temperature, pressure, and heat transfer distributions for $H/W_{in} = 3{:}1$ periodic segment presented in [6] are analogous to those presented in Sect. 8.2 for periodic segments with $H/W_{in} = 1{:}1$, 2:1 and 4:1.

Table 8.3 presents average values of $\overline{Nu}/Nu_0$ for different faces of the segment predicted in the work [4, 6]. These data testify that an increase in the aspect ratio from 2:1 to 3:1 entails lower rates of the heat transfer augmentation.

8.5.4 Two-Pass Ribbed Channel: Fluid Flow and Heat Transfer

Velocity and temperature fields. Fluid flow and heat transfer in the two-pass ribbed channel with $H/W_{in} = 3{:}1$ [6] are analogous to those discussed in Sect. 8.3 for $H/W_{in} = 2{:}1$. The velocity profile at the inlet to the bend shows acceleration of the flow near the inner divider wall, deceleration near the outer sidewall, and a large vortex resulting from the last rib upstream of the bend (like in Fig. 8.17). At the center of the bend, a Dean vortex emerges similar to that observed in the simulations and the experiments for the two-pass channel with $H/W_{in} = 2{:}1$, Fig. 8.16 [4]. Immediately upstream of the outlet pass, the flow shows a strong acceleration near the outer sidewall and a strong deceleration near the divider wall. Near the divider wall, a region of separation and recirculation with very low flow velocity emerges

Table 8.3 Average Nusselt number in the ribbed straight periodic segments with $H/W_{in} = 3{:}1$ and $H/W_{in} = 2{:}1$ as predicted in the simulations [4, 6]

	$\overline{Nu}/Nu_0$		
	Bottom	Rib	Total
Simulation for $H/W_{in} = 3{:}1$ [6]	1.72	1.94	1.79
Simulation for $H/W_{in} = 2{:}1$ [4]	1.83	2.12	1.93
Difference, %	−6.1	−8.3	−7.5

as a result of a sharp turn similar to that depicted in Fig. 8.17 for the two-pass channel with $H/W_{in} = 2{:}1$ [4]. Somewhat abated Dean vortex persists in the vicinity of the outer sidewall. Further downstream, a rib-induced large channel vortex arises and rotates in counterclockwise direction. After the third rib, flow tends to establish again a periodic pattern peculiar to sufficiently long straight ribbed channels.

Pressure drop and heat transfer. Relative pressure drop ΔP^* was calculated based on the difference of the average *static pressure* at the inlet and a cross section placed immediately at the end of the fourth rib in the outlet pass. With an increase in W_{el}/H the pressure drop curves for aspect ratios $H/W_{in} = 2{:}1$ to $H/W_{in} = 3{:}1$ first decrease and then level off to constant values (Fig. 8.34).

Increasing the aspect ratio from $H/W_{in} = 2{:}1$ to $H/W_{in} = 3{:}1$ caused a significant reduction of the pressure drop ΔP^*, which agrees with the data for the smooth two-pass channel in Fig. 8.30 [5]. However, the pressure loss in the channel with $H/W_{in} = 3{:}1$ are not so much lower in comparison with the channel with $H/W_{in} = 2{:}1$. As said in Sect. 8.4.4, the *static pressure*-based parameters ΔP^* and f in the channels with $H/W_{in} = 3{:}1$ and $H/W_{in} = 2{:}1$ [4, 5] are affected by the different ratios of the inlet and outlet cross sections.

Typical local heat transfer distribution in a two-pass channel with $H/W_{in} = 3{:}1$ look qualitatively analogous to Figs. 8.19 and 8.20.

Figure 8.35 depicts normalized averaged Nusselt numbers over the bend bottom (face *3*), and the bottom areas between the first and the second ribs (face *5*), the second and the third rib (face *6*), and the third and the fourth rib (face *7*) in the outlet pass. The highest heat transfer level is observed at the bend bottom; further downstream the Nusselt numbers decrease and finally level off. It can be seen from Fig. 8.35 that for smaller tip wall distances W_{el}/H periodicity tends to set on already

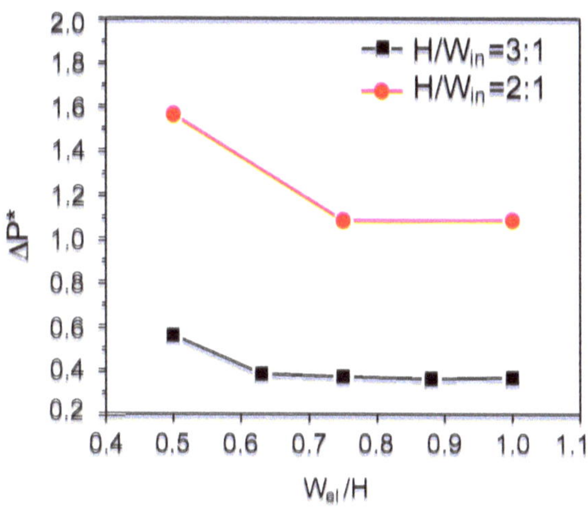

Fig. 8.34 Effect of W_{el}/H on the relative pressure drop ΔP^* for a two-pass ribbed channel. Data for $H/W_{in} = 3{:}1$—simulations [6]; data for $H/W_{in} = 2{:}1$—simulations [4]

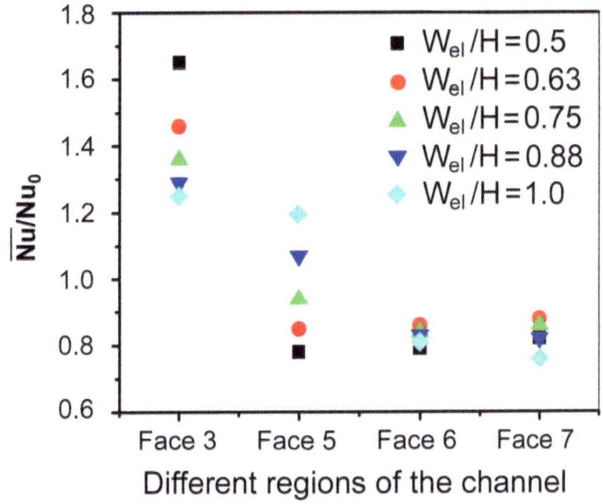

Fig. 8.35 Average Nusselt numbers (normalized by the Dittus–Boelter value Nu_0) in different regions of the two-pass ribbed channel with $H/W_{in} = 3{:}1$ and $H/W_{out} = 1{:}1$ [6]

after the first rib, but for larger tip wall distances the onset of periodicity drifts downstream and arises for the first time after the second rib [6].

Figures 8.36 and 8.37 represent a comparative analysis of the effect of the tip wall distance W_{el}/H on the average heat transfer over the bend bottom (face *3*) and tip wall (face *4*), respectively, for the channels with $H/W_{in} = 3{:}1$ and $H/W_{in} = 2{:}1$. Area-averaged Nusselt numbers were normalized with the Nusselt number value taken at $W_{el}/H = 0.5$. For $H/W_{in} = 2{:}1$, a trend of linear decrease is evident, while for $H/W_{in} = 3{:}1$ the Nusselt number diminishes nonlinearly and tends to level off for larger tip wall distances. It is worth noting that for larger W_{el}/H the curves for $H/W_{in} = 3{:}1$ lie higher than those for $H/W_{in} = 2{:}1$.

In Fig. 8.38, the data from Figs. 8.36 and 8.37 are normalized with the value Nu_0. For the channel with $H/W_{in} = 2{:}1$, the average Nusselt numbers are higher than

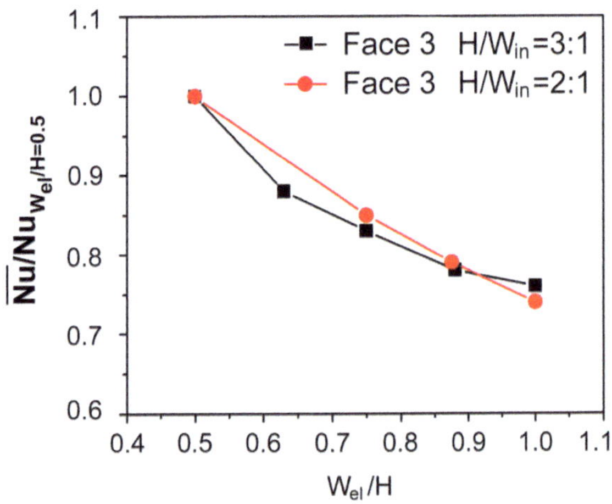

Fig. 8.36 Effect of the tip wall distance W_{el}/H on the average heat transfer over the bend bottom (face *3*) in the ribbed channels. Data for $H/W_{in} = 3{:}1$—simulations [6]; data for $H/W_{in} = 2{:}1$—simulations [4]

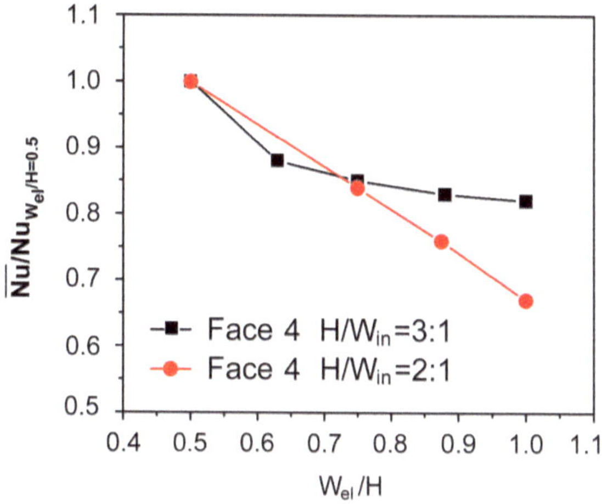

Fig. 8.37 Effect of the tip wall distance W_{el}/H on the average heat transfer over the tip wall (face *4*) in the ribbed channels. Data for $H/W_{in} = 3{:}1$—simulations [6]; data for $H/W_{in} = 2{:}1$—simulations [4]

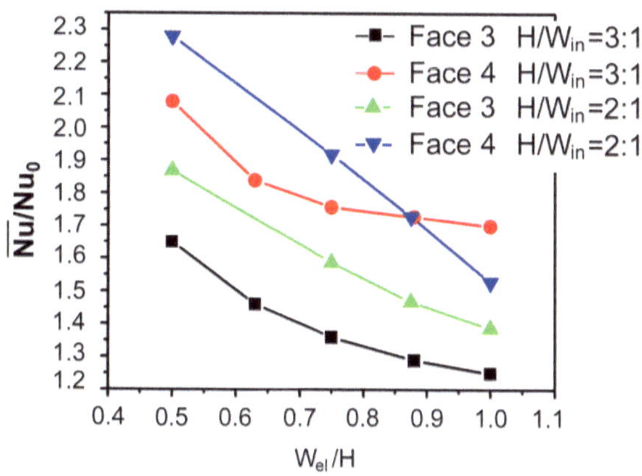

Fig. 8.38 Effect of the tip wall distance W_{el}/H on the average heat transfer over the bend bottom (face *3*) and tip wall (face *4*) in the ribbed channels. Data for $H/W_{in} = 3{:}1$—simulations [6]; data for $H/W_{in} = 2{:}1$—simulations [4]

those for $H/W_{in} = 3{:}1$, the trend being however changing to the opposite for face *4* at $W_{el}/H = 1.0$.

In Fig. 8.39, in the ribbed channel with $H/W_{in} = 3{:}1$, the Nusselt number over faces *5* + *6* + *7* (the outlet bottom) increases slowly together with W_{el}/H and apparently tends to an asymptotic constant value. For the outer sidewall *8*, the heat transfer rate behaves as a diminishing function of the tip wall distance W_{el}/H.

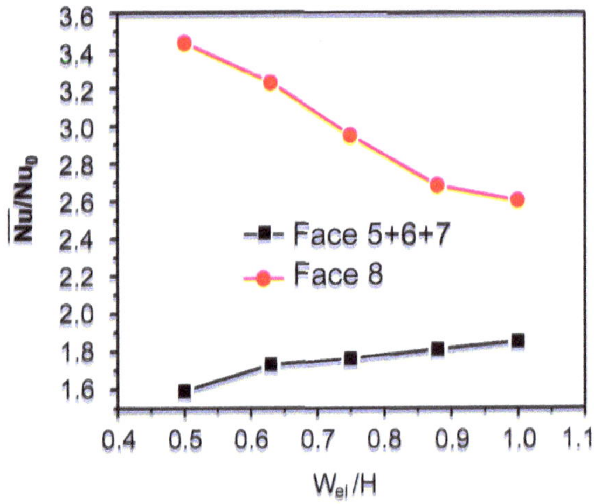

Fig. 8.39 Effect of the tip wall distance W_{el}/H on the average heat transfer over faces *5*, *6*, *7*, and *8* in the ribbed channel with H/W_{in} = 3:1 [6]

Both these trends can be attributed to the effect of the bend: for larger tip wall distances the area with the highest heat transfer shifts from the outer sidewall to the tip wall and bend bottom including the bottom upstream of the first rib comprised in face *3*.

Thus, the major conclusions drawn from the results presented above are as follows:

1. The overall *static pressure* drop decreases with the increase in tip wall distance and is noticeably less in the geometry with H/W_{in} = 3:1 than in the case of H/W_{in} = 2:1. This results from different ratios of the inlet and outlet cross sections of these two channels.
2. The average heat transfer rate over the tip wall and the bend bottom is smaller for the aspect ratio H/W_{in} = 3:1 than for the channel with H/W_{in} = 2:1.
3. For H/W_{in} = 3:1, the average heat transfer over the tip wall, bend bottom, and outer sidewall is a subsiding function of the tip wall distance. However, the heat transfer over the outlet bottom slightly increases and tends to flatten at a constant level.
4. 3D flow and heat transfer periodicity tends to set on already after the first or second rib in the outlet pass (depending on W_{el}/H).

References

1. Jenkins SC, Shevchuk IV, von Wolfersdorf J, Weigand B (2012) Transient thermal field measurements in a high aspect ratio channel related to transient thermochromic liquid crystal experiments. Trans ASME J Turbomach 134(3):Paper 031002
2. Jenkins SC, Zehnder F, Shevchuk IV, von Wolfersdorf J, Weigand B, Schnieder M (2013) The effect of ribs and tip wall distance on heat transfer for a varying aspect ratio two-pass ribbed internal cooling channel. Trans ASME J Turbomach 135(2):Paper 021001

3. Shevchuk IV, Jenkins S, von Wolfersdorf J, Weigand B (2007) Aerothermal efficiency of a ribbed channel in an internal gas turbine blade cooling system. Proceedings of the 7th ETC (Athens, Greece), pp 1041–1050
4. Shevchuk IV, Jenkins SC, Weigand B, von Wolfersdorf J, Neumann SO, Schnieder M (2011) Validation and analysis of numerical results for a varying aspect ratio two-pass internal cooling channel. Trans ASME J Heat Transfer 133(5):Paper 051701
5. Siddique W, El-Gabry L, Shevchuk IV, Hushmandi NB, Fransson TH (2012) Flow structure, heat transfer and pressure drop in varying aspect ratio two-pass rectangular smooth channels. Heat Mass Transf 48(5):735–748
6. Siddique W, Shevchuk IV, El-Gabry L, Hushmandi NB, Fransson TH (2013) On flow structure, heat transfer and pressure drop in varying aspect ratio two-pass rectangular channel with ribs at 45°. Heat Mass Transf 49(5):679–694
7. Cai L, Ota H, Hirota M, Nakayama H, Fujita H (2004) Influence of channel aspect ratio on heat transfer characteristics in sharp turn connected two-pass channels with inclined divider wall. Exp Therm Fluid Sci 28(6):513–523
8. Chen CC, Liou TM (2000) Rotating effect on fluid flow in a smooth duct with a 180-degree sharp turn. In: Proceedings of the 2000 ASME turbo expo (Munich, Germany): Paper GT-0228
9. Cho HH, Lee SY, Rhee DH (2004) Effects of cross ribs on heat/mass transfer in a two-pass rotating duct. Heat Mass Transf 40(10):743–755
10. Elfert M, Jarius MP, Weigand B (2004) Detailed flow investigation using PIV in a typical turbine cooling geometry with ribbed walls. In: Proceedings of the 2004 ASME turbo expo (Vienna, Austria) vol 3, pp 533–545
11. Han JC, Chandra PR, Lau SC (1988) Local heat/mass transfer distributions around sharp 180 deg turns in two-pass smooth and rib-roughened channels. Trans ASME J Heat Transf 110 (1):91–98
12. Hirota M, Fujita H, Syuhada A, Araki S, Yoshida T, Tanaka T (1999) Heat/mass transfer characteristics in two-pass smooth channels with a sharp 180-deg turn. Int J Heat Mass Transf 42(20):3757–3770
13. Hirota M, Fujita H, Cai L, Nakayama H, Yanagida M, Syafa'at A (2002) Heat (mass) transfer in rectangular cross-sectioned two-pass channels with an inclined divider wall. Int J Heat Mass Transf 45(5):1093–1107
14. Liou T-M, Chen C-C, Chen M-Y (2003) Rotating effect on fluid flow in two smooth ducts connected by a 180-deg bend. ASME J Fluids Eng 125(1):138–148
15. Mochizuki S, Murata A, Shibata R, Yang WJ (1998) Detailed measurements of local heat transfer coefficients in turbulent flow through smooth and rib-roughened serpentine passages with a 180 sharp bend. Int J Heat Mass Transf 42(11):1925–1934
16. Pape D, Jeanmart H, von Wolfersdorf J, Weigand B (2004) Influence of the 180° bend geometry on the pressure loss and heat transfer in a high aspect ratio rectangular smooth channel. In: Proceedings of 2004 ASME turbo expo (Vienna, Austria) vol 3, pp 685–695
17. Salameh T, Sundén B (2010) An experimental study of heat transfer and pressure drop on the bend surface of a U-duct. In: Proceedings of the 2010 ASME turbo expo (Glasgow, Scotland) 4 (Pts. A & B): 13–21
18. Schabacker J (1998) PIV investigation of the flow characteristics in internal coolant passages of gas turbine airfoils with two ducts connected by a sharp 180° bend. Ph.D. theses, EPFL, Lausanne, pp 1–149
19. Schüler M, Dreher HM, Neumann SO, Weigand B, Elfert M (2012) Numerical predictions of the effect of rotation on fluid flow and heat transfer in an engine-similar two-pass internal cooling channel with smooth and ribbed walls. Trans ASME J Turbomach 134(2):Paper 021021
20. Son SY, Kihm KD, Han JC (2002) PIV flow measurements for heat transfer characterization in two-pass square channels with smooth and 90 ribbed walls. Int J Heat Mass Transf 45(24):4809–4822

21. Xie G, Sunden B, Wang L, Utriainen E (2011) Parametric study on heat transfer enhancement and pressure drop of an internal blade tip-wall with pin-fin arrays. Heat Mass Transf 47 (1):45–57
22. Chen Y, Nikitopoulos DE, Hibbs R, Acharya S, Myrum T (2000) Detailed mass transfer distribution in a ribbed coolant passage with a 180° bend. Int J Heat Mass Transf 43(8):1479–1492
23. Iacovides H, Kelemenis G, Raisee M (2003) Flow and heat transfer in straight cooling passages with inclined ribs on opposite walls: an experimental and computational study. Experim Thermal Fluid Sci 27(3):283–294
24. Jang YJ, Chen HC, Han JC (2001) Computation of flow and heat transfer in two-pass channels with 60 deg ribs. Trans ASME J Heat Transf 123(3):563–575
25. Jang YJ, Chen HC, Han JC (2001) Numerical prediction of flow and heat transfer in a two-pass square channel with 90 ribs. Intl J Rotating Mach 7(3):195–208
26. Kirillov AI, Ris VV, Smirnov EM, Zaitsev DK (2001) Numerical simulation of local heat transfer in rotating two-pass square channels. Ann New York Acad Sci 934:456–463
27. Schüler M, Neumann SO, Weigand B (2009) Numerical investigations of pressure loss and heat transfer in a 180° bend of a ribbed two-pass internal cooling channel with engine-similar cross-sections. Proc IMechE Part A: J Power Energy 224(3):349–361
28. Cho HH, Wu SJ, Kwon HJ (2000) Local heat/mass transfer measurements in a rectangular duct with discrete ribs. Trans ASME J Turbomach 122(3):579–586
29. Han JC, Ou S, Park JS, Lei CK (1989) Augmented heat transfer in rectangular channels of narrow aspect ratios with rib turbulators. Int J Heat Mass Transf 32(9):1619–1630
30. Liu Y-H, Wright LM, Fu W-L, Han J-C (2006) Rib spacing effect on heat transfer and pressure loss in a rotating two-pass rectangular channel (AR = 1:2) with 45-degree angled ribs. In: Proceedings of the 2006 ASME turbo expo (Barcelona, Spain) 3 (Pts. A & B): 363–373
31. Chanteloup D (2003) Experimental investigation of heat transfer and flow characteristics in various geometries of 2-pass internal cooling passages. PhD Thesis, EPFL, Lausanne
32. Taslim ME, Liu H (2005) A combined numerical and experimental study of heat transfer in a roughened square channel with 45° ribs. Int J Rotating Mach 1:60–66
33. Tsang CLP, Gillespie DRH, Ireland PT, Dailey GM (2001) Detailed performance assessment of 16 % blockage interrupted ribs at 60 degree inclination in a square section turbine blade cooling passage. In: NATO RTA/AVT symposium on advanced flow management, Norway
34. Fu WL, Wright LM, Han JC (2006) Rotational buoyancy effects on heat transfer in five different aspect-ratio rectangular channels with smooth walls and 45 degree ribbed walls. Trans ASME J Heat Transf 128(11):1130–1141
35. Astarita T, Cardone G (2000) Thermofluidynamic analysis of the flow in a sharp 180 deg turn channel. Exp Thermal Fluid Sci 20(3–4):188–200
36. Han JC (1988) Heat transfer and friction characteristics in rectangular channels with rib turbulators. Trans ASME J Heat Transf 110(2):321–328
37. Han JC, Park JS (1988) Developing heat transfer in rectangular channels with rib turbulators. Int J Heat Mass Transf 31(1):183–195
38. Park JS, Han JC, Huang Y, Ou S, Boyle RJ (1992) Heat transfer performance comparison of five different rectangular channels with parallel angled ribs. Int J Heat and Mass Transf 35(11):2891–2903
39. Rallabandi AP, Yang H, Han J-C (2009) Heat transfer and pressure drop correlations for square channels with 45 deg ribs at high Reynolds numbers. Trans ASME J Heat Transf 131(7):Paper 071703
40. Cho HH, Kim YY, Kim KM, Rhee DH (2003) Effects of rib arrangements and rotation speed on heat transfer in a two-pass duct. In: Proceedings of 2003 ASME turbo expo (Atlanta, Georgia, USA) 5 (Pts. A & B): 433–442
41. Iacovides H, Kounadis D, Launder BE, Xu Z (2006) Experimental study of the thermal development in a rotating square-ended U-bend. In: Proceedings of 2006 ASME turbo expo (Barcelona, Spain) 3 (Pts. A & B): 647–655

42. Lee SW, Ahn HS, Lau SC (2007) Heat (mass) transfer distribution in a two-pass trapezoidal channel with a 180 deg turn. Trans ASME J Heat Transf 129(11):1529–1537
43. Lucci JM, Amano RS, Guntur K (2007) Turbulent flow and heat transfer in variable geometry U-bend blade cooling passage. In: Proceedings of the 2007 ASME turbo expo (Montreal, Canada) 4 (Pts. A & B): 159–167
44. Saha AK, Acharya S (2007) Turbulent heat transfer in ribbed coolant passages of different aspect ratios: parametric effects. Trans ASME J Heat Transf 129(4):449–463
45. Etemad S, Sundén B (2006) Numerical investigation of turbulent heat transfer in a rectangular-sectioned 90° bend. Numer Heat Transf A 49(4):323–343
46. Sewall EA, Tafti DK (2006) Large eddy simulation of flow and heat transfer in the 180° bend region of a stationary ribbed gas turbine internal cooling duct. Trans ASME J Turbomach 128(2):763–771
47. Viswanathan AK, Tafti DK (2007) Investigation of Detached Eddy Simulations in capturing the effects of coriolis forces and centrifugal buoyancy in ribbed ducts. Trans ASME J Heat Transf 129(7):778–789
48. Pape D (2004) Strömung und Wärmeübertragung in 180° Krümmern. Final Report on AG Turbo Project 2.4.10A, Institute of Aerospace Thermodynamics, Universität Stuttgart; Germany
49. Pape D, Jenkins S, von Wolfersdorf J, Weigand B, Schnieder M (2006) The influence of including a partially smooth section in the 2nd leg of an internally ribbed two pass cooling channel. In: Proceedings of the 2006 ASME turbo expo (Barcelona, Spain) 3 (Pts. A & B): 633–645
50. Ireland PT, Jones TV (2000) Liquid crystal measurements of heat transfer and surface shear stress. Meas Sci Technol 11(7):969–986
51. Kays WM, Crawford ME, Weigand B (2005) Convective heat and mass transfer, 4th edn. Mc-Graw-Hill. ISBN 0–07-246876-9
52. Kallinderis Y, Khawaja A, McMorris H (1996) Hybrid prismatic/tetrahedral grid generation for viscous flows around complex configurations. AIAA J 34(2):291–298
53. [FLUENT] ANSYS FLUENT User's Guide (2009) Version 12, ANSYS Inc
54. Baehr HD, Stephan K (2011) Heat and mass transfer, 3rd edn. Springer, Heidelberg
55. Kok KD (ed) (2009) Nuclear engineering handbook. CRC Press/Taylor & Francis Group, Boca Raton, London, New York
56. McAdams WH (1954) Heat transmission, 3rd edn. McGraw-Hill, New York
57. Maurer M, von Wolfersdorf J, Gritsch M (2007) An experimental and numerical study of heat transfer and pressure losses of V- and W-shaped ribs at high Reynolds numbers. In: Proceedings of 2007 ASME turbo expo (Montreal, Canada) 4 (Pts. A & B): 219–228
58. Xie G, Sunden B, Utriainen E, Wang L (2010) Computational analysis of pin-fin arrays effects on internal heat transfer enhancement of a blade tip wall. Trans ASME J Heat Transf 132(3): Paper 031901
59. Petukhov BS, Irvine TF, Hartnett JP (1970) Advances in heat transfer 6. Academic Press, New York
60. Han JC, Dutta S, Ekkad S (2013) Gas turbine heat transfer and cooling technology, 2nd edn. CRC Press/Taylor & Francis Group, Boca Raton, Florida

Chapter 9
Summary and Conclusions

This monograph presents results of the analytical and numerical modeling of convective heat and mass transfer in different rotating flows caused by (i) *system rotation*, (ii) *swirl flows* due to swirl generators, and (iii) *surface curvature* in turns and bends. Volume forces (i.e., centrifugal and Coriolis forces), which influence the flow pattern, emerge in all of these rotating flows. The main part of this work deals with rotating flows caused by system rotation, which includes several rotating disk configurations and straight pipes rotating about a parallel axis. Swirl flows were studied in some of the configurations mentioned above. Curvilinear flows were investigated in different geometries of two-pass ribbed and smooth channels with 180° bends.

Chapter 1 includes a mathematical description of the studied problems in the form of momentum, continuity, energy and convective diffusion equations in a vector form, Cartesian and cylindrical polar coordinates. Forces affecting the flow are also discussed in detail.

In Chap. 2, the aforementioned general mathematical description is customized for the rotating disk configurations. The chapter overviews in brief the existing mathematical methodology applicable to modeling the convective heat and mass transfer in such configurations, describes in detail the integral method developed by myself (referred to as "the present integral method" in this book), and gives a general analytical solution for turbulent boundary layer flow and heat transfer derived with the help of this method.

Chapter 3 represents a validation of the present integral method for the case of a single rotating disk. It was shown that the present integral method is essentially more accurate and enables modeling a wider range of the thermal boundary conditions than the methods of other authors. The novel analytical solution for temperature difference depending on two parameters provides a much better agreement of the Nusselt number with the experiments, and significantly expands possibilities for analytical predictions of heat transfer rates over a rotating disk subject to arbitrary thermal boundary conditions. Chapter 3 delivers also a critical overview of the most important experimental results for transitional flow, recommendations for estimation of average heat transfer of an *entire* disk, and briefly outlines some aspects of transient heat transfer over a single rotating disk.

I.V. Shevchuk, *Modelling of Convective Heat and Mass Transfer in Rotating Flows*, Mathematical Engineering, DOI 10.1007/978-3-319-20961-6_9

Chapter 4 outlines the solutions obtained with the help of the self-similar equations and the present integral method for the cases of (a) disk rotation in a fluid rotating as a solid body, (b) accelerating nonrotating radial flow, and (c) swirling outward radial flow in a cavity between parallel corotating disks. For laminar flow, the resulting novel approximate analytical solution based on the improved model for the modified enthalpy thickness deviates from the exact self-similar solution by not more than 3.1 %. An exact self-similar solution for orthogonal flow impingement onto a rotating disk testifies that in spite of the rotation, a solely impingement dominated regime of flow and heat transfer can emerge over the disk; the boundaries of this regime were pinpointed. Overall, the simulations correlate well with reliable experimental data for a stagnation point of a single impinging jet. For turbulent flow, the novel approximate analytical solution based on the present integral method agrees well with experiments. The present integral method demonstrates a higher accuracy at the expense of more accurate approximation of the radial velocity and temperature profiles in the boundary layer, and provides also a good match of the simulations with known experimental data for rotation cavities. For negative or approximately constant radial distributions of the wall temperature, negative Nusselt numbers (wall heat flux direction opposite to that in the source region) can emerge in the area of the Ekman-type layers.

In Chap. 5, self-similar solutions of the Navier–Stokes and energy equations were obtained for fluid flow in a conical gap for the configurations "rotating cone—stationary disk", "rotating disk—stationary cone", "corotating or contra-rotating disk and cone", and "nonrotating conical diffuser". The influence of the boundary conditions and various Prandtl/Schmidt numbers on the pressure, velocity, and temperature profiles, as well as on the Nusselt/Sherwood numbers was revealed.

Chapter 6 presents revised more accurate equations, which should be employed to recalculate the data for turbulent mass transfer for naphthalene sublimation in air to the conditions of heat transfer in air. Chapter 6 outlines also a novel methodology for simulations of temperature/concentration profiles for the Prandtl and Schmidt numbers much larger than unity. The present integral method further developed in this chapter enabled evaluating a relative thickness Δ of the thermal/diffusion boundary layers, which has not been performed by other investigators. It was demonstrated that the model with a decreasing function $\Delta(r)$ yields a new summand in the expression for the exponent at the Reynolds number, which determines functional dependence of Nusselt or Sherwood numbers. Consequently, theoretical relations obtained for the Nusselt and Sherwood numbers are in a good consistency with the selected empirical equations.

In Chap. 7, the commercial code FLUENT was used to simulate convective heat transfer in a pipe rotating about a parallel axis. Two factors were studied: (1) inlet angle of attack and (2) cross-section shape (circular/elliptic pipes). The elliptic pipes had (a) the same hydraulic diameter (i.e., 51.2 % increased cross-section area), and (b) the same cross-section area as that of the reference circular pipe and were installed radially (aligned with the radius of rotation) or circumferentially (perpendicular to the radius of rotation).

The heat transfer augmentation was observed only for the contra-rotating incoming air and the pipe. Elliptic pipes are preferable for heat transfer augmentation. In a cooling system with elliptic pipes obtained via morphing of the reference circular pipe and keeping the cross-section area unchanged, 12 elliptic pipes were packaged in a rotor in place of 12 circular pipes. More preferable circumferential elliptic pipes ensured 45 % of the total heat transfer augmentation in the entire cooling configuration, which is the highest among all the studied geometries. This was accompanied with 11.3 % smaller increase in pressure losses due to rotation than that in the configuration with 8 circumferential elliptic pipes with the enlarged cross-section (i.e., the same hydraulic diameter).

Chapter 8 is devoted to simulation and optimization of convective heat transfer in the varying aspect ratio two-pass internal ribbed cooling channels with 180° bends.

For a *periodic ribbed segment* of the channel, the averaged Nusselt numbers agreed well with experimental data. For $Re = 100{,}000$ and an angle of attack of 45°, most beneficial are the ribs with $e/D_h = 0.075$–0.125 for the aspect ratio $H/W = 4{:}1$, and with $e/D_h = 0.1$–0.15 for the aspect ratio $H/W = 2{:}1$ and 1:1. Rib heights exceeding the optimum entail a faster increase in the pressure loss and a very minor increase in the heat transfer rate.

In the *ribbed two-pass channel*, the bend geometry increases heat transfer because of the flow acceleration, impingement on the walls, and Dean vortices. The Nusselt number in the smooth channel with $H/W = 2{:}1$ increases almost 2 times in the bend and 2.6 times in the outlet pass. The Nusselt number in the varying aspect ratio channel reaches the levels of 2.3–2.4 in the bend. Relative pressure drop in the entire geometry behaves as a nonlinear function at W_{el}/W_{in} with a distinct minimum at $W_{el}/W_{in} = 1.75$, which overall conforms to experiments (deviations within 15 %). The aerothermal efficiency is maximal at $W_{el}/W_{in} = 1.0$–1.5. In the geometry with $H/W_{in} = 3{:}1$, the overall *static pressure* drop is noticeably smaller than in the case of $H/W_{in} = 2{:}1$. The average heat transfer rate over the tip wall and the bend bottom is smaller for the aspect ratio $H/W_{in} = 3{:}1$ than for $H/W_{in} = 2{:}1$.

To conclude, it was demonstrated in this book that the complex phenomena of fluid flow and convective heat transfer in rotating flows can be successfully simulated using not only the universal CFD methodology, but in certain cases by means of the integral methods, self-similar, and analytical solutions. The results of simulations presented in the book are in good agreement with experimental data available in the literature.

MIX
Papier aus verantwortungsvollen Quellen
Paper from responsible sources
FSC® C105338

If you have any concerns about our products,
you can contact us on
ProductSafety@springernature.com

In case Publisher is established outside the EU,
the EU authorized representative is:
Springer Nature Customer Service Center GmbH
Europaplatz 3, 69115 Heidelberg, Germany

Printed by Libri Plureos GmbH
in Hamburg, Germany